INTRODUCTION TO ENVIRONMENTAL STUDIES

JONATHAN TURK, Ph.D.

Naturalist

SAUNDERS GOLDEN SUNBURST SERIES IN ENVIRONMENTAL SCIENCE

1980   W. B. SAUNDERS / Philadelphia / London / Toronto

W. B. Saunders Company: West Washington Square
                          Philadelphia, PA   19105

                          1 St. Anne's Road
                          Eastbourne, East Sussex BN21 3UN, England

                          1 Goldthorne Avenue
                          Toronto, Ontario M8Z 5T9, Canada

Front cover illustration: Mountain Goats, Glacier National Park (Courtesy of John Barstow.)

Introduction to Environmental Studies   ISBN   0-7216-8919-1

Last digit is the print number: 9   8   7   6   5   4   3   2   1

PREFACE

In recent years the awareness of environmental problems has spread rapidly throughout our society. Manufacturers must be concerned not only with marketing and selling their products, but also with the control of pollution from their factories and with conservation of energy and materials. Land developers must recognize that land is limited and that soil, wildlife, and wilderness must be preserved. Builders may no longer construct homes and offices that are wasteful of energy. Scientists and engineers have become increasingly aware of environmental considerations. Homemakers, too, are faced with problems of pollution and rising fuel costs.

This book has been written in response to the need for environmental education by a wide variety of people. It attempts to provide an overview of this broad subject and to integrate the various social, economic, technical, and political issues. Interrelated problems of ecological disruptions, growth of human populations, land use, energy, nuclear power, food supplies, pesticides, air and water pollution, solid waste, and noise are all covered.

The text has been organized into study units:

Unit I : Introduction and Social Background
Unit II : The Biological Background
Unit III: Human Population
Unit IV: Resources and Energy
Unit V: Rural Land Use
Unit VI: Pollution

This arrangement should offer the instructor considerable flexibility. The text may be used for a full two semester course by taking advantage of all its features and teaching aids. Alternatively, with judicial choices of topics, it may serve for a one semester course or as a supplement for other classes to provide an overview of environmental problems.

Some specific features of the book include:

(a) Case Histories

A case history at the end of each chapter dramatizes the formal material. In addition, the reader will find personal interviews and experiences scattered throughout the text that serve to personalize and humanize the subject matter.

(b) Take-Home Experiments

At the end of each chapter, the reader will find a miniature "laboratory manual," which provides directions for some simple experiments. These experiments are designed to illustrate various topics in the chapter and to require no more, or very little more, equipment and supplies than are commonly

iii

available in the household. Some may be used as classroom projects. They should be fun to do.

(c) Problems and Questions for Class Discussion

There is a complete problem section at the end of each chapter. Some of the problems can be answered directly from the text material, while others involve additional thought. Following the problem sections there is a series of "Questions for Class Discussion." These questions do not necessarily have a "correct" answer but are meant to serve as a nucleus for group discussion of some of the important issues in the chapter.

(d) Chapter Summaries

Each chapter is provided with a summary to aid in easy review of the text material.

(e) Glossary

An extensive glossary of terms used in the text is provided at the end of the book.

(f) Use of the Metric System

Since metrication is both inevitable and (I believe) desirable, this book uses metric units extensively in the text, tables, and figures. In instances where British units are more familiar to many readers, both units are given. Measurements of heat are shown in calories, with the conversion factor to joules, the official SI energy unit. All the needed conversions are also provided in the Appendix.

Acknowledgments

I would like to thank my father, Dr. Amos Turk, for reading the entire manuscript and making many valuable suggestions.

I also owe special thanks to Mrs. Pearl Turk for assistance in proofreading, to Jan Logan for drawing many of the illustrations, and to my many friends at W. B. Saunders Company who put it all together.

Jonathan Turk

CONTENTS

Chapter 10
AGRICULTURAL SYSTEMS 196

Chapter 11
CONTROL OF PESTS AND WEEDS 220

Chapter 12
AIR POLLUTION 242

Chapter 13
WATER POLLUTION 266

Chapter 14

SOLID WASTES.. 284

Chapter 15

NOISE... 304

INTRODUCTION
TO
ENVIRONMENTAL
STUDIES

1

THE ENVIRONMENT — AN INTRODUCTION

1.1
THE NATURE OF ENVIRONMENTAL DEGRADATION

There are people who live in remote cabins in northern Canada and Alaska. They have few neighbors, and the environment in which they live has changed little since prehistoric times. Many men and women are farmers. They work with living plants and animals, but their environment is a controlled, human one. Crops are raised, animals are cared for, and unwanted pests are destroyed. Some people live in cities and work in offices and factories. Their surroundings consist mainly of objects made of concrete, brick, or steel. There are many regions of the Earth that people do not live in at all. Hikers and climbers cross high ridges and scale walls of ice and rock, but they do not live there. Deserts, oceans, and the great ice caps of the polar regions are mostly uninhabited.

Environmental science is the study of all these different places and habitats. It includes an examination of the air we breathe, the water we drink, and the sources of food that nourish us. It is also concerned with the appearance of our surroundings, and with the other living organisms — human and nonhuman — with which we share a place on Earth. Any action that makes the environment less fit for human life is said to be an **environmental degradation.**

People have been degrading their environment for many thousands of years. If you enter a cave that was inhabited by an ancient Indian family thousands of years ago, you can see large areas of the roof and walls that were blackened by smoke. Archaeologists digging in pre-Inca ruins along the coast of Peru have uncovered the remnants of garbage dumps that must have once been a source of odor and disease. But until recently, environmental damage was concentrated in a few small regions. Now, the danger is spreading over the entire planet. DDT is concentrated in the fat of penguins in Antarctica. Oil slicks appear in the central oceans. Synthetic chemicals have been detected in the high streams of the Rocky Mountains. Nearly all the rivers in the world are affected so badly that the water is unfit to drink. Scientists are issuing

3

that harm the environment. Many individuals waste large quantities of energy.

Environmental degradation is such a complex topic that we divide it into three main categories.

(a) Pollution.   Pollution is the contamination of the environment by the introduction of impurities. Smoke pollutes the air, sewage pollutes the water, junk cars pollute the land. We know for a fact that many parts of our environment are polluted. But the total effect of this pollution on human health and welfare is not completely understood.

(b) Depletion of Resources.   When you turn on your air conditioner or electric heater, you are cooling or warming the air around you, but you are not polluting it. However, the energy you use is probably generated at a power plant that burns coal or oil. The fuels must be extracted from the earth, and once they are burned they cannot be reused. The metal used to manufacture an air conditioner or heater is also mined from the earth. While these metals can be recycled when the appliance breaks down, often recycling is incomplete and the materials are eventually left rusting in a garbage dump.

It would seem reasonable to conserve our fuels and recycle our metals. However, many people do not regard depletion of resources as an immediate problem. Picture yourself walking along a road and seeing five automobiles pass you. The first is a small, old car, carrying four passengers. It seems to be in poor mechanical condition, and you note that a haze of blue smoke comes out of its exhaust pipe. The next four are late-model luxury cars, each carrying one passenger. The first car pollutes, but it conserves fuel. It is capable of driving many miles per gallon, and in addition there are four people in the car. Therefore, the number of passenger-miles per gallon is quite high. Let us assume that the other four cars do not pollute nearly so much. But they are certainly wasteful of fuel. The pollution from the first car makes you cough, which means that it affects you as soon as it is produced. The much greater depletion of the Earth's petroleum reserves caused by the other four, however, is

Ruins of an ancient Indian dwelling in Mesa Verde, Colorado. Some of the inside walls are still black from soot, indicating that pollution problems existed even in prehistoric societies.

warnings that human activities may upset weather patterns and change world climate.

Many different people and many different types of organizations contribute to the deterioration of our environment. Governments have permitted improper disposal of radioactive wastes. Some major oil companies ship petroleum products in poorly designed ships that spill oil into the ocean. Farmers often spread pesticides

(Reprinted with permission from *Yankee Magazine*, Dublin, N.H.)

easily forgotten, because there will still be gasoline for sale at the service station tomorrow.

Conservation is therefore often seen as a measure whose benefit will be realized later, perhaps only by our children or grandchildren. Unfortunately, many people are not particularly concerned with such long-range problems.

(c) Disturbance of the Natural Condition. If you are planning a trip next winter to Africa, and you need to know what clothes to pack, you must ask your travel agent. He or she will consult a book on world climate and tell you what temperatures and how much rain to expect. Of course, drought, rainy spells, and unseasonable cold or warm periods are all common. But, in general, climate is constant from year to year.

Recently we have begun to ask whether human activities can change world climate. Can the exhaust from high flying airplanes alter the atmosphere and harm living things? Will pollution of the lower atmosphere affect global temperature, rainfall, or agricultural productivity? Will the pollution of the oceans destroy its living organisms and upset the oxygen balance of the Earth? All of these questions have been explored by scientists, and they will be considered in more detail in appropriate sections of this book.

1.2
ENVIRONMENTAL PROBLEMS — AN OVERVIEW

There are no easy solutions in environmental science. There are no magic formulas that can suddenly reduce consumption and stop pollution. All the issues are complex. We recognize two basic approaches in the efforts to improve our environment. One approach can be stated as follows: Most major environmental problems are caused by twentieth-century technology. Therefore, the major problems should be solved by technology, and scientific answers must be found.

Another view is expressed in the following terms: People use the products of technology. It is people who drive cars, turn on electric lights, or litter the countryside with old beer cans. Therefore the problems are social and economic and we must search for social remedies to environmental problems.

In order to understand how these arguments apply in a real life example, let us look briefly at automobile pollution. Remember the mythical magic lamps and their genies? One had only to rub the lamp and speak one's wish, and it was done. The "fully equipped" automobile seems like a modern version of those magic lamps. A touch of the finger, hardly even a push, brings warming, or cooling, or music; it opens or closes windows, locks or unlocks the car doors, or even the garage doors. A touch of the toe brings a surge of acceleration, or a swift stop. Seated within these plush environments, people can move rapidly and comfortably from place to place. Consequently, cars are used by almost everyone who can afford them.

If you sit on a hilltop overlooking almost any city in the Western Hemisphere, you can see a brownish haze over the landscape. This haze is caused largely by automobile exhaust. It is called **smog** and is harmful to human health. The vehicles that produce smog also consume such large quantities of gasoline that petroleum reserves may soon be depleted. Is it possible to transport people from place to place comfortably and rapidly without producing smog or using large quantities of fuel?

Technical Solutions. We know that many people enjoy the privacy and convenience of the private automobile. But most of the vehicles in use today are inefficient. Therefore, a technical solution would be to design better, cleaner, more efficient cars. Then people could drive about as they pleased in an improved environment. Of course, it takes time to develop a complex device such as an automobile that is both efficient and free of pollution. If we will only be patient, say the people who favor technical solutions, the answer to pollution problems will be "delivered to our doorstep."

Transportation by moped uses much less fuel than transportation by automobile. (Courtesy of Columbia Manufacturing)

An example of an economic externality.

Social Solutions. Other people disagree strongly. They argue that we already have the knowledge to reduce pollution and consumption in the transportation industry. We need not wait for new technology. Small, compact cars travel two or three times farther on a gallon of gasoline than luxury cars do. Mopeds (motorized pedal vehicles) operate at up to 140 miles per gallon of gasoline and bicycles use no gasoline at all. Electric trolleys and trains can operate with low pollution and high efficiencies.

Recently, German engineers built a small automobile that is slow and uncomfortable but can travel 2566 kilometers (1595 miles) on a gallon of diesel fuel. All of these alternatives are less expensive than luxury cars. But day after day millions of people drive billions of miles in large, heavy, polluting, inefficient automobiles. In order to use the best available technology, people must change their habits and life styles. Therefore, social solutions are needed to solve transportation problems.

Many people believe that complex environmental problems will be solved only if both technical and social remedies are used. For example, if scientists develop more efficient transportation systems and if people use the best available technology, then we would conserve fuel and pollute less.

So you see, the problems we are studying have no single, simple solution. The issues are complex, and the answers are uncertain. This book will not provide instant remedies for our environmental problems. It will discuss the complexities of the issues and leave the reader to make his or her own decisions.

1.3
THE COSTS OF
ENVIRONMENTAL
POLLUTION — EXTERNALITIES

If you buy a new car, the price you pay accounts for the construction and maintenance of the factory, the cost of raw materials and labor, shipping charges, marketing expenses, and profit. There are many other costs, however, that are not accounted for.

Mining and manufacturing pollute the surrounding air and water. If the smoke from the mill darkens nearby houses, the price of a car does not include extra money to pay for more frequent painting. Water pollution kills fish and increases the cost of water purification. When a woman buys an automobile she doesn't pay commercial fishermen for the loss of income when the fish die. She doesn't help pay city water purification plants for the added costs of cleaning the water. If there is a loss of tourist trade because the lakes and streams are dirty, no one reimburses motel and restaurant owners. If the air and water pollution endanger people's health no one pays for increased medical bills or loss of work because of illness. There are many costs that are not paid for by the manufacturer but must be paid by someone. These costs are external to the price of manufactured goods. They are called **externalities.**

The total cost of all externalities is very high. As a very rough approximation, the costs of environmental damages to health, of depression of property values, and of losses of crops and materials are each somewhere between five and ten billion dollars per year. Of course, such estimates (especially those of damages to health) are subject to large errors, for they represent only known dangers and do not include possible long-term effects. In many cases the true price of externalities is higher than the cost that can be measured in dollars. If a person contracts cancer from an environmental poison we can calculate the total doctor and hospital bills and the value of lost work. But there is an added cost of human pain and suffering that cannot be calculated. Recreational opportunities, too, are changing. Fifty years ago, for instance, people on the south shore of Lake Erie fished, boated, swam, and explored the beaches and marshes. Today swimming there is unsafe, the number of tasty game fish has declined dramatically, the wild secluded beaches have disappeared, and boats ply polluted waters. People still find places to play. There are amusement parks, indoor swimming

Swimming in a pool is different from swimming in a natural environment.

pools, gymnasiums, and movie houses in the cities along the shore. But the destruction of unspoiled environments is a real loss to many people that cannot easily be calculated.

1.4
THE COST OF POLLUTION CONTROL

A large scale effort to clean the environment would benefit all of us in a great many ways, but it would be expensive. However, when the cost of pollution control is added up, people often neglect to balance that expense against the total cost of economic externalities. There is little doubt that pollution control is cheaper than pollution.

It is difficult to calculate the actual cost of pollution control. Some companies could not afford control equipment and would be forced out of business. Others could afford to reduce pollution only if they could raise the price of manufactured items. In some cases industries could recycle their wastes rather than discard them and pollution control would be virtually free. In many instances, plant managers have stated in court that it would be absolutely impossible to control pollution at a reasonable cost. When forced by the government to clean up their operations or close the plant, some of these same companies have installed efficient control equipment without going out of business. Careful estimates indicate that the costs needed to satisfy Federal environmental legislation were only about 1.1 percent of the gross national product (GNP) in 1973, that they will rise to about 1.5 percent in 1979, level off through 1983, and then decrease.* For example, suppose a manufactured item now costs $1000. If better pollution controls were used, and these controls cost 1.5 percent, then the price of the item would rise to $1015.

Some people claim that pollution control programs cause unemployment. This is a misleading statement. People must be hired to develop, build, and install control equipment. Of course the overall picture does not predict changes in individual industries. Obviously, a company that specializes in the manufacture of pollution control equipment will prosper; others will be harmed. Suppose, for example, that you work for a chemical company that pollutes the air and cannot afford the cost of control equipment. The plant closes down. Everyone now enjoys the clean air, but you are out of a job. If enough others like you feel that they are being treated unfairly, there will be vigorous opposition to programs of environmental control.

Many people have stated that energy conservation programs would put people out of work. This statement is also misleading. If people were hired to install solar heating units, build new mass transit systems, and work on similar projects, there would be many new jobs. If, on the other hand, we waste energy until, suddenly, we run out of oil and coal, then disaster will strike. Millions will lose their jobs and the economy will collapse.

---

*The sixth annual report of the Council on Environmental Quality, December, 1975.

1.5
SOCIAL APPROACHES TO POLLUTION CONTROL

Altruism

Altruism is devotion to the interest of others. If a person voluntarily acts to improve the environment, he or she is acting altruistically. If altruistic actions were taken by large numbers of individuals, environmental quality would be improved. For example, if everyone rode bicycles on any journey less than eight kilometers, automobile pollution would be reduced signifi-

The gross national product is the total market value of all goods and services produced in the country in a given year.

See Appendix for a discussion of the metric system.

Top view shows how fumes would pour from the smokestack of a steel furnace if pollution control devices were not installed. Bottom view shows operation with successful control. The design, construction, and maintenance of pollution control equipment not only preserves our environment, but provides jobs as well. (Photo courtesy of Bethlehem Steel)

cantly. Therefore, an altruistic act would be performed by a person who rode a bicycle when he or she would have preferred to drive a car. How have altruistic approaches succeeded? Many people do ride bicycles, take the bus, or drive compact cars. Consumers recycle paper, cans, and bottles even when they are not paid to do so. But in spite of much effort by a great many people, the environmental condition continues to worsen.

The problem is that not enough people behave altruistically. The great majority are not helping. Many people buy large cars even though they don't need the extra space. Others throw aluminum cans in the garbage because it is too much effort to bring them to the recycling center. Each person should act with care toward the environment, but we can't depend on altruism alone to solve our problems.

1.6
LEGAL APPROACHES TO POLLUTION CONTROL

Highway signs in Colorado read:

Most states have littering laws and a person can be fined or even put in jail for discarding trash onto the highways. Similarly, there are laws in sev-

eral states in the United States banning no-deposit bottles and aluminum beverage containers. National governments all over the world have enacted laws designed to reduce pollution. There are various types of anti-pollution laws.

One type of law simply **prohibits** a given act. The law against littering says you must not throw away any garbage onto the highway. If you do you will be arrested.

However, many problems are so much more complex than littering that an outright prohibition is difficult. Think of a city whose electricity is generated by a power plant that burns coal. Imagine further that the generating plant is smoky and smelly. Steam, soot, and poisonous gases are discharged into the air. If a government agency were to arrest the operators of the plant and to close it down, homes would be without electricity. People would be out of work. There are several approaches to complex problems of this sort. One approach is to order that the plant clean its operation within a given time limit. After, say, six months or a year the operators of the plant must install satisfactory pollution control equipment or be forced to shut down. Another possible approach is known as a **pollution tax.** Under this plan, a factory or electric generating plant would be charged for the pollution it produces. The polluters would either pay the tax and continue to pollute or clean the wastes and avoid paying the tax. If the tax were high enough, most industrialists would choose to reduce pollution. Although many environmentalists favor this type of law, problems do arise. Suppose that the operator of an electric generating facility chose to pay the tax and continue to pollute. The cost of the tax would be passed on to the consumer. Then a person who lived near the plant would breathe polluted air *and* pay higher electric bills.

Rationing is another possible approach to environmental improvement. Rationing was used a great deal during World War II to aid the war effort. People received coupons for a given amount of gasoline, rubber tires, meat, butter, and other items. You could purchase a rationed item only if

LOUIE

you had a coupon. Once your coupons were used up, you could not legally buy more even if you had enough money. Rationing was seriously considered in the United States in 1973 as a solution to the worldwide gasoline shortage. Under the plan proposed at that time, each driver would receive coupons allowing him to purchase 12 gallons of gasoline per week. Once that ration was used up the person would have to use mass transit or stay at home. This proposed plan was not put into effect.

Still another approach would be a **use tax.** Several use taxes have already been put into effect. For example, there is a four cent federal tax on every gallon of gasoline sold at the pump. Some people have suggested raising this tax. Then a commuter who normally drives to work would be faced with a decision. He could: (a) continue to commute in a large car and pay the tax; (b) join a car pool or buy a small car to use less gasoline and pay less tax; (c) ride a bus or train to work; (d) sell his house and move to a location near work; (e) remain in the same house and choose a new job located near his home. If the tax were high enough, many people would choose some alternative to use less gasoline.

Use taxes could also be levied on other nonrenewable resources such as metals, fertilizers, plastics, and other items.

Imagine that a factory in your neighborhood is polluting the air and water around you. What can you do? One possibility is to take its managers to court. In order to sue a manufacturer an individual or group must first have what is called **legal standing.** To

have legal standing you must be able to prove that you personally are harmed by the pollution. If you can show that the smoke from the factory has made you ill, you have legal standing. But a woman living in Chicago cannot sue an oil company in California for spilling oil in the Pacific Ocean simply because she is upset over the damage to the wildlife. However, if she owns waterfront land nearby, she can sue for damages to her property.

Sometimes a group is granted standing where an individual is not. Recent statements by Supreme Court judges suggest that perhaps valleys and mountains, like corporations, should be granted standing. Some cases are heard because they are con-

---

GASOLINE RATIONING

To ration or not to ration — that is the question. Pressure for rationing comes from many sides. Some of it is backed by cogent argument. Utility executives worry that they may not get enough fuel oil to run their generators. Manufacturers fear that they may not get enough energy and feedstocks to run their plants. Political pressure favors rationing as a means of keeping down the price of gasoline. Meanwhile, rationing kindles a gleam in the eye of intellectuals who think of coupons as a means of redistributing income.

Nobody can say now that we can avoid rationing. If the alternative is severe disruption of production or transportation, rationing may be the only answer. But the damage will be serious. The bruising battle over who gets how much; the bureaucratic nightmare of distributing coupons to tens of millions of people — snafus, forgeries, black markets — all this boggles the administrative imagination. The nearest comparison probably is not to wartime gasoline rationing, but to peacetime Prohibition.

By Henry C. Wallich
Newsweek
December 24, 1973

Can valleys and mountains, like corporations, having legal standing? (Photo of Mount Robson, highest peak in the Canadian Rockies)

sidered "public interest" cases. For instance, a factory spewing poisonous gases into the atmosphere might be sued by a town and legally declared a "public nuisance." Certain legal scholars declare that a pollution-free and healthy environment is one of the rights of the Ninth Amendment.

Court action may be effective, but it is often complex, slow, and expensive (see Section 1.8, Case History). Even in relatively simple cases a person or environmental group may be forced to fight a giant corporation. One court battle may cost half a million dollars or more and may require many years to complete. During this time the company may be allowed to continue to pollute. Despite these difficulties, legal action is sometimes effective. In many cases, individuals and communities have forced a manufacturer to close or alter its operation. Sometimes the company will fear the bad publicity of a court case and install pollution control equipment without going to court.

1.7
ULTIMATE PROSPECTS FOR ENVIRONMENTAL IMPROVEMENT

What type of life will people lead fifty or one hundred years from now? Pessimists predict a total collapse of civilization. Optimists see clean, efficient, pleasant, "space age" cities. Some people imagine a world where sickly people struggle to work wearing gas masks and ear plugs. Others see an environment that is neither cleaner nor dirtier than the one we live in at present.

Pessimists argue that the human condition will continue to deteriorate unless drastic laws are imposed. They believe that the skies will become clean only if the managers of polluting factories are jailed. Population will be controlled only if there is compulsory sterilization after a woman has had two children. The streets will be unlittered only if an individual caught throwing a cigarette on the ground is fined half a month's pay. On the other hand, optimists feel that the awareness of environmental problems has only just begun. We must have faith in the human genius to stay alive and belief that we will build a world that is increasingly pleasant to live in. It is difficult to make accurate predictions of the future. But surely one positive step is to understand relationships between us humans and the world we live in. The remainder of this book will provide a start toward this understanding.

1.8
CASE HISTORY: RESERVE MINING — A LENGTHY LEGAL BATTLE

Reserve Mining Company is one of the largest iron mining and processing corporations in the world. Every

Will the cities of the future be clean, efficient, and pleasant, or will they be polluted, dirty, unsanitary, pockets of extreme poverty?

Residential area of Boulder, Colorado. (Photo by Marion MacKay)

Thousands of homeless sleep on the streets of Bombay. (Courtesy of Sygma)

Aerial View of the Reserve Mining Company plant at Silver Bay, Minnesota. The white area in the photograph consists of taconite tailings poured into Lake Superior. (Wide World Photos)

Asbestos can be woven into a cloth. This cloth is flexible, yet does not burn. It is used for firemen's gloves, flameproof pipes, automobile brake linings, and other similar uses.

day the company digs nearly 100,000 tons of ore out of the ground. This ore contains iron mixed with less valuable minerals and rocks. Of the 100,000 tons mined, about 30,000 tons of concentrated iron ore pellets are produced and 70,000 tons of solid waste material are thrown away. Since 1955, the company has been dumping these wastes into the waters of Lake Superior.

The Reserve Mining operations are economically important to regions of northern Minnesota as well as to the country in general. The company employs 3000 people and supplies over 15 percent of the iron ore used in the United States. If the plant were shut down the local economy would be destroyed. Not only would the 3000 employees be out of jobs, but many of the business people in the neighborhood would also be out of work. There might be a national shortage of iron ore, and the price of steel could rise. That, in turn, would lead to increased inflation across the country.

The waste that is dumped into Lake Superior contains many different types of minerals. One of these is a fibrous material called **asbestos.** Some of these fibers float along the lake to Duluth, Minnesota, and to Superior, Wisconsin, and enter the drinking water in these cities. It is known that asbestos fibers in the air cause cancer in humans. Experts feel that it is likely that asbestos fibers in water also

The ore is crushed and then transported via a company railroad to the Silver Bay processing plant on the shores of Lake Superior. Nine to ten trains per day, each made up of 120 to 140 cars, are needed to transport the ore. (Photo courtesy of Reserve Mining)

cause cancer, but no one is sure. Since there is a strong possibility that Reserve is endangering the health of local residents, various government agencies have tried to force the company to stop the pollution. The legal battle has been long and complex.

In 1969 the federal government recommended that Reserve stop polluting Lake Superior. Nothing was done. Two years later the United States Environmental Protection Agency (EPA) ordered the company to develop a water pollution control program within six months. Reserve submitted a plan that was unacceptable, so the United States government took the company to court. The first court battle lasted two years. On April 20, 1974, a district judge ruled that the air and water pollution endangered the health of the people living in the area. He ordered the plant to shut down. Rather than clean up their operations, Reserve immediately appealed the legal case to a higher court. Two days later a court of appeals judge overruled the district judge and said that more time was needed to review the case. The plant was allowed to resume operations immediately. The company continued to dump 70,000 tons of waste into Lake Superior every day. One year later the higher court ruled that Reserve must stop water pollution "within a reasonable time." Nothing was done. In 1976, the court ordered that all water pollution must cease by July 7, 1977, unless "circumstances in the case had changed before that time."

By now seven years had gone by since the United States government had demanded that the pollution stop. During that seven years the mining had continued without any change in operation. After 1976 the company finally agreed to start construction of a water pollution control facility. This agreement was considered to be a "change in circumstances." In May, 1977, the court stated that as long as construction of the water pollution control facilities was proceeding normally the company would be allowed to continue to pollute until April 15, 1980.

The eight year legal battle is a classic example of a struggle between economics and the desire for a clean,

An aerial view of the Silver Bay processing plant that refines approximately 31 million tons of crude ore to produce 10.8 million tons of iron each year. Wastes from this plant are deposited into the waters of Lake Superior. (Photo courtesy of Reserve Mining)

healthy environment. In a sense nobody won. The government finally has forced Reserve to start construction of a waste disposal site that does not pollute Lake Superior. But from 1969, when the case first went to court, to 1980, when the construction should be complete, people will have lived in a polluted and perhaps poisonous environment for eleven years. The men and women who work at Reserve have held their jobs. Pay at the mine is high and most have been able to live

The iron ore that is processed at Silver Bay is mined with the aid of huge power shovels. Each shovel removes over ten tons of ore in a single scoop. (Photo courtesy of Reserve Mining)

a "good life." But the external price may be enormous. In the years ahead many may die of cancer contracted from mine pollutants.

An important and still unanswered question has been raised by the legal battle. In the future, should a large corporation be allowed to pollute for over a decade just because economic hardship would result if it were shut down and because it can afford the millions required for a lengthy court battle?

CHAPTER SUMMARY

1.1 THE NATURE OF ENVIRONMENTAL DEGRADATION

Environmental degradation is divided into three main categories: (a) **Pollution** is the deterioration of the quality of the environment by the introduction of impurities. (b) **Depletion of resources** is the consumption of valuable fuels, fiber, metals, or fertilizer. (c) **Disturbance of the natural condition** is a change in the global environment, such as a shift of world climate.

1.2 ENVIRONMENTAL PROBLEMS — AN OVERVIEW

Some people feel that environmental problems should be solved by searching for technical and scientific answers. Others feel that social and economic solutions are needed. Most probably, technical and social remedies must be used together.

1.3 COSTS OF ENVIRONMENTAL POLLUTION — EXTERNALITIES

In general, manufacturers do not pay the cost of the pollution they produce. These costs are called **externalities.** Externalities include environmental damages to health, lowering of property values, losses of crops and materials, and changes in recreational opportunities.

1.4 THE COST OF POLLUTION CONTROL

Pollution control will cost many billions of dollars, but pollution control is less expensive than pollution.

1.5 SOCIAL APPROACHES TO POLLUTION CONTROL

Altruism is the devotion to the interest of others. If a person voluntarily improves the environment he or she is acting altruistically. In general, such approaches have not worked well in our society.

1.6 LEGAL APPROACHES TO POLLUTION CONTROL

There are several legal approaches to pollution control. (a) Some laws prohibit pollution outright. (b) A **pollution tax** charges a company for the quantity of pollution emitted. (c) **Rationing** allows a person or company to purchase only a limited supply of a certain product. (d) A **use tax** places a tax on products or services such as gasoline, fuel oil, or electricity. (e) People can take a polluting company to court. The court may then order that the pollution be stopped.

1.7 ULTIMATE PROSPECTS FOR ENVIRONMENTAL IMPROVEMENT

Pessimists foresee an increase in all our problems, while optimists predict a bright, unpolluted future. Certainly a case could be made for both viewpoints, but perhaps the truth lies somewhere between these extremes.

KEY WORDS

Pollution
Externalities
Altruism

Pollution tax
Rationing
Use tax
Legal standing

TAKE-HOME EXPERIMENTS

1. Locate the site of the nearest aluminum beverage can recycling center in your neighborhood. Is it in a convenient location? If it is not, speak to the people who operate the center and discuss the possibility of opening a recycling operation in a nearby area. If there is no aluminum recycling center in your area, start one. How many

pounds of aluminum can you collect in a week? Where do you sell the aluminum? Report on your operation.

2. Think of a law (local, state, or federal) that would help improve the quality of your environment. Write your proposal to the town council or your state or federal legislator. Submit a copy of your letter and your reply to the class for discussion.

1. **Environmental disruptions.** What are the three categories of environmental degradation? Which types are most directly obvious to a casual observer? Which are the least obvious? Give reasons for your selections.

2. **Environmental disruptions.** Using your list of the three types of environmental disruptions from Problem 1, categorize each of the following: (a) Exhaust from high-flying aircraft upsets the balance of the upper atmosphere and may possibly change climate on Earth. (b) Exhaust from the same aircraft during takeoff creates a smoky haze near a large airport. (c) A business person flies to a distant city to talk to a partner when the meeting could have been easily conducted over the telephone.

3. **Environmental disruptions.** Give an example of pollution in your neighborhood. What is the source? What are the effects of the pollution? Discuss.

4. **Altruism.** List three changes that you could make in your daily life that would reduce consumption.

5. **Solutions to environmental problems.** Discuss the difference between technical and social solutions to environmental problems. List the arguments in favor of each.

*6. **Solutions to environmental problems.** Imagine that a fertilizer factory were polluting the air in a city. Would you recommend that the pollution be controlled largely by technical or by social remedies? Would your answer be different if the polluting factory were manufacturing disposable beverage containers? Discuss.

*7. **Solutions to environmental problems.** Electric generating stations are responsible for a large amount of industrial pollution. List some social remedies to the problems of pollution from this source.

8. **Vocabulary.** Define economic externality.

*9. **Cost of environmental control.** Suppose you were a reporter for your school newspaper and were assigned to compare the cost of stopping the pollution of the local river with the cost of economic externalities that the pollution produces. What sources of information would you seek out? Which category of costs would be more difficult to estimate? Why?

*10. **Economic externalities.** On February 26, 1972, heavy rains destroyed a dam across Buffalo Creek in Logan County, West Virginia. The resulting flood killed 75 people and rendered 5000 homeless. The dam was built from unstable coal mine refuse called "spoil." Engineers knew that the dam was unstable, but repair work was not done because it would have been expensive and would have raised the price of coal. (a) How would you relate the concept of economic externalities to this tragedy? (b) Suggest legislation designed to prevent future Appalachian dam disasters. (c) How will your legislation affect the price of coal and its competitive position in the fuel market?

11. **Economic externalities.** Discuss some of the economic externalities of the Reserve Mining Company's operation on the shores of Lake Superior.

12. **Economics.** Explain why the control of pollution may result in a net increase in the number of jobs. Give examples.

13. **Altruism.** Define altruism. Which of the following acts would be considered to be altruistic? (a) A poor person rides the bus to work because he cannot afford a car. (b) A wealthy person who owns a luxury car rides the bus to save world supplies of fuel. (c) A person collects aluminum cans and donates them to a local Girl Scout collection drive.

14. **Altruism.** Forty years ago, rag and paper dealers pushed hand carts through New York City buying rags and old newspapers from home owners. They made their living by reselling the scraps for

*Indicates more difficult problem.

manufacture of recycled paper. Were these people acting altruistically? Discuss.

15. **Altruism.** Some apartment rents include the cost of utilities. Others do not. What are the advantages of each system to the tenant? To the landlord? In which type of system would a person be more likely to turn the thermostat down? Discuss.

*16. **Usage of utilites.** About 10 percent of the gas consumed by a household kitchen stove with a pilot light is used to maintain the pilot light. Water that is simmering is just as hot as water that is boiling vigorously. An egg covered by half a pint of boiling water will harden just as rapidly as in a quart. Outline specific guidelines for a family of four that might cut its consumption of natural gas in half.

17. **Legal approaches to environmental control.** List four types of laws that could be enacted to control pollution. Discuss the relative merits of each.

18. **Damage suits.** How can legal suits against polluters be used? What are the drawbacks of depending on court action for environmental control?

19. **Legal standing.** Which of the following people or groups would be likely to have legal standing? (a) A person claims that air pollution from electric generating plants is changing world climate. He says that changes in climate might eventually upset agricultural productivity and raise the price of bread. (b) A group of commercial fishermen claim that water pollution from a nearby factory has killed the fish in a river. (c) A person states that the noise from a steel mill is disturbing his peace of mind. (d) A person living in Los Angeles complains that the Alaskan pipeline is destroying the wilderness quality of Alaska.

*20. **Court battles.** Imagine that there are ten factories in your neighborhood and each one is polluting the air. Imagine further that this pollution has made you sick. Discuss some of the complexities involved in suing the factories and forcing them to reduce pollution.

*21. **Legislation.** Discuss the following alternative proposals for reducing water consumption. (a) Tax all water usage at a constant rate per gallon. (b) Tax water usage at a progressive rate; that is, allow the tax to rise with increasing use. (c) Shut off all water from 2 to 4 P.M. (d) Shut off all water from 2 to 4 A.M. (e) Ration water at some reasonable level. What is the purpose of the legislation? What are the side effects? Can you propose better legislation?

22. **Public policy.** If you were the mayor of a small town, and a prosperous factory, the largest single employer in the town, were illegally dumping untreated wastes into the stream, what action would you recommend? What if the factory were barely profitable?

*Indicates more difficult problem.

QUESTIONS FOR CLASS DISCUSSION

1. How do you travel to and from school? Can you suggest ways for decreasing consumption in your personal transportation habits? Discuss with your classmates.

2. The cost of environmental control falls more heavily on some people than on others. Do you think it would be fair for the government to pay people who are hurt the most? Would you favor giving compensation to workers who lose their jobs because of environmental control rules? To companies whose costs for environmental cleanup are high? To stockholders whose investments are reduced in value? Do you feel that these payments might be opened to graft and corruption? What benefits to the environments would such payments provide?

3. When animals are slaughtered for human consumption, various wastes are produced. These include fat and bones from cattle, chicken feathers and entrails, blood, and unused fish parts. These wastes, when discarded, are a large source of water pollution. Some factories, called **rendering plants,** recycle these materials by converting them to tallow (which is used to make soap) and to animal feed products. However, many rendering plants discharge odorous pollutants into the atmosphere. Discuss the economic externalities of the recycling operation. If a rendering plant pollutes, should it be shut down? What if it cannot afford pollution controls? Should the government subsidize rendering operations or offer low term loans for pollution control? Discuss.

BIBLIOGRAPHY

An excellent general reference that covers most aspects of the field of environmental science is:
Paul R. Ehrlich, Ann H. Ehrlich, and John P. Holdren: *Ecoscience.* San Francisco, W. H. Freeman and Co., 1977, 1051 pp.

A good introduction to environmental economics is given by:
Donald T. Savage, Melvin Burke, John D. Coupe, Thomas D. Duchesneau, David F. Wihry, and James A. Wilson: *The Economics of Environmental Improvement.* Boston, Houghton Mifflin Co., 1974. 210 pp.

For an emphasis on natural resources refer to:
Ferdinand E. Banks: *The Economics of Natural Resources.* New York, Plenum Press, 1976. 267 pp.

Issues of public policy are covered by:
Daniel H. Henning: *Environmental Policy and Administration.* New York, American Elsevier Publishing Co., 1974. 204 pp.

Citations of legal cases appear in:
Jerome G. Rose (ed.): *Legal Foundations of Environmental Planning.* New Brunswick, N.J., Center for Urban Policy Research (Rutgers University), 1974. 318 pp.

A somewhat apocalyptic view is given by:
Robert L. Heilbroner: *An Inquiry into The Human Prospect.* New York, W. W. Norton & Co., 1974. 150 pp.

Two well-written and provocative books discussing economy and environment are:
Gerald Garvey: *Energy, Ecology, Economy.* New York, W. W. Norton & Co., 1972. 235 pp.
Edwin G. Dolan: *TANSTAAFL: The Economic Strategy for Environmental Crisis.* New York, Holt, Rinehart, & Winston, 1971. 113 pp.

Two engaging volumes on social policies for the environment are:
Garrett Hardin: *Exploring New Ethics for Survival.* New York, Viking Press, 1972. 273 pp.
Garrett Hardin and John Baden: *Managing the Commons.* San Francisco, W. H. Freeman and Co., 1977. 294 pp.

Reference books can never be quite current: For recent information on changes and advances in environmental studies refer to the following weekly and monthly journals:
Chemical and Engineering News. The American Chemical Society.
Environmental Science and Technology. The American Chemical Society.
Science. The American Association for the Advancement of Science.
Scientific American. Scientific American, Inc.

2
THE ECOLOGY OF NATURAL SYSTEMS

2.1
INTRODUCTION

Environmental science is a study of **systems.** A system is a collection of parts that act together in some way. Think of the human body. The heart, liver, brain, and stomach are all separate organs, yet none of them can act properly unless the entire system is functioning. The systems studied in environmental science vary widely in form and function. Yet they all share some similarities. Each one is far more than an independent collection of parts. If one part of a system is disturbed, the entire system is affected. If a single pipe carrying cooling water through a nuclear power plant bursts, various safety systems will automatically go into operation. It is even possible that the entire plant may be shut down. If the average temperature in a mountain stream rises a few degrees, the types of plants and animals living in the stream may be altered.

Not all changes affect a system equally. Sometimes a small disturbance is barely noticeable. You can cut a person's hair and he or she will still act normally. But in other situations, a seemingly small change can affect the system seriously. Throughout this book we will study systems. One important part of this study will be to observe how systems respond to changes.

No system is completely independent. A person must eat, drink, breathe, and eliminate wastes. All these activities involve exchanges with the environment. Similarly, a factory, which is a complex system, can operate only if raw materials are brought in and finished products are shipped out. Life on an island in the middle of the ocean couldn't survive without sunlight, rain, and an exchange of atmospheric gases.

An **ecosystem** is a system that includes plants and animals plus the physical environment in which they live. An ecosystem is defined to be nearly self-contained. Most matter is repeatedly recycled, and only small quantities flow into and out of the system. **Ecology** is the study of the flow of energy and materials within an ecosystem and the interactions between the plants and animals that live there.

Ecosystems differ widely with respect to size, location, weather patterns, and the types of animals and plants that live in them. A watershed in New Hampshire, a Syrian desert, the Arctic ice cap, and Lake Michigan are all distinct ecosystems. Common

A predator is an organism that eats other animals. Thus a coyote is a predator of mice, many types of birds are predators of insects, and lions are predators of zebras.

to them all is a set of processes. In each there are plants that use energy from the sun to convert simple chemicals from the environment into complex, energy-rich tissues. Each houses various forms of plant-eaters, predators who eat the plant-eaters, predators who eat the predators, and organisms that cause decay.

2.2
ENERGY RELATIONSHIPS WITHIN AN ECOSYSTEM

What is energy?

All systems need energy to function. Whenever something speeds up or slows down, or is heated or cooled, or whenever chemical reactions occur, exchanges of energy are involved. Energy is needed to accelerate a sports car, boil a cup of water, or build body tissue. Energy is released when a moving sports car slows down, water cools, or a log burns.

We must understand the difference between matter and energy. Coal, diamond, carbon dioxide, sugar, wood, metal, plastic, glass, muscle tissue, and soil are all forms of matter. Think of a lump of coal. Coal is mainly carbon. If we heat and compress this coal under extreme conditions, it will change to diamond. You may be surprised to learn that a diamond can burn. When diamond is burned, carbon dioxide is released. A plant can absorb this carbon dioxide and use it to produce sugar and other plant tissue. Sugar contains carbon. If the plant is buried in the mud for several million years, the tissue may be converted back to coal. The carbon may be used over and over again indefinitely. It may change form continually, but it is never lost or used up.

Matter can be recycled. Energy cannot. Although energy is never used up or lost, it may become unavailable for certain processes. A lump of coal can be burned in a steam locomotive. The heat from the fire drives the engine and supplies the energy to move the train. Suppose a person drove a train from Paris to Amsterdam. When the

A moving freight train has energy of motion.

This pile of coal can burn to produce heat. It has chemical potential energy.

A campfire is producing heat energy.

engine arrived in Amsterdam, the coal would be gone. But what happened to the energy? Could you somehow find it, save it, and use it to drive the train back to Paris? The answer is no. The heat from the coal was spread out into the environment. The air between Paris and Amsterdam was warmed slightly. But it would be impossible to reuse this heat to perform useful work. This very important concept will be discussed in more detail in Chapter 7. For now, remember that *materials can be recycled, but energy cannot.*

Energy lost to the environment as heat cannot be recycled to drive
the locomotive back to Paris.

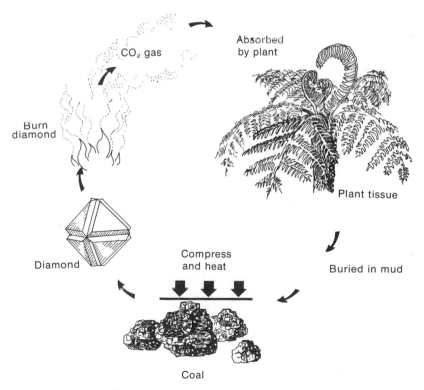

Carbon can be recycled over and over again.

Photosynthesis (fō′tə sin′ thi sis) is a compound word that comes from two Greek words. *Photo* comes from a Greek word meaning light, and *synthesis* is from a Greek word meaning to put together. In science, to synthesize is to form a chemical compound by altering or combining other compounds. So photosynthesis is a chemical synthesis initiated by light.

Photosynthesis

Since energy cannot be reused, ecosystems need a continuous supply of energy to survive. Almost all of the energy available to us on Earth comes from the Sun. Sunlight evaporates water that later falls as rain. The Sun heats the Earth unevenly, causing winds to blow. Sunlight strikes the leaves or needles of green plants. Plants are able to trap some of this energy and transform it into chemical energy. This process, known as **photosynthesis,** is illustrated in the equation below.

$$\text{carbon dioxide} + \text{water} \xrightarrow{\text{gives, in the presence of sunlight}} \text{sugar} + \text{oxygen}$$

$$6\ CO_2 + 6\ H_2O \xrightarrow{\text{sunlight}} C_6H_{12}O_6 + 6\ O_2$$

To understand the process of photosynthesis better, think of the following story. A farmer takes a small seed, say a pumpkin seed, and plants it in a pot of dry soil. She weighs the soil and the seed together. Then she waters her garden and lets the pumpkin plant grow. Huge vines spread and large orange pumpkins appear at the ends of the vines. But the soil does not disappear. If we dry the grown plant with the soil and weigh the entire system, we will find that it is much heavier than the original soil. But we know that the matter in the plant had to come from somewhere. The green leaves of the pumpkin trapped carbon dioxide from the air and water from the soil and combined these two compounds together to form sugar and other plant tissues.

What happened to the energy of the sunlight? It was trapped in the plant tissue. If we burn a pumpkin, it will produce light and heat, which is a way of releasing the stored chemical energy (see Fig. 2–1).

Carbon dioxide and water are low-energy compounds; they cannot be burned to produce heat. Sugar, the product of photosynthesis, contains stored chemical energy. We can burn sugar to produce heat. Carbon dioxide and water are released as byproducts.

$$\text{sugar} + \text{oxygen} \xrightarrow{\text{burns to produce}} \text{carbon dioxide} + \text{water} + \text{energy}$$

$$C_6H_{12}O_6 + 6\ O_2 \longrightarrow 6\ CO_2 + 6\ H_2O + \text{energy}$$

Respiration

Plants and animals use the energy stored in sugar and other plant matter to maintain body functions. This process is called **respiration** (res′ pə rā′ shən). Think of respiration as a slow, controlled burning. In fact, the equation describing this process is the same as the equation for a fire.

$$\text{sugar} + \text{oxygen} \xrightarrow{\text{respiration}} \text{carbon dioxide} + \text{water} + \text{energy}$$

$$C_6H_{12}O_6 + 6\ O_2 \longrightarrow 6\ CO_2 + 6\ H_2O + \text{energy}$$

We see that plants obtain their energy from the Sun. Animals eat the

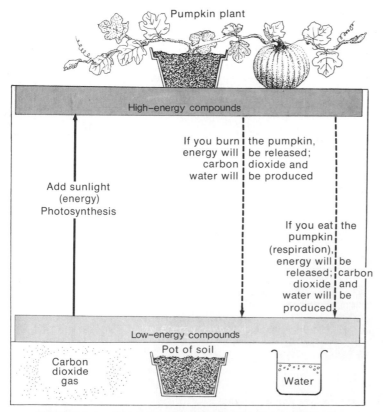

FIGURE 2–1.   Some energy relationships of a pumpkin plant.

plants and obtain their energy from them.

Energy Utilization in an Ecosystem

Consider an ecosystem that receives 1000 Calories of light in a given day (see Fig. 2–2). Most of this energy is not absorbed at all. Some is simply reflected back into space. Of the energy that is absorbed, most is stored as heat or used for evaporation of water. Plants also need energy to stay alive. (Of course, a person must eat even if he wishes to remain at constant weight. This energy is used to maintain proper body functions.) Plants, too, need energy to maintain their tissue. Some of the Sun's energy is used

Calorie (C) and calorie (c) are units of energy. 1 calorie (c) is the energy required to raise the temperature of 1 gram of water 1 degree Celsius. 1 Calorie (C) is the energy required to raise the temperature of 1 kilogram of water 1 degree Celsius. Therefore, 1 Calorie = 1000 calories. The large Calorie unit is also called a kilocalorie. (See Col. 1)

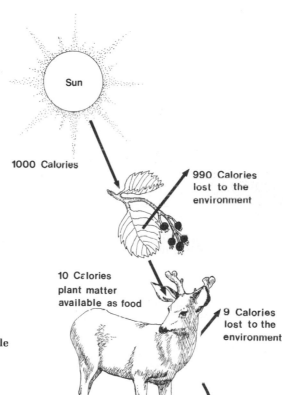

1000 Calories

990 Calories lost to the environment

10 Calories plant matter available as food

9 Calories lost to the environment

1 Calorie available as food

FIGURE 2–2. Energy flow through a simple food chain.

For the details of a study of the flow of energy in a forest ecosystem see James R. Gosz, Richard T. Holmes, G. E. Likens, and F. Herbert Bormann: "The Flow of Energy in a Forest Ecosystem." *Scientific American,* March 1978, page 93.

for this purpose. Of the original 1000 Calories, about 10 Calories are stored in the plant tissue as energy-rich material, which animals can use for food. What happened to the other 990 Calories? Where did they go? They are dispersed into the air in unusable forms just as the energy from a lump of coal that drives a locomotive is dispersed.

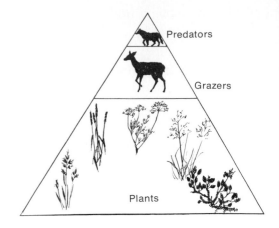

1000 Calories of sunlight $\longrightarrow$ 990 Calories lost to the environment + 10 Calories stored as plant tissue

Now suppose an animal, say a deer, eats a plant containing 10 Calories of food energy. The deer would use about 90 per cent of the energy to move about and maintain body functions. As a result, only about 1 Calorie would be stored in the form of body weight.

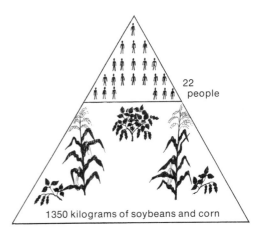

10 Calories of energy-rich plant tissue $\longrightarrow$ 9 Calories lost to the environment + 1 Calorie stored as animal tissue

A carnivore is an animal that eats the flesh of other animals. Carnivores can be large, like a tiger, or small, like a ladybug, or even a microscopic protozoan.

A mountain lion eating the deer is likewise inefficient in converting food to body weight, so the energy available to the carnivore is even less.

Food Pyramids

There is a finite amount of solar energy available to any ecosystem. Some of this energy is used by plants. Animals, too, need energy, but they cannot use sunlight directly. We learned above that the amount of usable energy *decreases* as it is transferred from sunlight to plants to animals. Therefore animals have *less* useful energy available to them than plants do. Because there is less energy available to animals, the total mass of grazing animals will generally be less than the total mass of plants in an ecosystem. Similarly, carnivores eat grazing animals, so there is a smaller mass of carnivores than grazers. These relationships are shown in Figure

Mass is a measure of the amount of matter. On the surface of the Earth, mass is proportional to weight.

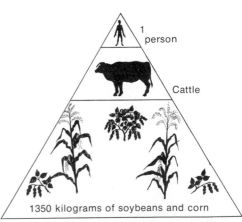

FIGURE 2–3. Food pyramids. *A,* In most ecosystems there is a greater mass of plants than grazers and a greater mass of grazers than predators. *B,* 1350 kilograms of plant matter can support 22 vegetarians. *C,* 1350 kilograms of plant matter can support only one person who eats meat.

2–3. These drawings are called **food pyramids.**

Suppose that a farmer wishes to plant a crop of soybeans and corn. The farmer can either eat the vegetables directly, or feed them to a cow

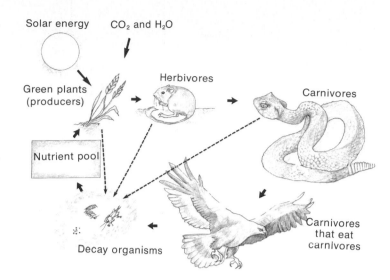

FIGURE 2–4. Simplified model of a terrestrial ecosystem.

and then eat the cow. It is obvious from the previous discussion that the Sun's energy is used most effectively if people eat plants. If vegetable crops such as soy beans and corn are eaten directly, 1350 kg of harvest can support about 22 people. If the plants are fed to cattle first, and the meat is eaten, only one person can be supported.

Food Webs

Natural ecosystems contain many different types of organisms. Plants trap the Sun's energy. Various types of grazing animals eat plants. Grasshoppers, mice, and deer are all grazers. These animals are eaten by predators. Praying mantises eat grasshoppers, owls eat mice, and mountain lions eat deer. In turn, some predators are hunted. Snakes eat praying mantises, and martens eat owls. Many species eat both plant and animal matter. These animals are called **omnivores.** Bears, rats, pigs, chickens, grouse, and people are all omnivores.

Of course, some individuals die and are not eaten right away. Instead their tissue remains to decay. The process of

A marten is a small mammal (about 75 centimeters, or 2½ feet, long). Martens live mostly in tress in the northern forests. They have a long glossy coat and a bushy tail. They hunt other small tree-living mammals and birds, such as squirrels, owls, and small land mammals such as rabbits, mice, etc.

Marten. (Courtesy of Fish and Wildlife Service, photo by David G. Allen)

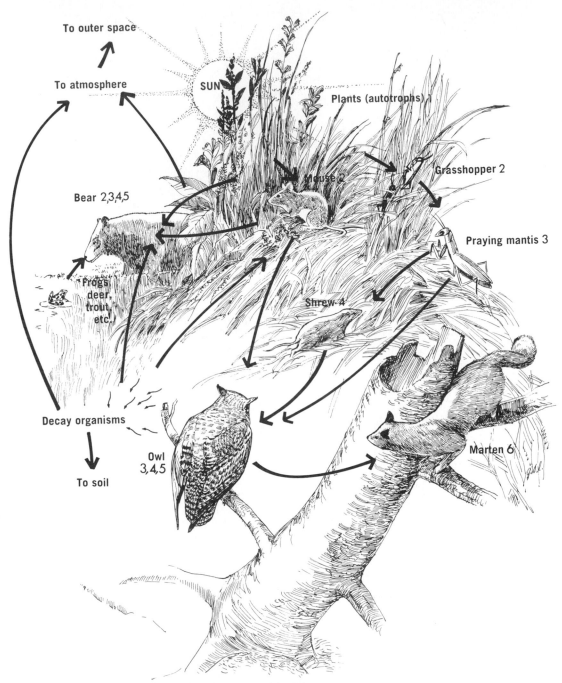

FIGURE 2-5. Simplified land-based food web. Arrows are in the direction of progressive loss of energy available for life processes. Numerals refer to levels of consumption.

decay is carried out by many widely varying types of species. These organisms, both plants and animals, are classified by function and are called **decay organisms.** As we shall see, ecosystems operate properly only if nutrients are recycled. Since decay organisms are responsible for recycling organic matter, they are essential to all natural systems. The complex patterns of consumption in a natural ecosystem are called **food webs.**

2.3
NUTRIENT CYCLES

Energy alone is not enough to support life. Imagine, for example, that some plants such as algae were sealed into a sterilized jar full of pure water and exposed to plenty of sunlight. The plants would use carbon dioxide and release oxygen during photosynthesis.

Soon the carbon dioxide would be used up and they would starve. If the experimenter added carbon dioxide, the plants still would not survive. They would starve for lack of other chemicals necessary for life. The life in the jar might continue, however, if some plant consumer (a snail, for example) were introduced, and the pure water were replaced by pond water to supply nutrients and decay organisms. Assuming that the right number of snails were added, the jar might now perhaps become a balanced, stable ecosystem. Nutrients would be continuously recycled.

Algae use water, carbon dioxide, sunlight, and dissolved nutrients to support life to build tissue. Snails eat the plants. Oxygen is one major waste product of algae and at the same time is essential for snails. In turn, snails release carbon dioxide, which is needed by the algae. When plants and snails die, they are eaten by decay organisms and the nutrients are recycled. We see that the nutrients in the sealed jar cycle from plants to animals to decay organisms and back to plants to be reused again. Thus, the community can exist indefinitely. In fact, sealed aquariums, in which life has survived for a decade or more, can be seen in some biology laboratories.

The Earth itself can be considered to be a sealed jar. Although it receives a continuous supply of energy from outside, it exchanges very little matter with the rest of the universe. Thus, for all practical purposes, life started on this planet with a fixed supply of raw materials. There are finite quantities of each of the known elements. The chemical form and physical location of each element can be changed, but the quantity cannot.*

Plants also digest sugar (respiration), releasing carbon dioxide and water, but they consume more carbon dioxide than they use. (See. Col. 1)

Carbon–Oxygen Cycle

We learned that plants synthesize sugars by combining carbon dioxide with water in the presence of sunlight. Oxygen is discharged as a by-product. But oxygen does not accumulate in the environment. Oxygen and sugar are consumed during respiration, and carbon dioxide and water are released as waste products. This completes the cycle, for carbon dioxide and water may be reused as raw material for new synthesis.

Recently, industrial activity has become an important factor in the carbon-oxygen cycle. People are burning large quantities of fossil fuels. When these fuels are burned, oxygen is consumed and carbon dioxide is released. For example,

Oxygen and carbon are also cycled by geological processes. Carbon dioxide dissolves in sea water and may at some time either become rock (such as limestone) or be utilized to form the skeletons of shellfish or coral. This loss is partially balanced by the action of inland water, which slowly dissolves deposits of limestone on land, releases some carbon dioxide, and carries other compounds to the sea.

$$\text{coal} + \text{oxygen} \xrightarrow{\text{fire}} \text{carbon dioxide}$$

$$C + O_2 \longrightarrow CO_2$$

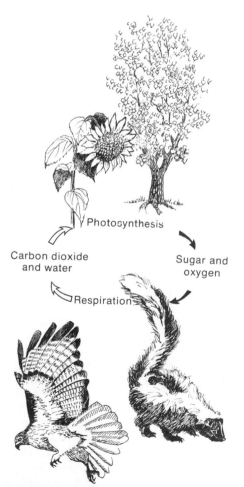

FIGURE 2–6. Biological carbon and oxygen cycle.

Photosynthesis

Carbon dioxide and water

Sugar and oxygen

Respiration

*This statement does not apply to radioactive elements.

It has been shown that the carbon dioxide concentration in the atmosphere has been rising steadily for the past 75 years. Many people believe that this added carbon dioxide may alter the temperature balance of the atmosphere and possibly affect life on Earth. This topic will be discussed further in Chapter 12.

Nitrogen Cycle

All plants and animals need nitrogen to survive. Nitrogen is one of the most common elements in our environment. It is four times more plentiful in the atmosphere than oxygen is. Yet the atmospheric nitrogen (present in the chemical form N_2) that surrounds us all the time cannot be used directly by most organisms. We inhale it with every breath, then exhale it. It is not assimilated by the body. Most plants and animals use so-called fixed nitrogen for growth. Ammonia (NH_3), nitrite (NO_2^-) and nitrate (NO_3^-) are "fixed" forms of nitrogen.
Atmospheric nitrogen (N_2) can be converted to fixed nitrogen in various ways. Certain bacteria and algae are able to fix atmospheric nitrogen. Some of these organisms live free in the soil, while others congregate in colonies on the roots of certain plants called **legumes.** Some common legumes are peas, beans, and alfalfa. Lightning and the action of sunlight on atmospheric nitrogen also help to synthesize fixed nitrogen compounds. In recent years, modern factories have been producing many millions of tons of nitrogen fertilizers.
Fixed nitrogen can be absorbed by plants and then passed into the food web. Animals eat the plants, and in turn are eaten. The nitrogen may now cycle for a considerable time. Organic decay might return it to the soil, or it might be digested and returned to the soil through feces or urine. Once in the soil, the nitrogen may reenter plant systems. If the plants are eaten, the nitrogen will be recycled back to animals. Because recycling is so efficient, an entire ecosystem can usually be supported by a small amount of nitrogen fixation. Eventually, fixed nitrogen is returned to the atmosphere. Denitrifying bacteria and fire are responsible for this process.

In agricultural systems, nutrient cycling is incomplete. A kernel of wheat does not return to the soil. Rather, it is loaded on a truck, shipped to a distant city, and eaten. The nitrogen-rich urine and feces of the person who eats the wheat is dumped into a sewer and is not generally returned to the field. Since grain crops cannot fix nitrogen themselves, farmers must fertilize soil. Nitrogen fertilization can be accomplished by any of three techniques. The simplest method is to fertilize the soil by recycling plant and animal wastes. Another technique is crop rotation. Legumes and grain are planted in alternate years in order to maintain soil nitrogen. Finally, chemists have learned to manufacture fertilizer by producing ammonia. Today, enormous quantities of ammonia are manufactured in this manner. Some people estimate that industrial fixation accounts for one third of the total annual production of nitrogen compounds on the Earth.

atmospheric + hydrogen $\xrightarrow[\text{processes}]{\text{industrial}}$
nitrogen

ammonia (a form of fixed nitrogen)

$$N_2 + 3\,H_2 \xrightarrow{\hspace{3cm}} 2\,NH_3$$

Atmospheric nitrogen (N_2) is plentiful but not usable by most plants and animals	It is converted to →	"fixed" nitrogen necessary for the growth of plants and animals

To "fix" means to make firm or stable. In this sense, a gas (which is not firm) can be fixed by binding it in some form of solid or liquid.

Mineral Cycles

Gases such as oxygen, carbon dioxide, and nitrogen are available in large quantities in the atmosphere. Minerals such as phosphorus, calcium, sodium, potassium, magnesium, and iron are generally less plentiful in most ecosystems.

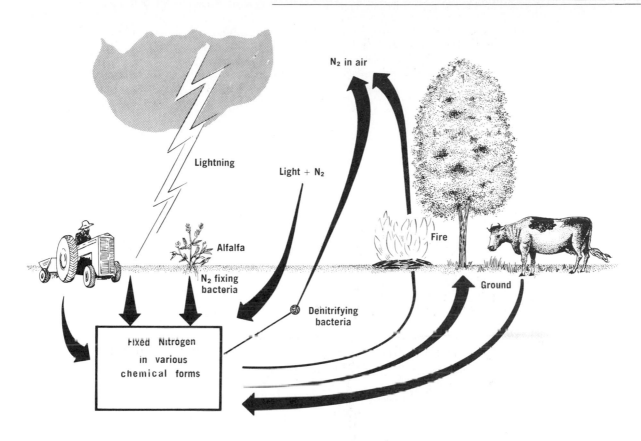

FIGURE 2–7. Pathways whereby nitrogen is removed from and returned to the atmosphere (nitrogen cycle).

The naturalist Aldo Leopold, in his book, *A Sand County Almanac*, writes a hypothetical story of a mineral atom, X. "X had marked time in the limestone ledge since the Paleozoic seas covered the land. Time, to an atom locked in a rock, does not pass.

The break came when a bur-oak root nosed down a crack and began prying and sucking. In the flash of a century the rock decayed, and X was pulled out and up into the world of living things. He helped build a flower, which became an acorn, which fattened a deer, which fed an Indian, all in a single year.

From his berth in the Indian's bones, X joined again in chase and flight, feast and famine, hope and fear. When the Indian took his leave of the prairie X moldered briefly underground, only to embark on a second trip through the bloodstream of the land.

This time it was a rootlet of bluestem that sucked him up and lodged him in a leaf that rode the green billows of the prairie June, sharing the common task of hoarding sunlight.... A forehanded deermouse cut the leaf in which X lay and buried it in an underground nest, as if to hide a bit of Indian summer from the thieving frosts. But a fox detained the mouse, molds and fungi took the nest apart, and X lay in the soil again, foot-loose and fancy-free....

One year, while X lay in a cottonwood by the river, he was eaten by a beaver.... The beaver starved when his pond dried up during a bitter frost. X rode the carcass down the stream.... He ended up in the silt of a backwater bayou, where he fed a crayfish, a coon, and then an Indian, who laid him down to his last sleep in a mound on the riverbank. One spring the river caved the bank, and after one short week X lay again in his ancient prison, the sea."

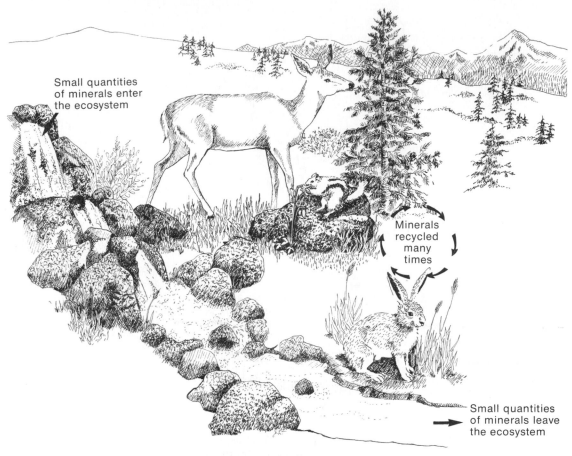

Small quantities of minerals enter the ecosystem

Minerals recycled many times

Small quantities of minerals leave the ecosystem

FIGURE 2–8. Mineral cycles.

For details of nutrients recycling in a natural ecosystem, refer to Chapter 4 of G. M. Van Dyne, *The Ecosystem Concept in Natural Resource Management*. (See references at the end of this chapter.)

Minerals enter a system after they have been dislodged from rocks by wind, ice, flowing water, and natural chemicals. They are removed by the action of flowing water. However, input and output rates are generally low. This means that life in an ecosystem depends on efficient recycling of nutrients. Each incoming mineral atom undergoes countless transformations from soil to plant tissue, from plant tissue back to the soil or to animal tissue and then to the soil, and around again and again before being washed out of the system.

2.4
MAJOR ECOSYSTEMS OF THE EARTH

The **biosphere** is the region including all the life-supporting portions of our planet and its atmosphere. This section briefly describes some of the major ecosystems of the biosphere.

Ocean Systems

The primary plants in all marine ecosystems are **phytoplankton** (fī′ tə plank′ tən). These are small, one-celled chlorophyll-bearing organisms suspended in the water. Phytoplankton are eaten mainly by small grazers called **zooplankton** (zŏ′ ə plank′ tən) (Fig. 2–9). These animals range in size from roughly 0.2 mm (0.008 in) to 20 mm (0.8 in) in diameter. There are few large grazers in the sea that would be comparable to land-based bison, deer, or cattle.

Many large omnivores and predators live in the sea. These species include fish, mammals, invertebrates, and reptiles. Waste products of dead and dying plants and animals sink downward toward the ocean floor. Many species live on the bottom and receive their food from above. Some of these are plants, others semimobile or immobile animals. Still others crawl or swim actively in search of food.

Nutrient cycles do not operate efficiently in the central oceans. Only the surface layers of the sea receive enough light for photosynthesis to occur. But nutrients fall downward to the bottom. Since most of the nutrients lie on the bottom, and the sunlight is on the top, relatively little life can survive in the central oceans. These regions have been likened to a great desert. In a few places, ocean currents carry nutrients to the surface providing food for plankton. Fish and other sea creatures that eat the plank-

A seal sits on the shore of the Pacific Ocean. (Photo courtesy of Amos Turk)

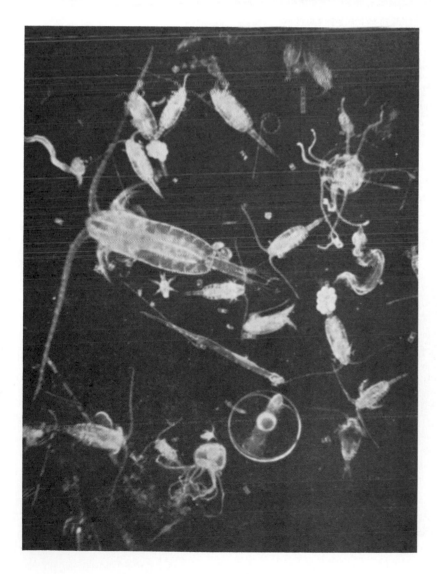

FIGURE 2–9. Living zooplankton. The shrimplike animals in the picture are various small crustaceans. Two tiny jellyfish with long tentacles are also seen (magnification 16×). (From A. Hardy: *The Open Sea.* London, William Collins Sons, 1966)

33

Salt marsh estuary system.

ton grow in abundance in these regions. Fish are also more numerous along the coast. Here, wave action carries bottom nutrients to the surface. Other nutrients are brought seaward by rivers, and a relative shallowness of the water helps to provide a series of rich ecosystems.

Estuary Systems

Coastal bays, rivers, and tidal marshes all lie next to the land. These regions are called **estuary** (es′ choo er′ ē) systems. Fresh water from rivers and streams brings a large load of nutrients to estuaries. Seaweed and marsh grasses grow in the shallow waters. Many fish live in these waters, and many more lay their eggs there because the newly hatched young can (a) take advantage of the high nutrient concentrations, (b) find shelter among the reeds and seaweeds, and (c) return easily to the open ocean when they grow to adult size. Therefore, estuaries are nursery grounds for many deep-water fish.

Alpine tundra in early spring in the Colorado Rockies.

Freshwater Systems

Lakes, ponds, and rivers are ecologically similar to ocean systems. Again, the food web starts with plankton and ends in large predators. Freshwater systems, unlike the ocean, are continuously fertilized by nutrients from the nearby soil. Because bodies of fresh water are shallower than the ocean, rooted plants, marsh grasses, lilies, and algae are much more important in the food web.

Terrestrial Systems

There are many different types of large stable terrestrial (land-based) ecosystems. Let us examine a few of the most important ones.

In the coldest regions, where the temperature is below freezing for much of the year, trees cannot survive. Low lying plants such as mosses, grasses, flowers, and a few small bushes cover the land. These ecosystems are called **tundra.**

Forests grow in warmer climates where there is ample rainfall. Some forests consist mainly of evergreens such as spruce and fir. Others are composed of broad-leafed trees like oak and maple. Near the Equator, tropical rain forests grow.

In the temperate areas where there is not enough rainfall to support forests, the stable ecosystems are **prairies** or **grasslands.** In areas, the virgin grasses of North America were as tall as a person. Elsewhere, they were short enough that the early scouts could "see the horizon under the bellies of millions of buffalo."

In very dry areas, where rainfall is less than 25 cm (10 in) per year, the ecosystems are **deserts.** Some deserts are barren; others are able to support some scrub brushes and cactus. They can be hot, as in Nevada, or relatively cold, as in eastern Washington.

2.5
ECOSYSTEMS AND NATURAL BALANCE

If we place a few square centimeters of grassy sod and a grasshopper together in a sealed aquarium, the grasshopper will eventually eat all the available grass and then die of starvation. But in a large prairie ecosystem, the grasshoppers do not eat all the grass. Studies have shown that under normal conditions 80 to 90 per cent of the plant matter falls to the ground uneaten. The 10 per cent to 20 per

A

B

C

A, Lightly grazed grassland in the Red Rock Lakes National Wildlife Refuge, Montana, with a small herd of prong-horns. *B,* Short-grass grassland, Wainwright National Park, Alberta, Canada, with herd of bison. *C,* A desert region in Arizona. (From Odum: *Fundamentals of Ecology,* 3rd ed. Philadelphia, W. B. Saunders Co., 1971)

cent that is eaten is shared by insects, birds, reptiles, rodents, large mammals, and other grazing animals. Only rarely does the population of grazers increase enough to eat most of the plants. Yet each individual animal tries to obtain as much food as it needs. How are ecosystems controlled? How can a natural meadow, lake, or forest exist in an orderly and relatively unchanged state for years?

Many opposing forces operate within a natural ecosystem. Animals eat and in turn are eaten. Fertility rates vary. Migration is common. Weather varies and climates change. Moisture and nutrients travel into and out of the soil. The net effect of all these events happening together is that in general most systems remain stable. If an ecosystem is disrupted, the system tends to regain balance. Such a tendency is called the balance

A redwood forest in northern California.

of nature, more formally **ecosystem homeostasis** (hō mē ə stā′ sis). For example, when there is drought in a grassland, plants do not grow well. The meadow mice that eat the grass become malnourished. When this happens, their fertility rate drops and their birth rate decreases. The hungry mice retreat to their burrows and sleep. They need less food and are less exposed to predators while they sleep, so their death rate decreases. Their behavior protects their own population balance as well as that of the grasses which are not consumed by hibernating mice.

Let us consider as another example the relative populatons of tigers, grazing animals, and grass in a valley in Nepal. Assume that the mountains surrounding the valley are so high that no animals can enter or leave the valley. One year rainfall is limited and mild drought conditions exist. Because water is naturally stored in pools, the drinking water supply is adequate. However, the grasslands suffer from the lack of rainfall. Therefore, food is scarce for the grazing animals, and many are hungry and weak. Such a situation is actually beneficial to the tigers because hunting becomes easier, and the tiger population thrives.

The following spring, when rains finally arrive, the grazing population is low. This permits the grass to grow back strongly because not much of it will be eaten. However, because the food supply for the tigers is low (and the grazers that are left are the strongest of the original herd), the tigers suffer a difficult time the summer after the drought. The third year there are fewer tigers, so the herd resumes its full strength and balance is achieved again.

Stable ecosystems, if examined superficially, do not seem to go out of balance at all. Actually, their homeostatic mechanisms work so well that slight imbalances are corrected before they become severe. If severe imbalances are prevented, it is very likely that the system will last for a long time.

A wonderful example of a stable system is a redwood forest. To the hiker entering the redwood forest on a hot summer day two characteristics of the environment are immediately obvious. It is cool, and there is a thick floor of spongy matter. The coolness, caused by the extensive shade of the tall trees, reduces water loss due to evaporation. The spongy floor is formed by a large bed of decomposed or partly decomposed matter such as fallen needles and rotting wood. This material holds the rainwater and thus serves as a reservoir for water and nutrients. Thus a mild drought does not kill the trees. In addition, a large amount of organic matter insures a steady functioning of the recycling system. The coolness of the forest offers still another advantage. Clouds passing over a cool air mass are likely to condense into droplets and fall as rain. Thus, the redwood forests tend to cause rain, restrict evaporation, and retain a substantial supply of water.

Redwood forests also protect themselves from fire. There is a large redwood forest on the west side of Route 1 as it runs through the town of Big Sur, California. Some time ago a tremendous fire swept through the forest. However, because the bark and the wood of the redwood tree are fire-resistant, the big trees were scarred but not killed. Because the forest environment was maintained, the forest community quickly grew back. If a fire of equal intensity had burned through a stand of pine, the trees would have died and many more years would have been needed for the forest to be rebuilt.

Not all ecosystems stay in balance all of the time. As we will see in the next chapter, imbalances do occur. Fires destroy the landscape, populations explode, and famine occurs in many systems.

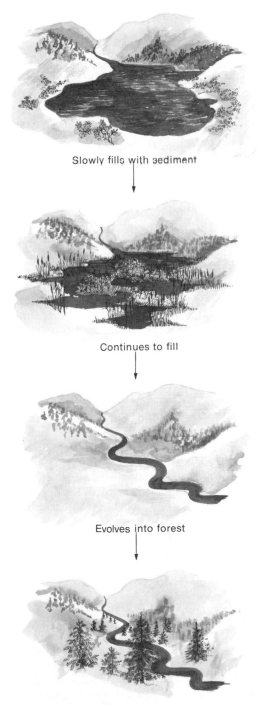

Slowly fills with sediment

Continues to fill

Evolves into forest

FIGURE 2-10. **Schematic view of one possible successional sequence.**

2.6
NATURAL SUCCESSION

Even well regulated systems change slowly with time. **Natural succession** is defined as the series of changes through which an ecosystem passes as time goes on. The **climax** is the final stage, the stage that is "unchanging." Of course, the word "final" is used with reservation, because the slow process of evolution changes everything. The composition of the climax depends on temperature, altitude, seasonal changes, and patterns of rainfall and sunlight.

Most systems that appear to be stable if we look at them for a few weeks or years are actually changing. A small lake is often used as an example of a stable ecological system. Plants, large animals, and microorganisms all exist in balanced relationships. Temporary imbalances occur and are adjusted, as in all natural systems. However, for most lakes the incoming streams and rivers bring more mud into the environment than is removed by the outgoing streams. The effect is small; it may not be noticed in one human lifespan. If you made a short-term study of the ecology of a lake, you probably would conclude that the ecosystem was in balance. But in time, the lake begins to fill up

A marsh evolving into a meadow. (From *North American Reference Encyclopedia.*)

A meadow evolving into a forest.

with mud. The vegetable and animal life changes. New plants appear that can root in the bottom mud and extend to the surface where light is available. The trout give way to carp and catfish. If an ecologist studies the lake at this stage he or she might again say that it was a stable system. Again, minor imbalances and adjustments could be observed, but the overall system appears stable. Yet mud continues to flow into the lake. In addition, it is common for the plants in the lake at this stage to produce more food than is consumed. Therefore, the bottom fills up with decayed plant matter. Eventually the lake may become so shallow that marsh grass can grow.

A marsh is also considered to be a stable system, but in most cases it gradually evolves into a meadow. Plant matter accumulates because growth is rapid and decay in the marsh systems is slow. The accumulated litter gradually fills in the marsh, which slowly becomes a meadow.

The meadow, too, may change. If the climate is right, trees can start to grow. First, shrubs appear, then quick-growing soft woods like birch, poplar, and aspen. The soft woods are replaced by pine; and finally, in what is called the climax, the pine is replaced by hardwoods.

We have shown an example of a system that succeeded from a lake to a marsh to a meadow to a forest. Not all systems, however, follow the same sequence. For example, the Great Plains that once stretched unfenced from the eastern slope of the Rocky Mountains almost to the Mississippi River never became forest because the rainfall wasn't sufficient. Instead the grasslands evolved into a climax system.

2.7
CASE HISTORY: THE EVERGLADES

The **Everglades** is a broad, swampy region of southern Florida (see Fig. 2–11). The Seminole Indians called it *Pay-hay-okee*, the river of grass. Indeed it is a large, shallow, slow moving river that starts from the shores of

Lake Okeechobee and flows approximately 160 kilometers (100 miles) into the ocean. The waters do not travel between well-defined river banks. Rather they flow through a series of swamps that is nearly 60 kilometers wide in places and slowly wind their way to the sea. One region of this swamp is covered mostly by low-lying plant matter called sawgrass.* Here and there in the sawgrass plain there are tree islands where dense stands of trees and bushes grow. These tree islands serve as nesting grounds and shelter for many species of birds, mammals, and reptiles.

South Florida has both a rainy and a dry season. During the rainy season the "river" level is high and the entire sawgrass plain is covered with about a meter (3 feet) of water. During the season of drought, some water flows in channels, but the sawgrass plain is dry.

Animals and plants have adapted uniquely to the Everglade system. During the wet season plant growth is quite rapid. The sawgrass grows in abundance. If this growing season were to continue year round, the Everland marsh would probably fill in with plant matter and succeed into a forest system. But in the winter time the water level starts to fall, and the grasses dry up. Almost every year, fires started by lightning or by people race across the plain. The timing of these fires appears to be crucial to the Everglade cycle. The sawgrass becomes dry and yellow during the start of the drought, even when there still may be a few centimeters of water left on the ground. If the grasses burn at this time the fire does little permanent damage to the swamp. The roots of the plants are covered with water and do not burn. The tree islands are usually safe from early winter fires. At this time there is still enough water so that the plants are damp and resistant to fire. The grass fires move rapidly across the plain and burn around the tree islands but do not consume them.†

---

*Actually, the sawgrass is not a true grass, but a type of sedge. Sedges look like grasses, but the structure of their stems is different.

†Occasionally large fires do destroy the tree islands.

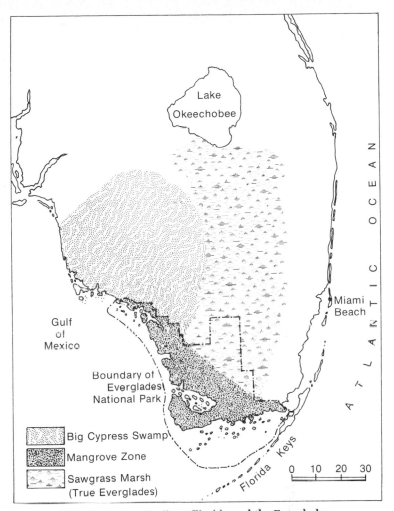

FIGURE 2–11. Southern Florida and the Everglades.

As the yearly drought continues, the waters recede still further. If not for another peculiar ecological adaptation, the swamp would dry up and most of the animals in it would die. It is the alligators that save the swamp. These animals scoop out large depressions in the marsh with their tails. Water collects in these "gator holes." When the plain dries up the fish seek the deep water of the holes to survive the drought. In fact, much of the aquatic life of the region becomes concentrated within the gator holes to survive until the next rainy season. Fish live and breed here. They also serve as food for the alligators as well as for predator birds and mammals.

We see that the ecology of the Everglades is complex. Plants and animals

The Everglades. This river of grass, or Payhay-Okee as the Seminoles call it, is a strange mix of temperate and tropical zones. Forest and jungle, fresh and salt water ecosystems survive together in the 2000 square miles that comprise the national park. The glades are home to rare species of wildlife, such as the crocodile, manatee, and wood stork, which are found nowhere else in the United States. (Courtesy of National Park Service)

Female alligator and young resting in a pool in the Everglades. (Courtesy of National Park Service, photo by Richard Frear)

Common egrets, wood storks, and white pelicans feeding in Everglades National Park. (Photo by George Laycock, Photo Researchers, Inc.)

have adapted to cycles of seasonal growth, fire, and drought. Alligators alter their own physical environment. In doing so they have also insured the continuation of other forms of animal life. The existence of all species is interconnected.

At the present time, civilization threatens this delicate cycle. Dairy farms, sugar cane fields, and orange groves lie just north of the Everglades. When farmers fertilize their crops, some of the fertilizers spill into local streams and eventually enter the sawgrass marsh. Sewage runoff from the cities also flows into the Everglades and fertilizes it. How can fertilizer harm an ecosystem? If the grasses and other plants grow too rapidly, the delicate relationships among plants and animals may be disturbed. Excess plant matter might choke the gator holes and fill the marsh with litter, algae, and weeds. Then the alligators, fish, and birds might die during the dry season. The plant-animal balance could be altered and the glades could be changed forever.

Pesticides also leak into the Everglades, killing plankton, fish, reptiles, and birds.

But perhaps the most serious threat to the marsh comes from water and flood control projects. The Everglades is dependent on a seasonal cycle of flood and drought. If the land did not flood during the wet season, the grasses and trees would not grow. If there were no drought, there would be no fire, and the marsh would slowly fill and die. But if the drought were too severe, the gator holes would dry up, and the animals would die. At the present time large quantities of water are pumped from Lake Okeechobee for irrigation and domestic use in the fields and cities of south Florida. Some of this water never returns to the Everglades. The United States Army Corp of Engineers has built a series of canals, levees, and water control gates in the marsh. They can reduce or even stop the flow of water to the southern Everglades. In fact, when a drought struck Florida from 1961 to 1965 most of the available water was pumped to nearby cities and farms, and millions of animals died. A public outcry forced the Water Control Bureau to reflood the glades. The threat, however, still exists.

The Everglades is a unique and fragile ecosystem. It is a product of millennia of co-evolution of plants and animals with their physical environment. If excess pollution or improper control destroys the swamp, it will most probably be gone forever. It would be very unlikely that it could ever be re-created.

CHAPTER SUMMARY

2.1 INTRODUCTION

A **system** is a collection of parts that acts together in some way. An **ecosystem** is a system composed of plants and animals plus the physical environment with which they interact. **Ecology** is the study of the flow of energy and materials within an ecosystem.

2.2 ENERGY RELATIONSHIPS WITHIN AN ECOSYSTEM

Energy is the capacity to do work or to transfer heat. Energy cannot be recycled; matter can.

Plants trap sunlight to build energy-rich sugars from carbon dioxide and water, which are energy-poor compounds. Oxygen is released as a byproduct. This process is called **photosynthesis.**

$$\text{Carbon dioxide} + \text{water} + \xrightarrow{\text{sunlight}}$$
$$\text{sugar} + \text{oxygen}$$

The energy stored in sugars is released during **respiration**.

$$\text{Sugar} + \text{oxygen} \longrightarrow$$
$$\text{carbon dioxide} + \text{water} + \text{energy}$$

If 1000 Calories of sunlight strike a plant leaf, only 10 Calories are converted to energy-rich tissues. An animal that eats this plant stores only 1 Calorie; the rest is released as heat. A predator that eats the grazing animal stores even less energy.

Food web is a description of the patterns of consumption in an ecosystem. Food webs are complex. There are always fewer meat-eaters than plant-eaters.

Food pyramid is a description of the

masses of plants and animals in an ecosystem. There is a smaller mass of grazing animals than plants, and a smaller mass of predators than grazers.

2.3 NUTRIENT CYCLES

Nutrients are continuously cycled and recycled within an ecosystem.

Carbon dioxide and oxygen are exchanged during photosynthesis and respiration.

Atmospheric nitrogen cannot be used by most plants and animals. It must be first "fixed" by specialized bacteria (many are found on the roots of **legumes),** lightning, the action of sunlight, or fertilizer factories.

Minerals are not generally plentiful in most ecosystems, so cycling of these materials is essential.

2.4 MAJOR ECOSYSTEMS OF THE EARTH

The **biosphere** is the region of the planet containing life-supporting systems.

The ocean is deep in places, but only the surface layers receive enough light for photosynthesis. **Phytoplankton** are the small, one-celled plants that are responsible for most of the photosynthesis in the ocean. **Zooplankton** are the primary predators in the sea. The central oceans are likened to a great desert.

Estuaries are protected regions of the sea that lie next to land. Estuaries support great quantities of sea life.

Freshwater systems are similar to ocean systems. Various terrestrial ecosystems include **tundra, forests, prairies,** and **deserts.**

2.5 ECOSYSTEMS AND NATURAL BALANCE

Ecosystem homeostasis — the balance of nature — is the sum of a great many actions that work together to maintain the stability of natural ecosystems.

2.6 NATURAL SUCCESSION

Natural Succession is the series of changes through which an ecosystem passes as time goes on. The **climax** is a final "unchanging" stage.

KEY WORDS

System
Ecosystem
Ecology
Energy
Photosynthesis
Respiration
Omnivore
Food web
Food pyramid
Nutrient cycle
"Fixed" nitrogen
Biosphere

Phytoplankton
Zooplankton
Estuaries
Tundra
Forest
Prairie
Desert
Ecosystem homeostasis
Natural succession
Climax

TAKE-HOME EXPERIMENTS

1. **Energy.** Weigh yourself in the early morning before breakfast. Now weigh all the food that you eat that day. Weigh yourself in the evening before you go to bed. What is the total weight of the food that you consumed during the day? Did you gain or lose weight or remain constant? Discuss.

2. **Energy.** Rub two sticks together vigorously. Feel the sticks with your hands. Are they warm? What happened to the work that you performed to rub the two sticks together? Could you find that work energy and reuse it? Discuss.

3. **Energy.** Build a compost pile by mixing together garbage, shredded newspaper, leaves, and other organic material. Pile the compost so that it is at least one meter high.

Let it sit for approximately ten days to two weeks. Measure the temperature of the outside air, and the temperature in the center of the compost pile. Which is hotter? Where did the heat come from?

4. **Nutrient cycling.** Take two shallow baking tins and cut a V-notch in the end of each as shown in the accompanying drawing. Fill one tin with plain dirt and one with grassy sod (if sod is not available, plant some grass seed in one tin and wait until the grass grows to maturity). Prop each tin up at a 30° angle. Then take two dry, clean measuring cups. Place one measuring cup under each tin so that any water flowing through the V-notch will spill into this container. Now make a watering can by punching small

holes in an old food can. Pour one liter of water through the watering can onto the plain dirt and another liter onto the sod. Collect the water that passes through the two V's. How much water was collected from each sample? Has any soil spilled over into into the measuring cups? Transfer the water and any soil into a frying pan. Evaporate the water slowly by placing the pan on low heat on the stove. Scrape out the soil from the two samples and see which is greater. (You could weigh each on a postal scale if you have one.) How much dirt and mud was collected from each sample? Using the results of your experiment discuss the role of plants in recycling minerals within an ecosystem.

5. **Decay organisms.** Obtain about a kg of fertile soil from a farm, woodland, or garden shop. Spread the soil carefully on a smooth piece of paper, and, using a magnifying glass, search for any living organisms. How many do you find? Draw pictures of them. If a microscope is available, place a small quantity of soil on a slide and examine

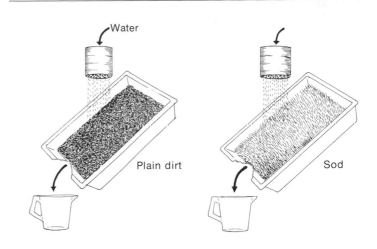

it. Can you see more organisms? What are they doing? How can the quantity and variety of life in your sample influence the quality of the soil?

PROBLEMS

1. **Systems.** What is a system? Are the separate parts of a system independent, or are they connected in some way? Discuss.

*2. **Systems.** The human body and automobile factory can both be classified as systems. Show how in each case a breakdown of one part can affect the whole. Show how a breakdown of some other part will not necessarily affect the whole. Briefly outline how each of these systems exchanges energy and raw materials with the outside environment.

3. **Vocabulary.** Define ecology; ecosystem; biosphere.

4. **Energy.** What is energy? Name two processes that require energy. Name two processes that release energy.

5. **Energy and matter.** Explain why matter can be recycled but energy cannot.

6. **Energy.** List five substances that are energy-rich. List five substances that are energy-poor.

7. **Energy and nutrient cycles.** We speak about nutrient cycles and energy flow. Explain why the concepts of nutrient *flow*. and energy *cycle* are not useful.

8. **Photosynthesis.** If you grow a pumpkin in a pot of soil, the pumpkin will become quite heavy, but the soil will loose only very little weight. Where did the material that was used to build the pumpkin come from? Discuss.

9. **Photosynthesis.** Discuss the energy changes that occur during photosynthesis. Where does the energy come from? Where does it go?

10. **Respiration.** Discuss the energy changes that occur during respiration. Where does the energy come from? Where does it go?

11. **Food web.** What is a food web? Discuss the major components of a natural food web. What is the importance of decay organisms in a food web? Discuss.

12. **Food web.** Outline a personal diet in which you are a herbivore; a carnivore; an omnivore.

13. **Food web.** Name an animal that is a grazer; a predator; a predator of a predator.

*14. **Energy.** Assume that a plant converts 1 per cent of the light energy it receives from the Sun into plant material, and that an animal stores 10 per cent of the food energy that it eats in its own body. Starting with 10,000 Calories of light energy, how much energy is available to a man if he eats corn? If he eats beef? If he eats frogs that eat insects that

---

*Indicates more difficult problem.

eat leaves? Of the original 10,000 Calories, how much is eventually lost to space?

*15. **Food web.** Sketch a diagram of a food web that primarily involves life in the air, such as birds and insects. Will you need links to terrestrial or aquatic systems?

16. **Food pyramid.** Imagine a large but isolated area with an adequate supply of plant food, equal numbers of lion and antelope, and no other large animals. The antelope eat only plant matter, and the lions eat only antelope. Is it possible for the population of the two species to remain approximately equal if we start with equal numbers of each and then leave the system alone? Would you expect the final population ratio to be any different if we started with twice as many antelope? Twice as many lions? Explain your answers. (Assume that lions and antelope have the same body weight.)

*17. **Nutrient cycles.** Give three examples supporting the observation that nutrient cycling hasn't been 100 percent effective over very long periods of time.

18. **Oxygen cycle.** Trace an oxygen atom through a cycle that takes (a) days, (b) weeks, (c) years.

19. **Nutrient cycles.** Why don't farmers need to buy carbon at the fertilizer store? Why do they need to buy nitrogen?

*20. **Carbon cycle.** The carbon dioxide concentration in the air just above trees varies considerably between night and day. From what you've learned about carbon, predict whether the atmospheric carbon dioxide concentration will be higher during the night or during the day.

21. **Nutrient cycles.** Certain essentials of life are abundant in some ecosystems but rare in others. Give examples of situations in which each of the following is abundant and in which each is rare: (a) water, (b) oxygen, (c) light, (d) space, and (e) nitrogen.

22. **Nitrogen cycle.** Describe three pathways whereby atmospheric nitrogen is converted to fixed forms that are usable by plants, and three pathways whereby fixed nitrogen is returned to the atmosphere.

23. **Ocean systems.** Discuss some similarities and differences between food webs in the ocean and on land.

24. **Estuary systems.** Give three reasons why estuaries are more productive than the central oceans.

25. **Terrestrial ecosystmes.** List four terrestrial ecosystems and describe the physical characteristics of each.

26. **Ecosystem homeostasis.** What does this term mean? Give two examples that show how homeostatic mechanisms operate.

*27. **Ecosystem.** Could a large city be considered a balanced ecosystem? Defend your answer.

*28. **Homeostasis.** Consider two outdoor swimming pools of the same size, each filled with water to the same level. The first pool has no drain and no supply of running water. The second pool is fed by a continuous supply of running water and has a drain from which water is flowing out at the same rate at which it is being supplied. Which pool is better protected against such disruptions of its water level as might be caused by rainfall or evaporation? What regulatory mechanisms supply such protection?

*29. **Ecology of soil.** Some fixed nitrogen is returned to the atmosphere by the action of certain soil organisms known as **denitrifying bacteria.** Do you feel that it would be wise to poison bacteria in the soil to conserve the fixed nitrogen supply? Justify your answer.

*30. **Ecosystem balance.** A sand dune ecosystem can survive flooding by salt water without suffering permanent damage, yet if a pine forest is similarly flooded it will not regain full productivity for many years. Is the pine forest less well balanced than the dune system? Discuss in terms of the homeostatic stability of ecosystems.

31. **Succession.** Define natural succession. What factors bring about changes in an ecosystem? What is the climax of an ecosystem? Cite three examples of a climax ecosystem.

*32. **Succession.** Imagine that a new island just arose in the South Pacific. Trace the succession that would be expected to occur. Estimate the time span required for the climax to be reached.

33. **Imbalance.** What do you think would happen to the Everglades if people built a set of dikes to ensure constant water levels all year round?

---

*Indicates more difficult problem.

1. Examine the site of a new construction project in your neighborhood. What type of system existed in that region before construction? Discuss the impact of the construction on local ecosystems. In your opinion has water balance been disturbed? What animal habitats have been disturbed? Has the stability of nearby ecosystems been endangered?

2. Carnivores use sunlight only indirectly and inefficiently. (See Section 2.2). Does this mean that an ecosystem would be "better off" without carnivores? Discuss.

3. More energy from the Sun is used to nourish a human being who eats meat than a human being who eats vegetables. Explain why this fact, by itself, is *not* an argument for or against vegetarianism.

BIBLIOGRAPHY

Three basic textbooks on ecology are:
Richard Brewer: *Principles of Ecology.* Philadelphia, W. B. Saunders Co., 1978. 278 pp.
Charles J. Krebs: *Ecology.* New York, Harper and Row, 1972. 694 pp.
Eugene P. Odum: *Fundamentals of Ecology.* 3rd Ed. Philadelphia, W. B. Saunders Co., 1971. 574 pp.

A periodical issue devoted in its entirety to "The Biosphere" is:
Scientific American. September, 1970. 267 pp.

Three books dealing with specific areas of natural ecology are:
R. Platt: *The Great American Forest.* Englewood Cliffs, N. J., Prentice-Hall, 1965. 271 pp.
B. Stonehouse: *Animals of the Arctic: the Ecology of The Far North.* New York, Holt, Rinehart and Winston, Inc., 1971. 172 pp.
G. M. Van Dyne, (ed.): *The Ecosystem Concept in Natural Resource Management.* New York, Academic Press, 1969. 383 pp.

A classic study of ecology and conservation as seen through the eyes of a naturalist is:
A. Leopold: *A Sand County Almanac.* New York, Sierra Club/Ballantine Books, 1966. 296 pp.

Two books that discuss the Everglades marsh system are:
Jean Craighead George: *Everglades Wild Guide.* U.S. Department of the Interior, National Park Service, 1972. 105 pp.
John Harte and Robert H. Socolow: *The Patient Earth.* New York, Holt, Rinehart and Winston, Inc., 1971. 364 pp.

3

ECOSYSTEMS AND NATURAL GROWTH

3.1
POPULATION GROWTH AND CARRYING CAPACITY

Grasshoppers living in a natural prairie do not generally eat all the grass. Of course, they eat some of it, but there is also enough left for other insects, as well as for rodents, birds, and large grazers. Even after all the animals have eaten their fill, some grass grows to maturity and goes to seed. Stalks and leaves fall to the ground uneaten and fertilize the soil. The system usually continues in a balanced and stable fashion. However, an individual grasshopper does not plan to build a stable system. An insect doesn't go on a diet to conserve food supplies. If an animal can find food it will eat all it needs. The grasshoppers do not eat all the grass because the *population* of the animals is controlled by environmental forces. In this chapter we will study the processes that control populations.

If the female of any species bred as often as possible, and if all the young survived and grew to maturity, the population of that species would grow very rapidly. For example, a bacterium can split into two bacteria in about 20 minutes. If enough food is available and there is no predation,

these two bacteria can grow into four after another 20 minutes. By the end of an hour the original bacterium will have become eight. By the end of a day and a half, a growing colony would have increased through 108 generations (36 hours at 3 generations per hour). Since each generation leads to a doubling of the number of individuals, the colony would consist of roughly 1,000,000,000,000,000,000,-000,000,000,000,000 (10^{33}) individuals. This number of bacteria could cover the entire surface of Earth to a uniform depth of 30 centimeters (1 foot). Such a growth pattern is known as a **geometric rate of increase.**

Most other organisms have longer generation times than 20 minutes. Therefore, their maximum growth rate is smaller than that of bacteria. Nevertheless, the breeding capacity of *all* species is large, and no ecosystem could support a geometric growth rate of any species for very long.

Refer to page 22 for definition of predation.

The geometric rate of increase is calculated as follows: Each generation leads to a doubling of the population. After two generations there are $2^2 = 4$ bacteria; after three generations there are $2^3 = 8$ bacteria; after 108 generations there are 2^{108} or 10^{33} individuals.

Imagine that a man and a woman have 12 children. Half of them are girls. Each girl marries and has 12 children, and like their parents, each girl has 6 boys and 6 girls. Assuming a like proportion in successive generations, the population from one married couple would grow as follows:

Generation	Number of People
1	2
2	12
3	72
4	432
5	2592
6	15,552

To calculate the number of people living in each generation, we start with the original couple — 2 people. If each woman bears 12 children, half of which are girls, we divide the population by 2, for only women bear children. Then we multiply by 12. In other words, we multiply by 6 for each generation.

Some combination of environmental pressures must, therefore, act to control the potential growth of every species. Examples of many of these environmental pressures are easy to observe. You swat a mosquito. A bird dies during a cold spell immediately following a winter rain. A puppy dies of diphtheria. You mow your grass before it goes to seed. A cat catches a rat. A bluejay chases a sparrow away from a crust of bread and the smaller bird loses a meal. Or you eat an apple and throw the core (including the seeds) into a garbage disposal unit. Other pressures are less easy to observe. Yet they are occurring all around us. A microscopic animal called a paramecium eats a bacterium. A fresh-water mussel eats a paramecium. A wild oat seed competes with a wild barley seed for a tiny, sheltered depression in the earth favorable for growth. A parasite infects a beetle. A seed fails to grow during a drought. An acorn rots during a particularly wet spring. Or a hailstone hits a caterpillar on the head. The sum of all the pressures that act together to inhibit the growth of a population is known as the **environmental resistance.**

If the population of any plant or animal becomes too large, then some pressure will generally act to control that population. For example, if there were too many grasshoppers in a field, each individual might have difficulty finding food, and some could starve or become weak. Predator populations would grow and predators would eat many of the grasshoppers. Perhaps disease would spread rapidly. All these pressures would act together to reduce the growth rate of the grasshoppers.

Every ecosystem can support a certain number of individuals of a given species. The **carrying capacity** is the maximum number of individuals of a given species that can be supported by a particular environment. The carrying capacity is not an absolute constant. It varies with food supply, weather, and other environmental factors. For example, more grasshoppers can live in a wheat field than in a short-grass prairie because there is more food in the wheat field. During a drought the carrying capacity will be lower than during a rainy season.

Most populations in most ecosystems remain stable most of the time. If you hike across the arctic tundra this summer or next, or ten years from now, the scenery will be relatively unchanged. Broad, hilly plains, covered by many flowers, stretch to the horizon. The thin layer of soil resting on icy ground is slippery and uneven, and in summer mosquitoes are abundant. Herds of caribou migrate across the land. Wolves, bears, foxes, wolverines, lynxes, and other animals all

Arctic tundra (with caribou). (Courtesy of Entheos Communications, Bainbridge Island, Washington, photo by Steven C. Wilson)

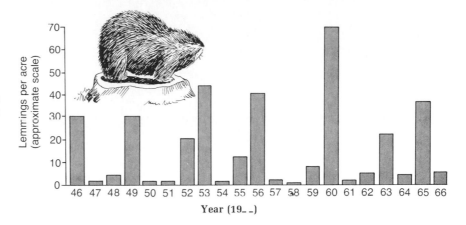

FIGURE 3–1. Lemming population cycles at Point Barrow, Alaska, from 1946 to 1966.

leave their tracks. Thus, over the long term, the system is stable.

But the populations of the plants and animals in the tundra change from year to year. For example one common arctic rodent is the lemming. The North American lemming population varies markedly in a three- or four-year cycle (see Fig. 3–1). One summer the population is extremely high. The next year the population rapidly declines.

For another year or two, the population recovers slowly. Then it skyrockets for a season and the cyclic pattern repeats itself.

Sometimes erratic changes of population occur even in old, seemingly stable systems. Earlier in this chapter it was stated that grasshopper populations generally do not grow beyond some stable carrying capacity. However, locust plagues (a locust is a type

---

The Tundra

"The bottom of the valley was soggy with water, which the thick moss held sponge-like close to the surface. This water squirted out from under his feet at every step. . . . He picked his way from muskeg to muskeg . . . across the rocky ledges which thrust like islets through a sea of moss. He crawled up a small knoll and surveyed the prospects. There were no trees, no bushes, nothing but a grey sea of moss scarsely diversified by grey rocks, grey lakelets, and grey streamlets. . . . As the day wore on he came into valleys where game was more plentiful. A band of caribou passed by . . . a black fox came towards him, carrying a ptarmigan. The man shouted. It was a fearful cry, but the fox, leaping away in fright, did not drop the ptarmigan."*

*From *Love of Life*, by Jack London.

Lemmings

There are several species of lemmings. Scandinavian lemmings are well-known for their spectacular marches to the sea. At periodic intervals, the populations of these rodents increase dramatically, and large hordes slowly march toward the ocean. Many are eaten along the way by large predators; others are killed by disease. When they reach the sea, most of those that remain plunge in and swim onward until they drown.

Locusts

"And the locusts went up over all the land of Egypt, and rested in all the coasts of Egypt: very grievous were they; before them there were no such locusts as they, neither after them shall be such.

For they covered the face of the whole earth, so that the land was darkened; and they did eat every herb of the land, and all the fruit of the trees which the hail had left: and there remained not any green thing in the trees, or in the herbs of the field, through all the land of Egypt." *Exodus*, Chapter 10, verse 14, 15.

A locust invasion in Morocco. (Courtesy of Food and Agricultural Organization of the United Nations, photo by Studios du Souissi)

of short-horned grasshopper) have occurred in many regions of the world since antiquity. This author witnessed a locust plague several years ago in central Turkey. It was harvest time and men and women were busy in the fields cutting stalks of wheat and collecting the grain. Suddenly a dark black cloud appeared on the horizon. Within about fifteen minutes the air around us was filled with flying insects. They smashed against our faces like hail stones, and people took shelter. I escaped to a nearby gas station. Soon the window was smeared with the bodies of grasshoppers that had flown into the glass. It was midday, but the room grew dark and we were forced to turn on the lights. After about an hour or two the locust cloud disappeared and I went outside. Farmers were walking slowly through their wheat fields, but the harvest had been eaten. The flowing meadows of grain were reduced to dirt. There was nothing left to eat.

Ecosystems are stable most of the time, but we must remember that sometimes stability is disrupted.

3.2
ENVIRONMENTAL RESISTANCE

Populations of plants and animals are controlled by a variety of factors.

Shortage of Nutrients, Light, and Space

If you study the population of plankton near the surface of the central oceans, you will find that although there is a lot of light, there are relatively few plankton. Of course, there is enough water to support life. Why, then, aren't there more plankton? The answer is that many nutrients are scarce in the deep sea. Nutrients from the land move slowly toward the central oceans, and many of these materials fall to the bottom of the sea where they are unavailable to the plankton that live near the surface. Thus, the populations of plankton are limited by a shortage of a few nutrients, even though there is plenty of water, light, and many mineral salts.

In some desert regions, many nutrients are relatively plentiful, but water limits the growth of plants and animals. In rain forests, plants are crowded together so densely that light and space become limiting. In general, the population growth will be slow if any of the essentials of life are scarce.

Climatic Factors

Temperature, wind, and wave action are other factors that limit population growth. Trees grow slowly in open, windswept areas, and they cannot grow at all in tundra regions where winters are particularly severe. Many ocean plants and animals do not live where wave action is intense. Polar bears do not migrate to the sunny south.

Biological Factors

The growth of organisms is limited by the presence of other organisms in the ecosystem as well as by the physical environment. Animals find food to eat and, in turn, are eaten by other animals. If food is scarce, the different organisms will compete for what is available. If wind, water, currents, waves, or changes in temperature make survival difficult, plants and animals will compete for shelter. Therefore, the environmental resistance also consists of pressures from competition and predation.

Competition. Competition arises when two or more organisms need a certain limited resource such as nutrients, water, sunlight, or living space. Often competition will result in the death of some individuals. Suppose that some pine cones blow onto a grassy meadow. As the trees grow, they will block out much of the sunlight, and many of the prairie plants will slowly die out. In addition, the needles that fall from the growing trees eventually change the soil chemistry, making conditions even more unfavorable for grasses and prairie flowers. Similarly, if two lizards struggle for a safe, cool burrow under a rock, the loser might find itself in the open where excess sunlight and exposure to predators may cause death.

Slugs are generally considered to be

The growth of trees in alpine regions is limited by weather conditions.

slow, deliberate animals. If you see one in the garden in the early morning, it will seldom be moving faster than a few centimeters per minute. But these animals fight viciously over food supply. If one slug is eating and another approaches, the first will often draw back like a snake and strike. A slug's mouth contains a small layer of filelike teeth. When it strikes it cuts its opponent's side with these teeth. The two animals will engage in battle. One may retreat, but they may fight to the death.

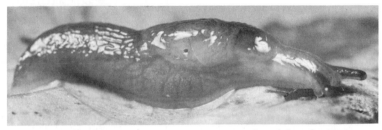

Land slug. (© N. E. Beck, Jr. from National Audobon Society)

Several species of barnacles and mussels compete for space in this intertidal zone in northern California.

Sea anemone. (© Arthur Ambler from National Audubon Society)

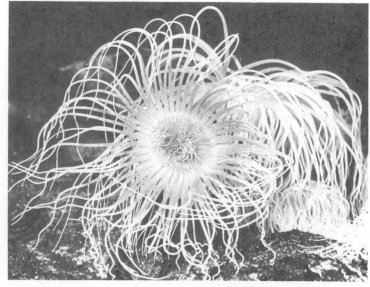

It is tempting to say that one species would be somehow "better off" if its competitors were eliminated. This conclusion is not always correct. In one region in northwest Washington, two animals, barnacles and sea anemones, compete for living space along the shore. A scientist thought it would be interesting to find out what would happen if all barnacles in a small area were killed or removed. When he did this, he found that the anemone population grew rapidly. But after about six months most of the anemones started to die. A careful study showed that these animals were dying from lack of water. In a natural system where anemones and barnacles live side by side, the barnacles provide shelter for the anemones so that they do not dry out from the hot summer wind. When the barnacles were removed in the test area, the anemones became exposed to the weather and died of exposure. Therefore, the competition that appeared to limit the growth of the anemones was actually necessary for their survival.

Predation. Predation is an interaction in which certain individuals eat others. Grasshoppers eat grass, grouse eat grasshoppers, hawks eat grouse. Of course, animals try to avoid being eaten. An insect flies away when a bird swoops after it. Deer run from wolves. Baboons join together to defend themselves from leopards. Many plants produce protective bark, spines, or chemicals to discourage predators. But predation is essential to the well-being of all ecosystems. If some organisms were not eaten by others, populations would grow so rapidly that ecosystems would become imbalanced. For example, many of the wolves and coyotes in the Kaibob Plains near the Grand Canyon were killed by ranchers in the early 1900's. As a result, the deer multiplied until they ate most of the available feed. Food became scarce and within a few years thousands of deer starved to death.

In the previous section we studied competition along the shoreline in northwest Washington. Predation, too, is essential for maintaining balance. Ecologists found fifteen different species of animals along one region of the coast. One of these species was a predatory starfish. The scientists removed all of the starfish from an experimental area. When the starfish

Starfish. (© Russ Kinne from National Audubon Society)

were removed, the populations of many of the grazing animals started to grow rapidly. Within a few months competition for food and space became intense. Animals starved, were displaced, and seven species became extinct in the experimental area. The balance of the ecosystem had become disrupted.

Predation is a delicate and fascinating process. I was fortunate enough to watch a lone timber wolf stalk a moose by a river in the northern Canadian forest. The wolf followed the moose along a riverbed, but did not approach too closely. The moose looked toward her predator frequently, but she never broke into a run. Instead she continued to feed and move slowly downstream. After about five minutes the wolf turned abruptly, trotted over a hill, and disappeared from view. Both animals knew that a single wolf was no match for a young, healthy cow moose. Studies have shown that most of the time old, crippled, sick, and very young moose are killed by wolves. A healthy adult is brought down only rarely by a wolf pack. Thus, only the strongest and smartest animals are left to reproduce. Over long periods of time, predation improves the health and vitality of the herd.

Artist's rendition of predaceous fish.

FIGURE 3–2. Grazers in Elk Island National Park. *A*, Moose; *B*, elk; *C*, American bison.

Parasitism is a form of predation. A parasite is a small plant or animal that consumes the tissue of a larger organism without killing it directly. Pine beetle parasites bore into the wood of pine trees and eat the sap of the living plant. (See the Case History at the end of this chapter.) Disease bacteria infect most kinds of plants and animals. Tapeworms may live in the intestines of mammals and eat the food that they find there. As parasites become strong, the host may eventually die.

3.3
NATURAL ECOSYSTEMS

Most natural systems are complex. Hundreds of thousands of different species live within a single valley, lake, or tidal region. In recent years, the complexity of many ecosystems has been threatened. Vast regions of the globe are being polluted, and this pollution is killing many plants and animals. What happens to an ecosystem when some of the species in it are killed? Of course, this question is difficult to answer. We learned in the last section that when one species of starfish was removed from an area, seven other animal species became locally extinct. Yet in other situations, removal of one species has been shown to have little effect. We do

know that most balanced ecosystems contain many different species. Diversity seems to be related to stability. For example, there are three species of large grazers in Elk Island National Park in Canada — moose, elk, and deer. Moose eat saplings and small bushes. With brush growth held in check, grasses find room to grow. Bison eat grass. Elk eat either leaves or grasses. In this way the community of grazers acts on the community of plants. If one species were exterminated, the balance would be disrupted.

A natural prairie contains many different species of plants. Bushes and perennial grasses do not die in winter but live for many years. They grow deep roots that collect water and nutrients from lower layers in the soil. On the other hand, annual plants sprout from seed every spring. They grow quickly, produce seeds, and then die in the fall. Annuals generally have shallow roots that lie mostly near the surface. During dry years there is so little water that many annuals die. However, the perennials, which use water from deep under ground, are able to live. These living plants hold the soil and protect it from blowing away with the dry summer winds. In years of high rainfall the annuals sprout quickly. Their extensive root

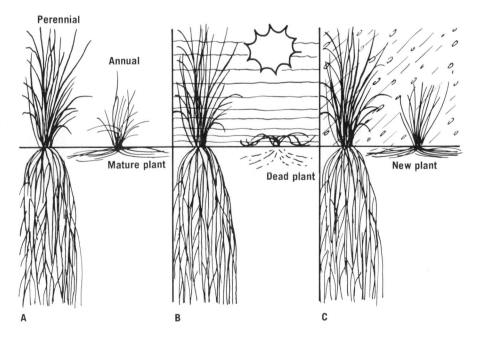

FIGURE 3–3. Root systems of prairie plants. *A*, Both annual and perennial plants grow in a prairie. *B*, In a dry year the annuals die but the perennials hold the soil. *C*, In a wet year all the plants thrive and hold the soil.

Perennial

Annual

Mature plant

Dead plant

New plant

A   B   C

The south-facing slope of this ridge is covered with grasses, cacti, and a few pine trees, while the north slope is densely wooded. (Photograph taken in Sunshine Canyon, Boulder, Colorado.)

systems spread rapidly and prevent soil erosion and water runoff. Thus, the soil is maintained in both dry and wet years.

In canyons in the Colorado Rockies ponderosa pine, juniper, grasses, and small cacti predominate on the dry sunny side, and blue spruce, douglas fir, and flowering plants predominate on the wetter, shady side. If a prolonged drought were to strike the canyons, many trees might die. But probably grasses and cacti would fill in the bare spots. The system would change but not die.

3.4
ECOSYSTEMS AND GLOBAL BALANCE

A lowland valley east of the Andes Mountains in Peru may be considered to be an independent ecosystem. Grasses, trees, and small shrubs support populations of insects, rodents, birds, and large grazers. Nutrient cycling, predation, and competition all contribute to ecosystem balance. But this system is not independent of all other systems. Ice melts in the high mountain glaciers and the water flows through upland tundra regions into the larger streams and rivers below. Normally only small quantities of nutrients will move downward toward the valley. If the high meadows are overgrazed, however, large masses of soil will erode and carry excess mud and silt downstream. Thus, the lowland valley ecosystem is dependent on the mountain systems. In turn this lowland valley "feeds" other regions further downstream. Gradually the rivers flow eastward into the jungles of Brazil and empty into the Amazon. The Amazon is the world's largest river. It drains an area nearly as large as the entire continental United States

The ecosystems of the Andes are interconnected with the jungles of the Amazon basin. (Bottom photo courtesy of the Brazilian Consulate)

and carries nearly 20 trillion liters of water per day into the Atlantic Ocean. Along with this water, many tons of mud, silt, and nutrients are carried out to sea. The river is so massive and powerful that people sailing in the ocean have tasted fresh water and seen mud from the Amazon before they could see land. This nutrient-laden water serves as a rich food supply for millions of ocean fish that live near the mouth of the river. Thus, the ecosystems of the Andes are connected with the fisheries of the southern Atlantic Ocean.

But the connection does not end here. Vast regions of the ocean are interconnected by currents — salt water rivers that flow within the ocean basins. One of these currents flows by the mouth of the Amazon. The moving ocean waters collect some of the river nutrients and carry them northward. These nutrients are slowly mixed with others and carried into the northern hemisphere to the coasts of Newfoundland and Labrador in Canada. Plankton grow well in these rich northern waters and many types of fish live in abundance. These fish are partly dependent on the melting glaciers in Peru for their food. Such connections are far-flung, but they do exist.

There are a great many other ecosystems that are dependent on each other even though they lie thousands of kilometers apart. For example, the entire Arctic region does not receive enough solar energy to support much

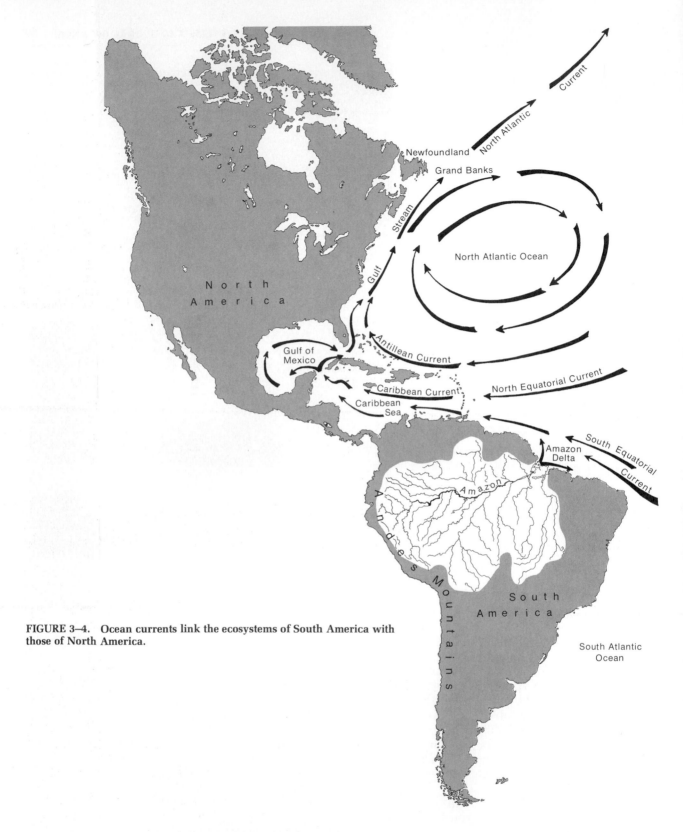

Gulf Stream

Newfoundland

Grand Banks

North Atlantic Current

North Atlantic Ocean

North America

Gulf of Mexico

Antillean Current

Caribbean Current

North Equatorial Current

Caribbean Sea

Amazon Delta

South Equatorial Current

Amazon

Andes Mountains

South America

South Atlantic Ocean

FIGURE 3–4. Ocean currents link the ecosystems of South America with those of North America.

life. But warm winds and ocean currents from the south carry enough heat to the tundra to help melt the winter snows and give the grasses a chance to grow.

In a larger sense, all ecosystems on this planet are interconnected. Before life existed on Earth, the atmosphere contained mostly nitrogen, ammonia, hydrogen, carbon monoxide, methane,

Nitrogen, $N \equiv N$

Ammonia,

$$\underset{H \quad H \quad H}{\overset{N}{\diagup | \diagdown}}$$

Hydrogen, $H-H$

Carbon monoxide, $C \equiv O$

Methane,

$$H - \underset{\overset{\displaystyle |}{H}}{\overset{\displaystyle H}{\overset{|}{C}}} - H$$

Water,

$$\underset{H \qquad H}{\overset{O}{\diagup \diagdown}}$$

Chemical formulas.

and water vapor. Gaseous oxygen was rare. This atmosphere would be poisonous to most plants and animals that exist today. Many scientists feel that the tiny one-celled plants that first appeared on Earth gradually began to produce excess oxygen. For millions of years the oxygen concentration in the atmosphere rose slowly. Large plants and animals did not

evolve until the smaller organisms had produced enough oxygen to change the chemistry of the atmosphere.

At the present time the atmosphere contains about 20 percent oxygen. If the concentration were to rise to 25 percent, fires would burn uncontrollably across the planet. If the carbon dioxide concentration were to rise by a small amount, plant production would increase drastically. Yet, for millions of years, neither of these disasters has occurred. How is the balance maintained? The answer appears to be that it is maintained by living systems themselves. It appears that the net effect of the gas exchange among trillions upon trillions of plants and animals is that the chemistry of the atmosphere remains stable. One scientist, J. E. Lovelock, compared the entire biosphere to a living creature. He called that creature by the Greek name for Earth, **Gaia.** If all life on Earth were to die, oxygen would once again become a rare gas. The atmosphere would once again be poisonous to complex plants and an-

Even if Lovelock's theory is correct, there appears to be enough oxygen in the atmosphere so that even if most plants on Earth died, animals could breathe for hundreds or even thousands of years.

The Earth consists of a delicately interconnected set of ecosystems. (Courtesy of NASA)

Dry conditions on the eastern slope of the Colorado Rockies encourage growth of species of yucca.

A woman cutting dead timber from a stand of ponderosa pine killed by pine beetles.

The native pine trees in many Rocky Mountain regions were cut by miners during the gold rush days to build log cabins.

imals. If this theory is correct, it states that a large biological catastrophe such as the death of the oceans could eventually upset the atmospheric balance. In turn, life on Earth could be threatened.

3.5
CASE HISTORY: THE PINE BEETLE EPIDEMIC ON THE EASTERN SLOPE OF THE COLORADO ROCKIES

The pine beetle *(Dendroctonus ponderosae)* is a natural parasite of the ponderosa pine. It bores holes into healthy trees and sucks the sap. Certain species of fungi infect the beetle holes and further upset the tree's life support system. If enough beetles attack a given tree, the tree will die.

At the time this is being written (1978), the ponderosa pine forests in several Rocky Mountain regions are

being destroyed by the pine beetle. Our case history will concentrate on the mountain ecosystems near Boulder, Colorado. The epidemic has been so severe that in some places in this area nearly all of the mature trees have died. The pine beetle is not new to these hills. It has been a part of the ecosystem for many years. Why, then, did this epidemic occur only recently?

The eastern slope of the Colorado Rockies has a relatively dry climate compared with other forest systems. Only about 45 cm (18 in) of rain falls annually, and the dry air and the hot sun of the region encourage rapid evaporation. Small cacti and yuccas grow on exposed hillsides. The trees tend to thrive best in gullies or shaded areas where water is more plentiful and the summer sun less intense. About 100 years ago, before gold and

Miners also cut trees to supply timbers for mine supports. (Courtesy of State Historical Society of Colorado)

A healthy ponderosa pine.

silver miners settled in large numbers in these regions, there were only a few trees on the mountainsides. The Colorado gold rush brought many settlers to the area. These miners cut nearly all the trees that existed, and they used the wood to build cabins and to timber mines. Within a few years the hillsides were virtually stripped. Only the low-lying brush, grasses, and cacti remained.

But we must remember that a great many pine cones must have lain on the ground. As the seeds sprouted, the young trees started to grow in a favorable environment. With the large trees gone there was little competition for space, sunlight, and water. Therefore, the saplings were healthy, and new forests began to grow quickly. Since several young trees require less space

and water than is needed by one large tree, these new forests grew to be much denser than the original system. By the 1960's, the young trees had grown to a mature size. The ecosystem appeared to be healthy. Yet, since the new forests were much denser than the original ones, competition for vital nutrients, especially water, was much more keen. The available moisture was now distributed among many more trees, so all the trees suffered from a lack of water.

If a beetle attacks a strong, well-watered tree, the tree will produce enough sap to force the beetle out of its burrow. Thus a healthy tree can repel a beetle attack, just as a healthy person can avoid infection even if there are many flu viruses in the air. On the other hand, a weak tree, living with inadequate supplies of water, cannot produce excess quantities of sap. Therefore, it is more likely to be infected by beetles. In the dense weakened forests of the present day, there are large areas where there are no trees healthy enough to repel a beetle attack. The beetles have bored successfully into the wood of many trees, and the forests are dying.

What will happen in the future? Of course, we can't predict with certainty. In some regions all the trees may die, leaving a grassland similar to that left by the miners and loggers a hundred years ago. In that case, the cycle may repeat itself. A new dense forest may grow, only to be killed by beetles a century from now. Then a new barren grassland would be left. In such a situation, true ecological balance will not be achieved. Instead, there would be a cycle of rapid growth followed by death. On the other hand, perhaps most, but not all, of the trees will die. The death of the weak individuals may reduce the competition for water. If that happened the remaining trees might become healthy enough to survive the beetle attack. The beetle population would then diminish and a stable forest system would regenerate.

The infestation of beetles continues, and it may be many years, perhaps decades or centuries, before we will be able to determine whether the forests will recover their original balance.

3.1 POPULATION GROWTH AND CARRYING CAPACITY

The **carrying capacity** is the maximum number of individuals of a given species that can be supported by a particular environment. Most populations in most ecosystems remain stable most of the time. But population cycles and erratic disruptions do occur.

3.2 ENVIRONMENTAL RESISTANCE

Populations of plants and animals are controlled by shortage of nutrients, by extremes of climate, by competition, and by predation. (Parasitism is a form of predation.)

3.3 NATURAL ECOSYSTEMS

Most stable systems are diverse. Diverse systems are generally flexible enough to adjust to changing conditions.

3.4 ECOSYSTEMS AND GLOBAL BALANCE

Distant ecosystems are interconnected via rivers, ocean currents, and wind systems. In a larger sense, all the plants and animals on Earth are interconnected via the oxygen–carbon dioxide cycle.

Geometric rate of increase
Environmental resistance
Carrying capacity
Competition

Predation
Parasitism
Gaia

1. **Population growth.** Buy or bake a loaf of bread that contains no preservatives. Eat most of the bread, but save one slice. Place this slice in an open dish in a quiet place in the room. Within a few days, mold will start to grow on the bread. You are asked to measure the growth of the mold as a function of time.

Take a piece of wax paper and draw a series of horizontal and vertical lines on it with one centimeter spacing between them. You now have a grid with a series of squares that are each one centimeter on a side. When the mold starts to form, place the grid over the bread and use it to estimate (a) the total surface area of the bread and (b) the surface area covered by mold. Remove the paper. Repeat this measurement once a day for ten days. Draw a graph showing the growth of the mold population as a function of time.

2. **Natural ecosystems.** Have the class survey the region near your school and classify the ecosystems in your neighborhood. Some common classifications might be: forest or woodland, grassland (defined here as a meadow or prairie that is not regularly cared for by people), beach-dune system, salt water marsh, swamp, desert, city park, or agricultural field. Have one member or group of the class study each type of system. Each person or group is then asked to make a detailed report on a four square meter section of the chosen ecosystem. It is important to choose the four square meter study area at random. One crude way of random choice is to throw a ball into the air. Then designate the place where the ball lands as the center of your study area. Measure out four square meters around this center. Count the number of different plant species in your area. Can you identify which plants are annuals and which are perennials? What area of bare soil is exposed? Do you feel that your study area is too small, or too large, for an accurate representation of the entire system? Comment on the general condition of the entire system (your four square meters plus the surrounding area). Compare your results with those of your classmates.

3. **Interspecies interaction.** In this experiment you will observe and record one or more cases of competition and predation. The observations can be made in the field or in the laboratory. There is no limitation to the type of study that can be conducted. Two examples are given below, but you are encouraged to use your own imagination.

(a) Set a bird feeder in a convenient location

and keep the feeder well stocked with bread and seeds. Observe the behavior of the birds. Do some individuals chase others away? Do some species of birds dominate the feeder? Describe your observations. If possible, take photographs, and bring them to class.
(b) Take a slow walk in the woods or in a park. Can you observe any instances where two plants appear to be competing for light, space, or water? Can you prove that competition is occurring, or would a further experiment be necessary? Can you observe the growth of any plant parasites? Are any insects present on the plants? Can you see the insects eating the plants or eating each other? Describe your observations. If possible, take photographs, and bring them to class.

PROBLEMS

1. **Definitions.** Define carrying capacity. Explain why there is a carrying capacity for every species within an ecosystem.

2. **Geometric growth rates.** Imagine that the population of a given species quadrupled (increased by multiples of four) every 10 years. If there were 10 individuals in 1950, how many would there be in 1960, 1970, 1980, 2000? Draw a graph showing the number of individuals as a function of time. (Plot time on the horizontal, or X, axis, and population on the vertical, or Y, axis).

3. **Carrying capacity.** The worldwide human population has been increasing continuously for the past few hundred years. Can this trend continue indefinitely? Are human populations subject to the constraints of a worldwide carrying capacity? Explain.

4. **Natural balance.** Discuss the statement, "Natural systems are perfect because they are always in harmonious balance."

*5. **Lemming cycles.** It is mentioned in the text that lemming populations vary on a three- to four-year cycle. Do you think that the populations of other species of the ecosystem would also cycle? Discuss.

6. **Environmental resistance.** List four components of the environmental resistance. Give examples of each.

*7. **Shortage of nutrients.** The central oceans could be made more productive if they were fertilized. Do you think it would be a good idea to fertilize the oceans? Discuss.

*8. **Physical components of the environmental resistance.** Many mountain streams flow rapidly down deep hillsides, then move slowly through upland meadows. In which region would you expect to find more rooted water plants? Discuss.

*9. **Environmental resistance.** The physical components of the environmental resistance (shortage of nutrients and climate) are sometimes closely associated with the biological components (predation and competition). Discuss this statement and give examples to support it.

10. **Competition.** An organism may compete with another organism of the same species or with organisms of other species. Explain and give examples.

11. **Competition and predation.** Competition and predation may be harmful to one individual and at the same time helpful to the entire system. Discuss and give examples.

*12. **Predation.** One ecologist stated that predators live on capital while parasites live on interest. Explain. Is this true from an individual or from a community standpoint? Explain.

*13. **Predation.** Would you expect a buffalo herd to be healthier after years of being hunted by men with bows and arrows or by men with guns? Explain.

14. **Predation.** Explain why predation is essential to the health of an ecosystem. Is predation beneficial to the health of each individual? Discuss.

15. **Competition and predation.** Categorize each of the following as competition, predation, parasitism, or none of the above: (a) A mistletoe sucks the sap from a pine tree on which it grows. (b) A paramecium eats a bacterium. (c) An elephant steps on an ant. (d) Two trees growing side by side reach out for light. (e) A wolf pack hunts and kills a young moose. (f) A wolf pack eats a moose that had died of other causes.

16. **Ecosystem stability.** Which do you expect to be better able to survive a

*Indicates more difficult problem.

drought: a cornfield or a natural prairie? Explain.

17. **Natural systems.** In recent years many species have been driven to extinction across the globe. Discuss, in general, how the extinction of a species may upset an ecosystem.

18. **Natural systems.** Discuss the role of plants and animals in regulating the atmosphere.

1. Select a nondomestic plant or animal with which you are familiar and discuss the primary components of the environmental resistance that affects it.

2. Choose a large ecosystem in your neighborhood with which you are familiar. What rivers or streams enter and leave the system? What are the prevailing winds in the area? Do they bring warm air or cold air to the area? Discuss the linkages between your system and distant systems.

BIBLIOGRAPHY

The three basic textbooks cited in the bibliography of Chapter 3 all relate to this chapter as well. Several books dealing specifically with population biology are:

Arthur S Boughey: *Ecology of Populations.* 2nd Ed. New York, Macmillan Co., 1973, 182 pp.

Arthur S Boughey: *Contemporary Readings in Ecology.* Belmont, Calif., Dickenson Publishing Co., 1969. 390 pp.

Kenneth E. F. Watt: *Ecology and Resource Management.* New York, McGraw-Hill, 1968. 450 pp.

Edward O. Wilson and William H. Bossert: *A Primer of Population Biology.* Stamford, Conn., Sinauer Assoc., 1971. 192 pp.

Two comprehensive books dealing with the distribution of species on islands and continents are:

Robert H. MacArthur: *Geographic Ecology — Patterns in the Distribution of Species.* New York, Harper & Row, 1972. 269 pp.

Robert H. MacArthur and E. O. Wilson: *The Theory of Island Biogeography.* Princeton, N.J., Princeton University Press, 1967.

4

THE EXTINCTION OF SPECIES

4.1
WHAT IS A SPECIES?

No two individual organisms are exactly alike. You are different from your brother, your sister, or your best friend. But there are groups of organisms that have common characteristics. Plants are basically different from animals. Animals can be divided into a great many subgroups. Marine jellyfish form one distinct class that is different from all others. Insects form another class. Animals with backbones form a broad classification. We recognize that animals with backbones can be further subdivided into classes such as reptiles, birds, and mammals. Of course, there are many different orders of mammals such as rodents (mice, rats, squirrels), cetacea (whales, dolphins, porpoises), and primates (monkeys, apes, people). How far can we keep classifying?

Individual plants or animals breed only with plants and animals that are quite similar. They do not breed with dissimilar ones. Therefore, it is convenient to classify organisms that breed together into discrete groups. A **species** is defined as a group of plants or animals that breed together but do not breed with members outside the group. For example, one look tells us that all elephants are alike in many ways. But Indian elephants are slightly different from African elephants. These two groups of animals do not

The African Elephant (*above*) and the Asian Elephant (*below*) are recognizably similar. But they represent two different species. The Asian elephant is smaller, has smaller ears, and the females of the species do not grow tusks. African and Asian elephants do not mate with each other. (Both photographs taken by Mark N. Boulton, Photo Researchers, Inc.)

67

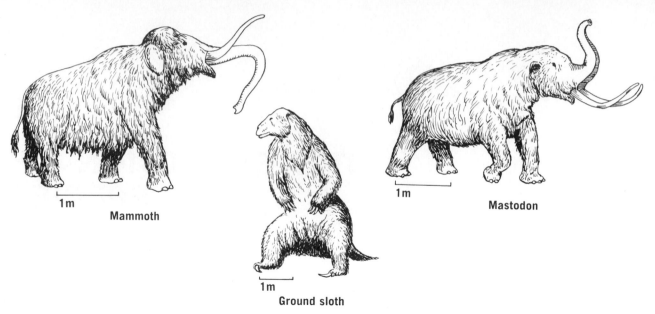

Mammoth

Ground sloth

Mastodon

Extinct North American animals.

mate with each other. Therefore, they form separate species.

When organisms mate, many of the characteristics of the parents are often passed on to the infants. Tall people tend to bear tall children. Swift antelopes have a high probability of giving birth to swift baby antelopes. Character traits are mixed within a species because members of the species breed together. But traits are not passed from species to species, since members of different species do not mate. Therefore each species forms a unique group. If a species is destroyed, it can never be re-created.

4.2
EXTINCTION OF SPECIES

Since the beginning of life on this planet, new species have continually been formed and existing species have continually been driven to extinction. About 500 million years ago, enormous numbers and varieties of primitive sponges populated the seas. For 30 or 40 million years these sponges dominated the oceans, from pole to pole. Then, in a relatively short period of time, most of the species of sponges became extinct. Somehow the environment had changed so that the sponges could not survive.

In more recent times, the rise and

fall of the dinosaurs was certainly one of the most outstanding events in the history of the Earth. Small dinosaurs first evolved some 225 to 250 million years ago. For perhaps 25 to 50 million years they slowly developed and grew in numbers and variety until they dominated the Earth. The dinosaurs reigned for 100 million years, and then they all died.

Of course, the period following the extinction of dinosaurs was not empty of life. During this time, the previous-

Top, Triceratops. *Bottom,* Stegosaurus.

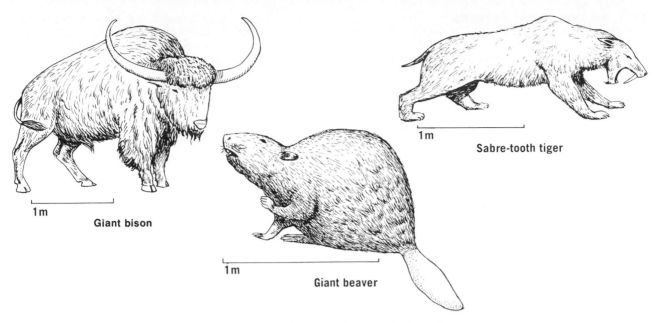

Giant bison

1 m

Giant beaver

1 m

Sabre-tooth tiger

1 m

Extinct North American Animals (*Continued*).

ly slow evolution of mammals accelerated. Many species, including human beings, arose. The next significant wave of extinctions occurred about 10,000 years ago. Before this time, mammoths, mastodons, camels, wild pigs, giant ground sloths, giant long-horned bison, woodland musk oxen, tapirs, bear-sized beavers, and many different kinds of now-extinct deer roamed the North American continent. They were hunted in part by sabre-toothed tigers, giant jaguars, and dire wolves. All these animals have become extinct. Mass extinctions occurred at the same time in what are now South America, northern Asia, Australia, and Africa.

What caused these extinctions? There are many uncertainties. Perhaps changes in climate had an effect. It is believed that primitive people also played some role in the drama. All the grazing animals that became extinct were hunted by early humans. The extinctions occurred just about the time that these people developed advanced hunting techniques such as throwing spears, shooting arrows, and setting fires to drive animals over a cliff.

For the next five to six millennia, very few additional species became extinct. Then suddenly, in recent times, a new age of species destruction has begun. Extinction rates started to increase rapidly in the last half of the nineteenth century, just about the same time that the industrial revolution started. Today many more animal and plant species are seriously endangered. Unless current trends reverse, extinction rates will continue to accelerate in the near future.

Hunting techniques of early humans. Artist's conception of how the Tule Springs site, Nevada, may have looked 12,000 to 10,000 years ago. (Courtesy of John Hackmaster)

More than half of the known extinctions over the past two thousand years have taken place in the last sixty years. Nearly 10 percent of these extinctions have taken place in the last fifteen years.

The United States Department of the Interior has published a list entitled "Endangered and Threatened Wildlife and Plants." (*Federal Register*, Vol. 42, No. 135–July 14, 1977; amended Feb 23, 1978.)

TABLE 4–1 PAST AND PRESENT
EXTINCTION RATES

Year Span	Average Number of Years Required for Extinction of One Mammalian Species
1–1800	50
1801–1850	25
1851–1900	1.6
1901–1950	1.1
1951–2000	>1

4.3
WHY ARE SPECIES BEING DESTROYED TODAY?

We know that climate has been relatively constant in recent years. Therefore it wasn't any change of weather that expelled grizzly bears from California and passenger pigeons from Wisconsin. We also know that devastating disease epidemics have not occurred recently in wild animal populations. Vegetation hasn't changed. We know, in short, that people are the major agents in the extinction of species today. The following paragraphs discuss four major destructive mechanisms: (a) destruction of habitats; (b) introduction of foreign species; (c) extermination of predators; and (d) hunting by people for food or fashion.

Destruction of Habitat

In recent years the ax, the chainsaw, the tractor, and the bulldozer have destroyed many natural ecosystems. Some organisms have been killed outright. Many more have died because their homelands and food supplies have been destroyed. In North America, thousands of species of prairie grasses and flowers are disappearing as the prairies are plowed to make way for farms. One of the reasons that California condors cannot survive is that their territories have been altered by ranchers, hunters, road builders, developers, and miners.

In Asia, many species of plants and animals, including the Bengal tiger, are endangered as the jungles in which they range are being cut. The list of destruction of habitats goes on and on. Current reports estimate that over one hundred species of large animals face extinction, along with tens of thousands of species of spiders, mites, crustaceans, insects, and plants. Most of these are endangered, at least in part, by loss of habitat.

Habitat is not just a specific place. It is the sum total of all the conditions necessary for life. If an animal migrates from one place to another, the migration route is part of its habitat. The entire route must be preserved for the animal to live. Pollution may destroy a habitat without changing its physical appearance. In particular, pesticides have led to the decline of many species of animals. Chapter 12 discusses how pesticides used against insects are often destructive to birds and other animals.

The problem of destruction of habitat is just one aspect of a broader problem of **land use.** For example, there are a few old climax forests left in the Southern United States. Some people believe that the areas should be maintained as wilderness. People could hike and camp there, and the habitat of endangered species such as the trailing arbutus and the ivory-billed woodpecker would not be destroyed. On the other hand, many loggers feel that the valuable timber should be cut. They argue that the wood is needed to build new homes. Some farmers feel that the region could be converted to fertile farmland. Developers eye the land as a potential site for factories, stores, or suburban housing.

In recent years commercial interests — logging, farming, urban development, and mining — have destroyed natural ecosystems across the globe. Today, vast sections of the Amazon jungle in Brazil are being cut for farmland. Strip miners destroy wheat fields in the United States and Canada in their search for coal. Farmers in East Africa are plowing the plains that once supported many unique forms of wildlife.

Land use experts are attempting to plan world systems so that people can live in greater harmony with their environment. It is a difficult problem. It

A

B

A, Trailing arbutus. (Photo by Alvin E. Steffan from National Audubon Society) *B*, Giant white pine tree — a reminder of the primeval American forest. (From Rutherford Platt: *The Great American Forest.* © 1965. Published by Prentice-Hall, Inc., Englewood Cliffs, New Jersey) *C*, Ivory-billed woodpecker (Photo by James T. Tanner from National Audubon Society) *D*, Whooping crane. (Photo by Allan D. Cruickshank from National Audubon Society)

C

D

71

Important grazers that have evolved on different continents.

is hard to tell a hungry hunter that animals should be kept for future generations. Large corporations have much power and often look toward short-term profits rather than long-range stability. People need food, space, energy, and recreation all at the same time. With careful land planning, many endangered species could survive. Without careful planning, many will become extinct.

Introduction of Foreign Species

In prehistoric times, migration of species from continent to continent or even from watershed to watershed often required thousands of years and in many instances did not occur at all. Consequently, each continent has given birth to various native species.

Bison are native to the North American prairie. Kangaroos live on the prairies of Australia, and elephants graze on the African savannas. Many plants also originated in specific areas. Corn first appeared in North America, wheat in the Middle East, and rice in East Asia.

During modern times, many species have migrated as free passengers on trucks, trains, automobiles, airplanes, and boats. Many migrating organisms have been unable to compete in foreign lands and they have died. But some have succeeded so well in their new environment that they have disrupted the ecological balance and have endangered other species. For example, the American chestnut tree used to forest much of the Atlantic coast in the United States. Now the chestnuts are seriously endangered. A parasitic tree fungus imported from China has killed most of the American trees. Both American and Chinese trees had developed resistance to the parasites that grew on their own continent. But they were not resistant to foreign parasites. Fifty years after the Chinese fungus arrived, all large American chestnut trees on the East Coast had died. Perhaps some young trees will survive and produce a new breeding stock for the species, but the outcome is uncertain.

There are many other examples of this type. The Japanese beetle, imported from the Orient, feeds on many crops, such as soy beans, clover, apples, and peaches. The American vine aphid, imported to France from the United States, was responsible for destroying three million acres of French vineyards. In fact, over half of the major pest insects in many areas are imported.

Chestnut blight. (From Odum: *Fundamentals of Ecology,* 3rd ed. Philadelphia, W. B. Saunders Co., 1971.)

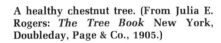

A healthy chestnut tree. (From Julia E. Rogers: *The Tree Book* New York, Doubleday, Page & Co., 1905.)

derstand the true complexity of the ecosystem.

The coyote is not now an endangered species, so why use this particular example here? The reason is that in the case of the coyote there is still time to initiate a program of balanced control. Thus we may be able to avoid a crash extermination campaign that would later be followed by a crash conservation program to save this species.

From
THE VILLAGE BLACKSMITH
by Henry Wadsworth Longfellow
Under a spreading chestnut tree
  The village smithy stands;
The smith, a mighty man is he,
  With large and sinewy hands;
And the muscles of his brawny arms
  Are strong as iron bands.

Extermination of Predators

Humans have always shared the planet with other predators. During the ages of co-evolution, delicate and curious relationships have developed. In general, large predators avoid attacking humans. This does not mean that people are never killed by other predators, but just that the total number of attacks has been relatively small. Yet many people think that they have to kill predators such as lions and wolves in order to secure safety for humans.

A second major source of conflict between predators and people is competition for food. Wild carnivores do kill game and domestic stock. Consequently, hunters and farmers have often attempted to exterminate predators. But elimination of all predators does not mean more food for humanity. As an example, let us examine the relationships among ranchers, grass, sheep, and coyotes in the Western United States and Canada. Coyotes' diets include many different items. They eat field mice, rabbits, pack rats, prairie dogs, sage hens, and bodies of lambs that have died of natural causes. They also kill young lambs. If the coyotes were exterminated, the populations of many rodents would increase. Jack rabbits, mice, and prairie dogs would flourish and eat great quantities of grass. There would be less food available for the sheep. On the other hand, a large coyote population would kill many sheep. Therefore, farmers would be wise to control coyote populations, but not to destroy all of them. Unfortunately, mass poisoning programs are being proposed by many ranchers who may not un-

For a detailed account of interactions between predators and humans, see Roger A. Caras: *Dangerous to Man.* New York, Holt, Rinehart and Winston, Inc., 1975.

Hunting for Sport and Fashion

It seems sad and entirely inexcusable that people should kill endangered species for sport or fashion. Yet this is exactly what is happening. Conservationists estimate that between 500 and 700 wild giant sable antelopes live today in west Central Africa. These animals have a set of magnificent curved horns, sometimes five feet long. These horns may easily be the animal's downfall. Hunters come from many parts of the world to shoot a big bull and bring home a trophy. Many hunters and tourists who are unable or unwilling to shoot a giant sable purchase their trophy from native hunters and return home with an expensive wall decoration. Hope for the survival of this species is slim.

The snow leopard is one of the few large predators living in the high country of the Himalayas. Leopard skins are considered to be elegant, and snow leopard skins are warm as well. As a result, there are fewer than 500 snow leopards left alive.

These two examples were chosen out of several hundred cases where sport and fashion hunters are responsible for large-scale killing of animals. Other well-known examples include Nile crocodiles for handbags and shoes, many jungle cats for furs, and several species of birds for their elaborate feathers.

4.4
HOW A SPECIES BECOMES EXTINCT — THE CRITICAL LEVEL

The pressures discussed in the preceding section reduce the numbers of

Coyote hunting a mouse. (Photo by Jen and Des Bartlett, National Audubon Society, Photo Researchers, Inc.)

Mountain lion cornered by bounty hunters in Colorado. (Photo by Carl Iwasaki, *Life Magazine*)

Giant sable antelope. (Courtesy of Richard D. Estes)

plants and animals of the species. But they rarely eliminate the last few surviving individuals. What factors lead to the final extinction of a species? Or, in other words, how does an endangered species become an extinct species?

Let us consider the case of the passenger pigeon. In the late 1880's, approximately two billion of these birds flew over the North American continent in flocks that blackened the skies. Commercial hunters killed millions for food and many more for fun, for the species was thought to be indestructible. As hunting pressures increased, the pigeon populations naturally suffered. By the early 1900's market hunting was no longer profitable. Yet thousands of pigeons survived. Then suddenly they all vanished—poof. Ducks, geese, doves, and swans had all been hunted and they survived in reduced numbers. Why did the passenger pigeons die out?

There appears to be a **critical level** for every population. As soon as the population of a species is reduced below the critical level, the species is no longer expected to recover. Even if the pressure is lifted, the species may well be on a one-way street to extinction. Suppose, for example, that hunters reduce the population of a species. Later a law is passed that prohibits hunting of the endangered species. If the critical level has already been reached, the species might die out even though the animals are no longer disturbed by humans. The critical level is different for each species. There are several reasons for these differences:

(a) Failure of Reproduction. When animals face certain types of stress, many of them fail to reproduce normally. For example, female rabbits reabsorb unborn fetuses into the bloodstream during drought. Many animals such as the Javanese rhinoceros have never bred successfully in captivity. Even people sometimes suffer from failure of reproduction. For example, an abnormally high percentage of American males became impotent during the economic depression of the 1930's. We just don't know

The passenger pigeon—a lesson learned too late.

enough about animal behavior to predict how a certain stress will affect fertility of a given animal. But we do know that in certain cases, destruction of range and pressures from hunters alter the reproductive potential of the remaining animals. If a species that is already hard pressed suddenly stops reproducing, extinction may result.

(b) Ecosystem Imbalance. If many plants and animals in an ecosystem are killed, the ecosystem may become imbalanced. A severe imbalance may then lead to the extinction of a species. The original North American population of a few billion passenger pigeons must have supported a large and varied population of predators. When commercial hunters slaughtered a significant number of pigeons, they did not shoot a proportional number of predators. It is possible that the pigeon was exterminated because the ratio of predator to prey was so unfavorable.

(c) Reproductive Isolation. If there are only a few individuals in a large area, members of the opposite sex may have a hard time finding each other. The Case History at the end of this Chapter discusses the wholesale killing of blue whales. These animals live in a huge area in the Antarctic, South Atlantic, and Indian oceans. If only a few individuals remain alive, they may never get together to mate.

(d) Bad Luck. When the population of a species becomes low, bad luck can strike the final blow. An unlucky example is the stellar albatross. The albatrosses were hunted heavily, but nevertheless a few flocks survived and bred. Then, in 1933 as one flock was nesting peacefully on an offshore island, a volcano erupted, killing most of the adults and all of the young. Again, in 1941, another volcano erupted and destroyed a second nesting population. The species survived for another 20 years, and reproductivity had just begun to accelerate when the last sizable flock was caught in a typhoon and destroyed. The future of the species is in question.

(e) Inbreeding. (mating between closely related individuals such as brother and sister or cousin and cousin) poses another danger to the existence of reduced population. Biologists have shown that, in general, unions between brother and sister or cousin and cousin have a high probability of producing weakened offspring. In a small population such mating has to occur quite frequently because there are so few individuals to mate. This inbreeding increases the probability that a species will become genetically weak and perhaps extinct.

4.5
SPECIES EXTINCTION AND THE LAW

In recent years many people have become increasingly aware of the need to preserve endangered species. But many others have not. In 1969 the hides of 113,069 ocelots, 7934 leopards, and 1885 cheetahs were imported into the United States. Many more hides were shipped to Paris, London, and Tokyo. Conservationists realized that legal action was necessary. In 1973 delegates from every major nation in the world except the People's Republic of China gathered to discuss the problem. They agreed to outlaw all trade of any part of any plant or animal that was endangered. In the same year the United States passed the Endangered Species Act. This law states that it is illegal to import or transport endangered species or products of endangered species. Furthermore, it is illegal to destroy the habitat of any endangered species. Canada, Great Britain, Taiwan, Kenya, and several other nations have also passed similar legislation.

These laws have certainly reduced

Leopard skins and other hides on sale in a street market in Srinagar, India. I asked the head of game management in Srinagar if it was legal to trap leopards and sell their skins. He informed me that the animals are endangered and carefully protected. I then asked about the hides for sale in the street. He told me that sometimes the law is difficult to enforce.

The partially completed Tellico Dam. This project has been halted by concern for the continued existence of the Snail Darter. (Courtesy of the Tennessee Valley Authority)

Snail darter.

the wholesale slaughter of endangered species. But a law on paper must be enforced if it is to save animals' lives. At John F. Kennedy Airport in New York City custom officials confiscate about $300,000 worth of products of endangered species every year. This collection includes leopard skin coats, hunting trophies, crocodile shoes, and many other items. No one knows how many more products are smuggled in. As another example, certain species of whales are currently endangered. Yet Japanese and Russian whalers continue to hunt these animals (see the Case History, Section 4.7).

In the United States, several court battles have challenged the Endangered Species Act. In 1967, workers began construction of a large dam called the Tellico Dam in Tennessee. Six years later, when the Endangered Species Act was passed, the dam was nearly complete. Construction had already cost $103 million. The dam was to provide enough electrical power to heat about 20,000 homes in the region.

In August, 1973, a zoologist studying the rivers in the area discovered a small species of fish that he called the snail darter. This rare little animal eats snails that live along the bottom of a few small streams. The snail darter is found nowhere else in the world. If the Tellico Dam is completed, the habitat of the darter will be destroyed and the species will become extinct. A United States District Court ruled that since the Endangered Species Act states that it is illegal to destroy the habitat of an endangered species, the construction of the dam must be halted. This court order saved the snail darter and left an incomplete pile of concrete in the river. The Tennessee Valley Authority (TVA), builders of the dam, took the case to the Supreme Court. In June, 1978, the Supreme Court upheld the decision of the lower court. Further construction of the dam was forbidden. The majority opinion of the court stated that, "The plain intent of Congress in inacting this statute was to halt and reverse the trend toward species extinction, whatever the cost." But the case is not closed. There is a possibility that Congress may pass a law to change the Endangered Species Act so that projects such as the Tellico Dam may be completed.

4.6
THE NATURE OF THE LOSS

Many people rise to the defense of "majestic species" such as the bald eagle or the Bengal tiger. But what about the lowly snail darter? This fish wasn't even discovered before 1973! What is its importance in the scheme of things? In a larger sense, why worry about the extinction of species at all?

There are several reasons to worry. In some ways the most compelling arguments are the aesthetic and religious ones. Different individuals may express their feelings in different ways. To some, species and the wilderness should be preserved simply because they exist. Others may say that humans have no right to exterminate what God has created. Still others may mention the great enjoyment they get from visiting an untouched wilderness area. There would be less richness, variety, and fascination of life on this planet if plants and animals were destroyed.

A second reason is concerned with development of medical and biological sciences. Many important discoveries have depended on various plants and animals as experimental subjects. Many different species have been used in these experiments. When Thomas Hunt Morgan started his famous studies of genetics, he chose the fruitfly, *Drosophila*, not because the life cycle of these insects was of particular interest in itself but because the animals were easy to study.

Wild plants and animals are often essential to agricultural scientists. Domestic corns are susceptible to various diseases. Maize is a member of the corn family. Wild maize does not produce nearly as much food as corn does. But these plants are more resistant to disease. By crossing domestic corn with wild maize, people have been able to develop high yielding, resistant crops. In the United States natural maize is often considered to be a weed and is poisoned. The loss of this species would be a severe blow to modern agriculture.

Scientists believe that there may be as many as 800,000 different species of plants. Of these only a few dozen species provide the major food supply for humans. A few more are cultivated for drugs. How many other plant spe-

Giraffe on the African Plain. (Photo courtesy of Amos Turk)

cies may be useful? In a recent study* it was learned that a great many rare plants have at one time been used for food or medicine. Agriculturalists are studying some of these. There is hope that new and efficient crops and medicines may be developed.

The same situation has arisen with animals. In Africa, many native grazing animals have been killed to leave room for cattle. But cattle often contract various tropical diseases and die. The native animals have natural immunity to many of these diseases. In certain regions, farms produce less meat in a given land area than natural prairies do. Yet people are destroying the natural prairies to make room for cattle.

Still another reason for saving species from destruction concerns the richness and variety of life on the planet. The Earth is constantly changing. Climates change, environments change, products and foodstuffs change. Traditionally, species of plants and animals have always evolved along with their changing environment. If a mountain range uplifted, certain organisms evolved to live at higher altitude. If climate changed,

Drosophila magnification about 6×. (Courtesy of Dr. E. Novitski, Department of Biology, University of Oregon)

*See Siri von Reis Altschul: "Exploring the Herbarium," *Scientific American*, May, 1977, p. 96.

Koko signs "visit" in requesting to have the gorilla Michael join her in her room for a play session. (Photo by Ronald H. Cohn, Ph.D., courtesy of the Gorilla Foundation)

many organisms adapted. Our world is now developing and changing rapidly. Most scientists feel that if our systems are to remain stable, plants and animals must evolve to meet the change. But evolution is slow. If we allow species to become extinct now or in a few short years, the total global potential for adaptation and change will be greatly diminished. The stability of future systems may be endangered.

Traditional biology has taught that we cannot assume that animals have "human" feelings. According to this belief it is improper to say that a dog or a monkey is jealous, angry, happy, or content. A dog may bite a person, but we cannot assume that a dog bites because it is *angry*. A series of experiments with chimpanzees and gorillas has made people wonder whether this idea of animal emotion is really true. Monkeys and apes do not have well-developed vocal cords. Therefore, they cannot speak. But several of these animals have recently been taught various types of sign language. For the first time in history, a human

being can talk directly to a member of another species. The results have been significant. These animals express fear, love, and reason. They remember the past and project into the future.

In one case, a gorilla, Koko, was taking a bath. After washing herself the ape tore the sponge to shreds. Her trainer, Penny, yelled at her and told her not to do that again. Penny shook the sponge in the ape's face and asked "What is this?" The ape was silent for a moment and then said, "Trouble." Then she reached out and embraced Penny. Another time a photographer was asking Koko to pose for a picture. Koko didn't like the photographer and signed "Me angry, me bite." (She was calmed down and did not bite the photographer.)

Our entire moral and legal system is based on the assumption that animals do not feel emotions as humans do. But now we know that at least some animals not only feel emotions, but they can express them as well. Once we accept this fact, we must recognize that an animal has a right to live, just as a human does.

4.7
CASE HISTORY: THE BLUE WHALE

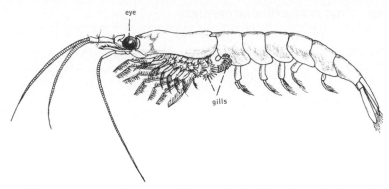

Krill (Euphausia superba) are small (5 cm) shrimplike animals that provide the major food source for most Antarctic whales.

Antarctica is a cold, frozen continent, covered with ice and having almost no vegetation. But, the southern oceans that surround this continent are rich, fertile, and teeming with life. In fact, life is more abundant in the Antarctic Ocean than in many central tropical oceans. Deep-sea currents carrying valuable nutrients rise in the lower latitudes and fertilize the surface layers of the southern ocean. The nutrient-rich waters support large populations of phytoplankton, the primary producers of the ocean. In turn, there are a great many zooplankton that feed on the tiny plant species. The zooplankton population consists, in part, of various species of small, shrimplike animals known as **krill.** An individual krill is only a few centimeters long. But these animals group together in enormous schools that may be ten meters thick and cover a surface area of several kilometers.

Schools of krill this size naturally represent a significant food supply. In fact the schools are large and plentiful enough to feed the largest animal that has ever lived on this planet — the blue whale. A blue whale is about 30 meters (100 ft) long and weighs 90,000 kg (100 tons). It has no true teeth, but rather a set of elastic, horny plates called **baleen** that forms a sieve-like region in the whale's mouth. A blue whale feeds by swimming through a school of krill with its mouth open, allowing the krill to pass through the baleen but excluding any larger objects.

About the turn of the century, it was estimated that there were approximately 200,000 blue whales in the world. In 1920, over 29,000 whales were killed. The blubber was boiled down to oil to be used directly as a lighting fluid or as a raw material for the manufacture of cosmetics, medicinals, lubricants, shoe polish, paint, and other products. Some of the meat was eaten by humans; some was thrown away. In later years some was ground into food for dogs and domestic mink. Whaling was a dangerous but profitable business.

Whale populations have been destroyed by hunting pressures. In nature, small animals usually have a higher reproductive rate than larger animals. For example, some female mice can give birth to about twenty-five infants a year. Insects lay several thousand eggs a year. Both these animals have evolved to live with powerful predators. Many young are produced. Most of them die before they mature. But large animals, and especially large mammals, generally bear only a few young. The infants are cared for, and predators are less numerous, so most of the infants live to maturity. This cycle is part of the natural balance. If many babies were born, and most of them lived, they would quickly overpopulate their range and upset their ecosystems. For millions of years blue whale populations were stable. Few predators attacked the whales. Birth rates and death rates were low. Then suddenly whalers sailed south and killed many of the animals. Whales are defenseless against harpoons armed with high explosives. Furthermore, dead whales (unlike dead mice or dead insects) are not rapidly replaced by newborn infants. As a result, the whale population started to decline rapidly. In the middle 1950's biologists warned the whaling communities that blue whales would be threatened with extinction if current practices continued.

The wholesale slaughter of the whales was not only disruptive to natural ecosystems it was economically foolish as well. If whalers would practice reasonable conservation efforts and allow a large breeding population

Skeleton of an Atlantic right whale. Note the numerous sievelike baleen in the whale's mouth. The blue whale is a plankton eater and also has baleen rather than teeth. (Courtesy of the American Museum of Natural History)

A Finback whale in the Sea of Cortez in Mexico. (Photo by Jen and Des Bartlett, Photo Researchers, Inc.)

to survive, they would be able to harvest several thousand individuals a year. On the other hand, if conservation was not practiced, the breeding population would be destroyed. Then, in a very short time, there would be no more whales. Both whales and whalers would suffer. Unfortunately, no nation or group of nations controls the open ocean. Each owner of a whaling vessel is able to do as he pleases. Often, the owners choose to kill whales as quickly as possible before other whalers catch them. They ignore long-range economic or ecological consideration. So the slaughter continues.

The International Whaling Committee (IWC) was established in 1946 to regulate whaling practices. The Committee ignored the recommendations of most knowledgeable marine biologists and set limits that greatly exceeded the levels that would allow whales to survive. The whalers ignored even these generous guidelines and killed more whales than the IWC recommended. Predictably, the whale population declined and the total catch decreased. In 1964–65 only twenty whales were harvested, and in 1974–75 the catch dropped still further. With the blue whale population nearly extinct, whalers turned to hunting other, smaller species. In turn, some of these may become endangered in the near future.

At the present time, the blue whale population is estimated at a few thousand. Some experts feel that the critical level may have been reached already and that the blue whale, largest of all animals ever to live on this planet, is destined for extinction. At the same time, the total yield from whaling has declined to well below the potential productivity of the southern ocean. As the result, the world community has directly suffered from the slaughter of the whales. It is now illegal to import products of the blue whale into many nations, such as the United States, Great Britain, and Canada. The problem is that people from a few nations, including Japan and Russia, continue to hunt Antarctic whales. In a sense, the sea belongs to all of us, the people of Earth. But at present, international maritime law cannot be enforced. A few people are becoming wealthy by hunting whales, and everyone else is suffering. If current practices continue and species of whales become extinct, we will all suffer a great aesthetic and scientific loss. In addition, a productive region of the Earth's surface will produce less protein and fewer raw materials for the human population.

CHAPTER SUMMARY

4.1 WHAT IS A SPECIES?

A **species** is a group of plants or animals that breed together but do not breed outside the group. Character traits are mixed within a species but are not passed from species to species.

4.2 EXTINCTION OF SPECIES

Species have evolved and become extinct for hundreds of millions of years. It is believed that people were responsible for many extinctions approximately 10,000 years ago. Recently rates of extinction have, once again, been high.

4.3 WHY ARE SPECIES BEING DESTROYED TODAY?

People are the major agents of species extinction today. The major mechanisms of extinction are: destruction of habitat, introduction of foreign species, extermination of predators, and hunting for sport and fashion.

4.4 HOW A SPECIES BECOMES EXTINCT — THE CRITICAL LEVEL

When a population of a species falls below a certain number, called the **critical level,** the species may easily become extinct even though many mating pairs remain alive. Critical levels operate because:

(a) When animals face certain types of stress, many of them fail to reproduce normally.

(b) Often the pressures that act on one species of an ecosystem do not affect the entire ecosystem equally, and severe imbalance can result which might lead to extinction.

(c) The population density may decline to a point where members of the opposite sex have a hard time finding each other.

(d) When the population of a species becomes low, that population is particularly subject to bad luck.

(e) Inbreeding poses a genetic danger to the existence of reduced populations.

4.5 SPECIES EXTINCTION AND THE LAW

Many national and international laws prohibit further destruction of species. However, poachers break laws. In the United States conflicts have arisen between those who want to protect rare habitats and those who favor further industrial development.

4.6 THE NATURE OF THE LOSS

The loss of a species is: (a) an aesthetic loss; (b) an injustice, from an ethical or religious viewpoint, if it results from human actions; (c) a loss of plants and animals that may have a potential scientific or economic interest or value; (d) a decrease of genetic variety on the planet. In addition, people must realize that animals have feelings and we should regard life in high esteem.

KEY WORDS

Species
Land use
Critical level

Inbreeding
Krill
Baleen

TAKE-HOME EXPERIMENTS

1. **Habitats.** Prepare a list of the various types of environments that exist within a 10 km radius of your home. Typical entries might be: residential; commercial; public park; woodlot; farm; lake; stream or river; undeveloped mountain. What types of wild plants and animals live in each of the sections? Discuss the effect of civilization on species distribution in your area.

2. **Foreign species.** Go to local farmers or nursery owners and ask them to list the major insect pests in your region. How many of these pests are native to your area? How many have been imported from foreign lands?

PROBLEMS

1. **Definitions.** Define species. Explain why a species cannot be re-created after it has become extinct.

2. **Extinction of species.** Discuss the statement, "People are, and have always been, the primary agents in the extinction of species."

3. **Animal extinctions.** Discuss the role of primitive people in the extinction of mammals approximately ten thousand years ago. Do you feel that people, alone, caused these extinctions? Defend your answer.

4. **Species extinction.** How do we know that people are the primary cause of species extinction today? Discuss.

5. **Species extinction.** List the four major factors leading to species extinction in modern times. Discuss each briefly.

6. **Habitat destruction.** List the major factors which lead to habitat destruction in the modern world. What factors would you list as being most disruptive to wild animals? Explain.

*7. **Habitat destruction.** Elk normally feed in the high mountains in summer and travel to lower valleys during the winter months. Imagine that a developer was planning to build a housing complex in a mountain valley in Montana. The developer claims that since only 10 percent of the elk's annual range is being destroyed, the herd will not be seriously affected. Do you agree or disagree with this argument? Defend your position.

*8. **Introduction of foreign species.** Australia has been separated from Asia, Europe, and Africa for millions of years, and consequently many species of plants and animals unique to this continent have evolved. When European settlers immigrated to Australia in the nineteenth century, they brought with them many new species of plants and animals, such as sheep, dogs, rabbits, and various cereal grains. In many cases these new organisms displaced native ones. Many native species have recently become extinct. Suggest some reasons why these substitutions have occurred.

*9. **Environmental impact of major construction.** When ships pass from ocean to ocean through the Panama Canal they are raised through a series of locks, sail across a freshwater lake, and then are lowered through a second series of locks. Passage through the canal would be easier if a deep trench were dug to connect the Atlantic and Pacific oceans directly. Would such a canal affect the survival of aquatic species? Discuss.

10. **Predators and people.** Explain why elimination of coyotes might actually reduce the number of sheep that could be raised on a western range.

*11. **Predators and people.** Many Alaskan natives believe that people should be allowed to poison wolves and hunt them from airplanes. They argue that wolves kill moose and thereby reduce the number of moose available for human consumption. Do you agree with this argument? Discuss.

---

*Indicates more difficult problem.

12. **Critical level.** Define critical level. Give an example.

13. **Critical level.** Discuss how the critical level might differ greatly from species to species. In your answer, briefly explain how each of the following factors could affect different species differently: (a) reproductive failure, (b) ecosystem imbalance, (c) ability of males to find females, (d) luck, and (e) inbreeding.

14. **Endangered species and the law.** Discuss some legal protections for endangered species. Are the laws 100 percent effective? Explain.

15. **Justification for the preservation of species.** Discuss the importance of wild grasses in the modern world.

16. **Blue whales.** Some people believe that plants and animals should be managed for maximum value to people. Imagine that you were not concerned at all with the welfare of whales, but wanted to increase the total harvest of food for humans. Would you recommend that whalers kill many blue whales in the next few years? Discuss.

QUESTIONS FOR CLASS DISCUSSION

1. The Serengeti Plain in East Africa is one of the last homes of large herds of migratory animals. Wildebeests, zebras, gazelles, giraffes, elephants, rhinoceroses, lions, jackals, cheetahs, leopards, and many other animals live there much as they have lived for tens of thousands of years. Conservationists believe that the plains should be maintained as a game preserve so that the animals may live in peace. They state that there should be no human interference in the Serengeti, except tourism. The argument continues that tourists who visit the region will spend enough money to support the native people in the area.

The Masai tribespeople have lived and hunted in the Serengeti long before governments attempted to regulate the region. They believe that they should be allowed to continue to live as they have for generations. However, with recent advances in modern medicine, the population of the Masai has been increasing drastically. If they are allowed to hunt and farm the Serengeti they might overpopulate it and destroy the ancient ecosystems.

Some game management scientists think that the wild herds should be harvested for meat. They feel that with careful management thousands of animals could be shot yearly, and eaten. The herd would be reduced, but not eliminated. And tourists could still visit the region. Discuss the merits of each of the various positions — the conservationists, the Masai, and the game management scientists.

2. Prepare a classroom debate. Have one side argue that the Tellico Dam should be completed, thereby destroying the habitat of the snail darter. Have the other side argue that the dam construction be stopped permanently. One important issue in the debate is: Where do you draw the line? Are bald eagles more worthy of being conserved than snail darters? Why or why not?

BIBLIOGRAPHY

Four books which deal specifically with the destruction of animal species are listed below:
Roger A. Caras: *Last Chance on Earth.* New York, Schocken Books, 1972. 207 pp.
Kai Curry-Lindahl: *Let Them Live.* New York, William Morrow & Co., 1972, 394 pp.
H. R. H. Prince Philip, Duke of Edinburgh, and James Fisher: *Wildlife Crisis.* Chicago, Cowles Book Company, 1970. 256 pp.
Vinzenz Ziswiler: *Extinct and Vanishing Animals.* New York, Springer-Verlag, 1967.

Two valuable books which deal more specifically with ecosystem alterations are:
David W. Ehrenfeld: *Biological Conservation.* New York, Holt, Rinehart and Winston, Inc., 1970. 226 pp.

Charles Elton: *The Ecology of Invasions by Animals and Plants.* New York, John Wiley & Sons, 1958. 181 pp.
A detailed, fascinating, and advanced discussion of the Pleistocene extinctions is given in:
Paul S. Martin and H. E. Wright, Jr. (eds.): *Pleistocene Extinctions.* New Haven, Yale University Press, 1967. 453 pp.

An interesting book that discusses the relationship between wild animals and humans is:
Roger A. Caras: *Dangerous to Man — The Definitive Story of Wildlife's Reputed Dangers.* New York, Holt, Rinehart and Winston, Inc., 1975. 422 pp.

Overpopulation is causing serious problems throughout the world. In New Delhi, India, many homeless people sleep in the streets.

5

GROWTH AND CONTROL OF HUMAN POPULATIONS

5.1
INTRODUCTION

Think of the place where you grew up. Picture in your mind how it looked when you were young and what it looks like now. If your child-hood home was a suburb or a small town, some of your favorite secret wooded areas may no longer exist. Today, these woodlands are the drive-ways of large apartment complexes or the sites of huge shopping centers. Destruction of scenic, quiet areas is

View of Harlem River High Bridge looking northwest toward Manhattan from the Bronx, 1861. (From the collection of the New York Historical Society)

one of the effects of rapid population growth. Ask your parents to compare the childhood appearance of the place in which *they* grew up to its appearance now. Their memories will probably provide sharper contrasts than your own. If your parents are from New York City, they may remember farms in the Bronx. Californians will remember vast expanses of uninhabited land. Southerners will recall small cities and very rural areas. If you were to ask your great-grandparents to recall their childhood environments, they might tell you about wagon trains, buffalo herds, and Indian children riding ponies across the prairies. In other regions of the world, rapid population growth not only has changed the appearance of the land, but has also been a factor that has caused misery and starvation. In fact, most of the major environmental problems in the world today are related to population size.

Dramatic changes in local populations have occurred since prehistoric times. Humans first evolved about 100,000 years ago. During the early years of human life, the total population of the Earth must have varied widely. In some years there were more deaths than births, so the human population decreased temporarily. By the first century A.D., however, the world population had already established its present pattern of almost uninterrupted growth. At the time of the discovery of America, there were about one-quarter billion people alive on Earth. In 1650, about a century and a half later, world population had doubled to one-half billion. In another 300 years, world population multiplied fivefold to 2.5 billion persons. During the 1950's the population increased almost another one-half billion. In 1978, world population was approximately 4.5 billion persons. Today there are more people in China than there were people on Earth in 1650. Or consider yet another comparison. More than two thirds of all people born since 1500 are alive today. Figure 5–1 shows a schematic graph of world population size since the emergence of human beings.

A glance at the curve shows that world population growth is becoming more and more rapid. Indeed, Figure 5–1 may well cause the reader to panic. If world population continues to grow at its current rate, very soon there will be too many people for the Earth to support. If poverty and starvation exist now, how can an even larger population be fed and cared for? Destruction of land, depletion of natural resources, production of wastes, and pollution of the air and water can all be expected to increase with increased population.

What will actually happen in the

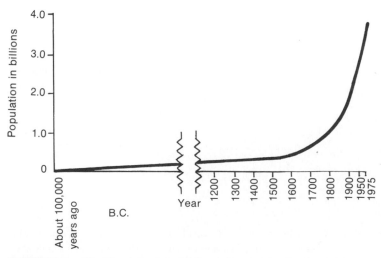

FIGURE 5–1. World population size from emergence of *Homo sapiens* to 1975.

future? If the present rate of world population increase were to continue, there would be ten people for every square meter of the Earth's surface in less than 700 years. But we know that it is impossible for people to stand elbow to elbow on this planet. One square meter of the Earth's surface cannot feed, clothe, and shelter ten people. Therefore, the world population cannot continue to increase at its present rate for long. In fact, it is obvious that the *world's population size cannot grow forever*, not even very slowly. What factors will eventually control growth? Optimists predict that populations will be controlled by calm, rational decisions made by families and nations. Other people believe that famine, war, and disease will eventually limit the number of people on the planet.

5.2 PREDICTING FUTURE POPULATIONS

It is not easy to predict how human populations will change. The problem is that complex political, scientific, and social changes affect population growth. In 1900, a forecaster may not have guessed that World War I would have killed millions in Europe. In 1920, many people did not foresee that new drugs would drastically reduce death from infectious diseases such as pneumonia. In 1950, hardly anyone predicted that many women in most developed nations would decide to bear fewer children than their mothers did.

"Excuse me, sir, I am prepared to make you a rather attractive offer for your square." (Drawing by Weber; © 1971 *The New Yorker Magazine*, Inc.)

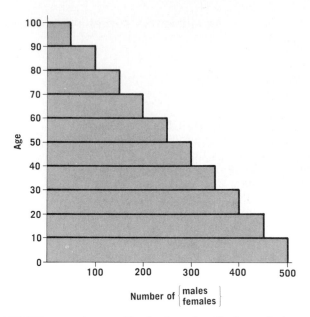

FIGURE 5–2. Age-sex distribution of an ideal population.

Demography is the statistical study of human populations. A demographer examines the structure of existing populations and tries to predict future growth rates. How does a population grow? Imagine that there is a small town that we will call Peopleville. One thousand people live in the town. In one year 17 babies are born, 9 people die, 3 move into the town, and 1 moves away. At the end of the year the population will be:

```
  1000 people originally in the town
+   17 babies born
−    9 deaths
+    3 people moving into the town
−    1 person moving away
  ─────────────────────────────────
  1010 people at the end of the year
```

This calculation tells us what *has* happened in the past but gives us little clue as to what *will* happen in the near future. A demographer who wants to predict the population a year later will have to estimate the number of births, deaths, immigrations, and emigrations. More information is needed to forecast these events. Obviously, if there are a great many young men and women in Peopleville, there is a high probability that there will be many births. If there are many very old people, we might expect more deaths in the near future. A demographer studies the ages and sexes of people in a population in order to predict future trends.

Figure 5–2 is an **age-sex distribution** for an imaginary population. This graph tells us that there are about 500 infants and children between the ages of 0 and 10. Many children die at an early age. In the population in Figure 5–2 there are only about 400 people between the ages of 10 and 20. The probability of dying decreases for young adults, so there are 350 people between the ages of 20 and 30. Very old people have a high probability of dying soon, so the population declines rapidly at the top of the graph.

Let us use this graph to predict future population growth. We see that there are more babies than teen-agers, more teen-agers than young adults, and more young adults than older people. Women in the age group 17 to 40 are most likely to bear children. In five years the large population of young teen-agers will become adults, and bear children. In ten years the even larger population of children will grow up and bear more children. Therefore, from the graph in Figure 5–2 we can predict that the population will increase rapidly in the near future.

Do all population distributions look like the one in Figure 5–2? No, not at all. Each nation has its own individual distribution, because the number of men, women, and children is different for each country. Figure 5–3 shows the age-sex distribution for three nations: India, Norway, and West Germany. We see that the curve for India has a very broad base. There are a great many young children in the population. Therefore, we would expect that the population would expand rapidly during the next generation. In Norway, there are approximately an equal number of people in each age group between 0 and 30. Ten years from now there will be about the same number of men and women in the reproductive age group as there are now. Most probably, the population will remain fairly constant in the near future. In West Germany, there are *fewer* infants and children than young adults. Ten years from now there will be fewer people in the reproductive age group than there are at present. Therefore, it

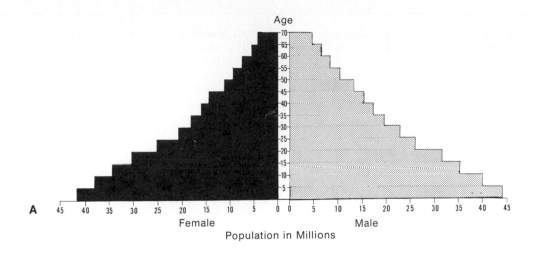

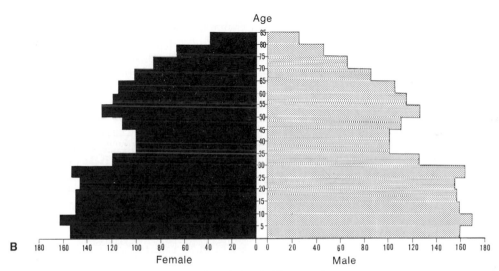

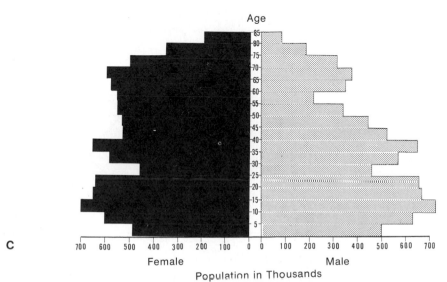

FIGURE 5–3.　Age-sex distributions for three nations. (Source: *U.N. Demographic Yearbook, 1975*) *A,* India — a rapidly expanding population. *B,* Norway — a stable population. *C,* West Germany — a declining population.

The developed nations of the world are those that are most heavily industrialized. The United States, Canada, all of Europe, USSR, and Japan are among the developed nations. The developing nations are those that are less heavily industrialized. India, all of Africa, Southeast Asia, and most of South and Central America are developing. One third of the world's population lives in the developed nations but consumes 85 percent of the global resources.

seems likely that the population will eventually decline.

5.3
THE DEMOGRAPHIC TRANSITION

For many thousands of years, human population grew slowly. Suddenly, in the fifteenth century, population began to rise rapidly. What happened? Have women been bearing more and more children during the past 500 years? The answer is no. In fact, the average number of births per family has become smaller. The change has occurred because people live longer, on the average, than they used to.

When nutrition is poor, water unclean, and infectious disease common, relatively few people live to adulthood. Many children are born, but many die. In some societies, half of all live-born infants do not reach their fifth birthday. As modern principles of health care are introduced into a society, death rates start to fall.

In the developed countries, the death rates have been dropping steadily for over a century. During the past one hundred years, clean water has become increasingly available, milk has been pasteurized, and new drugs have reduced death from disease. In addition, improved agriculture and more humane distribution of food have continued to aid in improving children's health. Therefore, death rates have declined gradually. During this time, people have come to understand that it is possible to raise a family without having many babies. Consequently, the birth rate has been declining for almost a century. Since birth rates and death rates have *both* declined, the populations have remained relatively stable.

$$\text{(a)} \quad \text{Birth rate in year X} = \frac{\text{Number of live children born in year X}}{\text{Midyear population in year X}} \times 1000$$

$$\text{(b)} \quad \text{Death rate in year X} = \frac{\text{Number of deaths in year X}}{\text{Midyear population in year X}} \times 1000$$

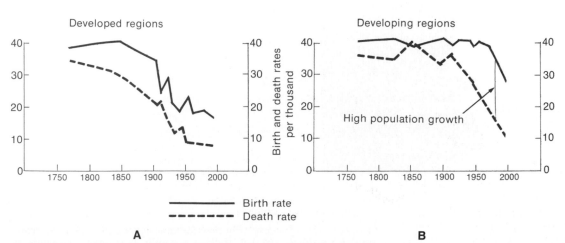

FIGURE 5–4. Birth and death Rates in developed and developing regions.

TABLE 5–1.   SOME TYPICAL BIRTH AND DEATH RATES BEFORE, DURING, AND AFTER DEMOGRAPHIC TRANSITION*

	Birth Rate/1000	Death Rate/1000	Rate of Natural Growth/1000
Very high birth and death rate			
Afghanistan	49.2	23.8	25.4
Niger	52.2	25.5	26.7
High birth rate, moderate death rate			
India	42.9	16.9	26
Morocco	46.2	15.7	30.5
Moderate birth rate, low death rate			
Argentina	22.7	9.4	13.3
Spain	18.2	8.0	10.2
United States	14.8	8.9	5.9
Low birth rate, low death rate			
Switzerland	11.7	9.0	2.7
Netherlands	12.9	8.3	4.6

*Source: UN Population and Vital Statistics Report, January 1, 1978.

Poor countries of the world have not experienced such gradual change. For centuries, women bore large families, and some half of the children died. Then in the 20th century modern medicine arrived suddenly from the rich nations. Health care improved dramatically in a relatively few years. Death rates dropped very quickly. People's patterns of behavior in the developing nations have not had time to adjust. Birth rates are still high. With high birth rates and lower death rates, population has increased rapidly.

Demographers summarize the population change of a country in the following way. Societies with primitive medical care are characterized by both high birth and high death rates. Since the difference is small, there is little or no growth. When modern medicine is introduced death rates among children drop. Birth rates, however, remain relatively constant. The combined effect of these two trends — falling death rates and constant birth rates causes population to grow rapidly. After some time, people get used to the fact that fewer of their children will die young, and birth rates drop. Therefore, the developed countries have low growth because birth and death rates are both low. This series of changes from high birth and death rate to low birth and death rate is known as the **demographic transition.** Table 5–1 lists some countries at various demographic stages.

5.4
ZERO POPULATION GROWTH

Most of the major environmental problems in the world today are at least partially related to population size. Think of it this way. An explorer traveling across the western parts of

Recent research has indicated that this picture of the demographic transition may be oversimplified. In several European countries, for instance, birth and death rates declined simultaneously and at nearly the same rates in the nineteenth century. In France, examination of records of births and deaths indicates that in many parishes the decline in birth rate actually *preceded* the decline in death rates. Further study of transition is an area for research in demography.

North America two hundred years ago would see few other people. In the evening it would have been a simple matter to pitch camp and go to sleep. Tree limbs and brush provided more than enough firewood. Clean drinking water was abundant. The rivers were teeming with fish, and game was plentiful on the prairies and in the woodlands. If a hunter shot a deer or caught a fish, the populations of game were not disrupted. Land, fuel, and food were all plentiful. Today, the situation has changed. Imagine that you visit Glacier International Park on the United States–Canadian border. So many people visit this region that simple actions such as walking through the woods or setting up tents have begun to upset the ecosystem. Therefore specific camping areas and hiking trails have been set aside. People are crowded together. It is illegal to gather firewood, for fuel is scarce. There is a danger that streams might become polluted, so safe drinking water must be piped into the camping areas. Hunting and fishing are carefully regulated. Even if everyone acts with concern toward the environment, the presence of so many people threatens the forest and mountain ecosystems,

The same problems exist throughout the world. Every person needs space, fuel, water and food. Waste products are produced. If population is sparse and people are careful there will be more than enough for everyone. On the other hand, if there are too many people, it becomes a major problem to provide even simple necessities.

The problem of overpopulation is easy to recognize but extremely difficult to solve. Birth rates change slowly. *Even if, by some miracle, birth rates were to be reduced today, human population would continue to increase for several decades.* It is important to understand why populations cannot stabilize quickly.

Suppose that each woman in a region gave birth to exactly two children. On the average, more boys are born than girls, so that in the next generation there would be fewer women capable of bearing children. Some children die before they grow up. Therefore, the total population would eventually decline. The **replacement level** is the average number of children that each woman must bear to maintain a constant population — that is, to achieve **zero population growth.** The replacement level varies with child death rates. In the United States, population would eventually stabilize if each woman bore an average of 2.1 children. In many developing nations, the replacement level is about 4.5.

The birth rate of a population depends on: (a) how many babies are born to each woman, and (b) how many women of reproductive age there are in the population. If there are many young women, we would expect the birth rate to rise temporarily, even if each woman has only one or two children. In the United States in 1976, each woman, on the average, was expected to give birth to 1.8 children. Yet the population continued to grow. How could that be? There was a large number of women in the most active childbearing age (17 to 40). There were even more females in the pre-childbearing age (0 to 17). In fact, approximately 70 per cent of all women were in the reproductive age group or younger. Thus, there were a lot of potential mothers in the population. Middle-aged women continue to bear babies. Young girls and teen-agers will soon grow up and have babies of their own. Therefore, the population will continue to rise. Even if women continue to bear an average of only 1.8 children, the population in the United States will continue to increase well into the 21st century. Only later will the population stabilize and decline.

The situation for a country such as India is even more discouraging. Right now approximately 40 percent of the population in India is under the age of fifteen. Even if women began to bear children at the replacement level *immediately,* the population will rise until it is 1.6 times as great as it is now. No one expects the birth rate to drop overnight. An optimistic estimate is that the birth rate in India

Of course, no woman bears 2.1 children. What this means is that 100 women would bear 100 × 2.1 children, or 210 children.

might be expected to gradually decline to the replacement level in thirty years. If this happens, the population of India will grow to about 2.5 times its present size. Somehow, all these people must be fed, clothed, and cared for.

5.5
LAWS AND SOCIAL ATTITUDES CONCERNING POPULATION GROWTH

In the past, most nations have been in favor of population growth. The Declaration of Independence of the United States accused King George III of England of trying to limit population. Even the ancient codes of Hammurabi of Babylonia and Emperor Augustus of Rome contained provisions encouraging births. Only since World War II has the desirability of population growth been seriously questioned by many governments. Today different nations have different policies toward growth. Poor countries with high birth and death rates are most concerned with raising the standard of living. The governments try to improve medical care. This improvement leads to decreased death rates and a rapidly increasing population.

Some countries with a rapidly expanding population are trying to encourage a reduction in births. Governments issue statements of public policy, distribute propaganda posters, and offer information on methods of family planning. Some governments even offer supplies for contraception. In recent years, several nations have found that these methods do not lead to a rapid enough decline in birth rates. As a result, stronger methods have been adopted. In Singapore, for example, the state provides paid maternity leave from a job for two births but not for any more. Also, only the first four children in a family have access to first-rate primary education. Laws of this type are effective, but they do have their faults. What happens to the fifth child in a family? He or she didn't ask to be born. Yet the child who is denied an adequate education will find it very difficult to hold a job in the future.

He has endeavored to prevent the population of these States: for that purpose obstructing the Laws for Naturalization of Foreigners; refusing to pass others to encourage their migrations hither, and raising the conditions of new Appropriations of Lands. (From The Declaration of Independence of the United States.)

Ultimately, however, it is family decisions, not government policies, that determine birth rates. Why do some couples bear no children at all, and others choose to bear ten or even fifteen? Of course, we can't give any definite answer to that question. Some people may find that children provide companionship. Parents may see a continuation of themselves in future generations. It could be that some people simply love children and want to have young ones around the house. Or that babies prove masculinity of the father or femininity of the mother. Or that they provide social recognition. In some societies children help support a family and provide old age security for the parents. In other societies, women may be active in a career and thus choose to bear small families. Tax or welfare structures may affect family size. Couples may decide to have a child to save a marriage or to maintain social or religious customs. Sometimes pregnancies are unwanted.

In general, middle-class people tend to have smaller families while poorer people tend to have larger families. With fewer mouths to feed and backs to clothe, well-to-do families become wealthier. Children receive the benefits of a good education. They can then find jobs when they grow up. A positive cycle develops. Small families promote wealth and wealth promotes small families.

Many social workers try to encourage poor people to have small families. They argue that if there are only a few children, then each one will have a better opportunity in life. Response to these arguments has been slow. Customs and habits do not change overnight. Birth rates are still high in many areas of the world.

Children in a developed society.

Joan P. Mencher discusses this problem in an article entitled, "Socioeconomic Constraints to Development: The Case of South India." (*Transactions of the New York Academy of Sciences,* 1973, pp. 155–167).

In recent years, many new reliable and inexpensive birth control methods have been developed. These devices obviously have had an effect on birth rates. But at the present time the birth rates of many nations are still increasing. Problems arise with birth control. Often women in developing societies become pregnant when they use birth control techniques carelessly. In some regions, women mistrust family planning programs and will not participate. One author reports that in southern India poor people are often mistreated in medical clinics where they obtain birth control devices. She says, "To have a baby does not require contact with hospital people, but to *avoid* having a baby requires contact with maternity assistants, doctors, etc., all of whom tend to treat the poor and low-caste people as 'animals.' " Many people avoid these clinics.

You can see that birth rates are influenced by many factors. Social attitudes and family decisions are important. Economic position and education often affect the reasons why people choose to have or not have children. Other factors such as child labor laws and employment for women influence people's decisions. The availability of medical aid and birth control are also important. Obviously, it is not easy to change patterns of population growth.

Family Planning clinic in India.

Children in a developing society.

5.6
CASE HISTORY: TAIWAN

Taiwan (formerly called Formosa) is an island about 150 km off the southeast coast of mainland China. Its area, 36,000 km², is about the size of Delaware and Maryland combined. Nearly half of the island is mountainous, and few people live there. The remainder is densely populated. The total population is slightly over 15,000,000 persons, or three times the combined populations of Delaware and Maryland.

Figure 5–5 shows the population growth of Taiwan over the past 400 years. Observe that from 1600 to 1900

Terraced farms in the mountains of Taiwan. (Photo courtesy of the Chinese Information Service)

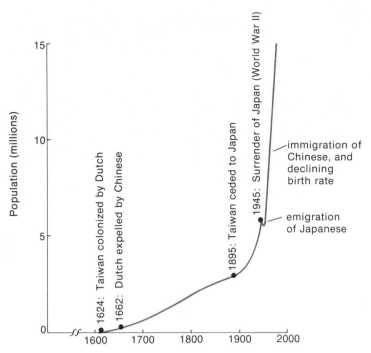

FIGURE 5–5.   A demographic history of Taiwan.

the population rose slowly. Then a rapid increase occurred. After World War II the Nationalist and Communist Chinese struggled for the control of the mainland of China. When the Communists won, the Nationalists es-

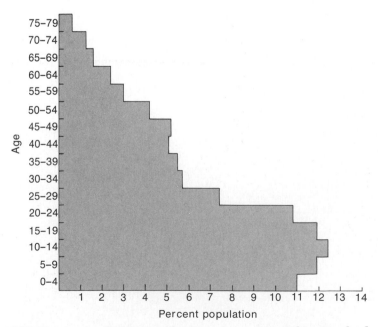

FIGURE 5–6.   Age distribution of Taiwan; Jan. 1, 1976. (*China Yearbook, 1976*)

tablished a government on Taiwan. During the period from 1947 through 1964 more than one million people migrated from the mainland to Taiwan. At the same time, the death rate on Taiwan declined sharply. By 1964, the population had exceeded 12 million. By 1975, the population had reached 15 million. Thus, the population of Taiwan increased fivefold in the 70 years from 1905 to 1975! (see Fig. 5–5).

How has this extreme population pressure been felt? Taiwan has not experienced a food shortage. In fact, it is one of the few nations of the world that are still net exporters of food. The nation has maintained one of the highest standards of living in Asia. The large supply of cheap labor has provided Taiwan the manpower for rapid industrialization. The people are among the healthiest in Asia. The infant mortality rate is low, the life expectancy high, and the level of nutrition more than adequate. Literacy rates are among the highest in the developing world. Clearly, the effects of rapid population growth in Taiwan have not been disastrous.

However, rapid growth has led to an age-sex distribution (Fig. 5–6) with many children. In 1968, 42 percent of the population was under 15 years old. In the developed countries, the comparable proportion was 22 to 31 percent. The government has found it expensive to educate all of these children. Moreover, the population has grown so rapidly that there has been a shortage of physicians, dentists, nurses, and health technicians. Other social costs of rapid population growth have been apparent. Farms have been broken up into tiny plots as they have been divided among many children. Cities have grown noisy and polluted. Housing is inadequate and streets are congested with traffic. These changes have led to a conviction both in the government and among the general public that population size should be limited.

The religion of the island is a mixture of Buddhism, Taoism, and Confucianism. Confucianist teaching values large families, specifically urging families to have at least one son. However, none of the religions specifically forbids family planning. Taiwan, then,

High-rise apartments were introduced to Taipei in the 1960's. Today there are hundreds of apartment buildings in the city. (Photo courtesy of the Chinese Information Service)

This balloon, flying over Taipei, says "Childbirth needs planning for a happy family!" (Photo by Daniel Turk)

A district health station in Taipei with a poster urging family planning. (Photo by Daniel Turk)

is an example of a country ripe for an active and successful family planning program. The population growth has been very fast, so that the problem of population is obvious. The people are literate, healthy, and well fed. All these factors are important for the success of family planning. In spite of this, the initial introduction of family planning met with resistance. During the period from 1920 to 1950, the government's official position was to encourage growth. Population during these years was checked by three natural regulators — disease, famine, and war. Suddenly, around 1950, the situation changed dramatically. Political stability, improvements in agriculture, and modernization of medicine and public health measures led to a jump in the population growth. In 1950, responding to this sudden spurt in growth, a government commission issued a pamphlet on birth control. Official reaction was unfavorable. The leaders of the government had just been involved in several years of bitter war with the Communists. They viewed the pamphlet as a Communist plot to weaken the military by limiting the number of future soldiers.

In less than a decade, however, the governor of Taiwan became convinced that a reduction in population growth would be good for his country. It would not reduce the size of the army for at least twenty years and would be a stimulus to the economy. A national program for teaching women about contraception was begun.

In 1962, the Population Council began to study the best way to approach family planning in Taiwan. The study concluded that in Taiwan the best method would be to make visits to individual homes to teach about contraception. In 1968 the government officially started a program for population control. The program has enlisted the help of many government agencies and private organizations. Field workers and private doctors assist under government contracts. Family planning is promoted by the mass media, including radio and television spots, slide shows in movie theaters, posters in buses, newspaper releases, even advertisements on matchboxes. Field workers visit new mothers in their homes to offer free family planning assistance.

The program has been remarkably effective. The rate of population growth has fallen from 3 percent in 1958 to 1.9 percent in the mid-1970's. The government hopes to reduce the rate still further.

CHAPTER SUMMARY

5.1 INTRODUCTION

World population has increased rapidly in recent years. However, we know that population cannot increase forever, for the Earth can support only a limited number of people.

5.2 PREDICTING FUTURE POPULATION

Demography is the statistical study of human populations. In order to predict future populations, a demographer must know the **age-sex distribution.** If there are many young men and women in a population, we would expect a fairly high birth rate in the future. If there are many old people, we would expect a high death rate.

5.3 THE DEMOGRAPHIC TRANSITION

Societies with primitive medical care have high birth and high death rates. Populations remain stable. When modern medicine is introduced, death rates drop, but birth rates remain high for a while. The population rises very rapidly. Later, birth rates fall, and the population growth slows down. This sequence of changes is known as the **demographic transition.**

5.4 ZERO POPULATION GROWTH

Most of the major environmental problems in the world are related to population size. The **replacement level** is the average number of children that each woman must bear so that **zero population growth** is eventually reached. If there are a high percentage of young women in a society, the population will grow temporarily even if women bear children at the replacement level. Zero population growth will not be realized until the age-sex distribution of the population stabilizes.

5.5 LAWS AND SOCIAL ATTITUDES CONCERNING POPULATION GROWTH

Many complex and interrelated factors combine to affect birth rates. It is not easy to change present patterns quickly.

KEY WORDS

Demography
Age-sex distribution
Demographic transition

Zero population growth
Replacement level

PROBLEMS

1. **Growth.** Explain why human population size cannot grow forever.

2. **Age-sex distribution.** What is an age-sex distribution? How is it useful for predicting future population growth?

3. **Age-sex distribution.** Examine the four age-sex distributions in Figure 5–7. For

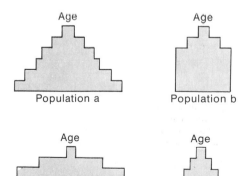

FIGURE 5–7. Hypothetical age-sex distributions.

each figure, predict whether the population will: (a) rise very rapidly; (b) rise slowly; (c) remain constant; (d) decrease.

*4. **Population distribution.** How would you expect each of the following to affect population growth? Consider which age groups are most likely to be affected by each event and how the event affects population change: (a) famine; (b) war; (c) lowering of marital age; (d) development of an effective method of birth control; (e) outbreak of a cholera epidemic; (f) severe and chronic air pollution; (g) lowering of infant mortality; (h) institution of a social security system; (i) economic depression; (j) economic boom; (k) institution of child labor laws; (l) expansion of employment opportunities for women.

5. **Demographic transition.** What is a demographic transition? How does it arise? Compare birth and death rates before, during, and after demographic transition.

6. **Demographic transition.** Name four factors that have been responsible for reducing death rates in developed nations. How have these factors affected developing nations differently from developed nations? Discuss.

7. **Demographic transition.** Explain why populations in many developing countries have increased rapidly in recent years.

---

*Indicates more difficult problem.

8. **Replacement level.** What is the replacement level? If women in the United States bear children at the replacement level, will zero population growth be realized immediately? Why or why not? Discuss.

9. **Zero population growth.** Would it be practical to stop population growth in India immediately with an active family planning program? Discuss.

10. **Population growth.** List and discuss briefly an economic, a political, and a social factor that is important in determining birth rates.

11. **Population growth.** Is it likely that birth rates in the developing nations will decrease to the replacement level within the next five years? Discuss.

12. **Social policies.** In the United States, families are allowed an income-tax deduction for each child. In most developed countries, each family is allotted an annual grant for each child. How do you think these tax laws affect family size?

13. **Social policy.** If you had the responsibility of discouraging population growth, would you consider reducing income tax deductions if a family has more than four children? Reducing health benefits? Whom would such policies harm?

QUESTIONS FOR CLASS DISCUSSIONS

1. How large was the family that your grandparents grew up in? Your parents? How many children are there in your family? How many children do you think an average family in a country such as the United States should have? Compare your answers with the answers of your classmates. Discuss in terms of changes in population growth rates.

2. Discuss the effects of population growth in your community. How do these problems compare with the problems of population growth in developing nations?

3. Rapid introduction of medical care in poorer countries has led to rapid population growth. In many cases, this population growth has, in turn, led to starvation and misery. Do you feel that people should stop sending medical aid to developing nations? Or should they send more aid? Discuss.

BIBLIOGRAPHY

This chapter has introduced demographic techniques for analyzing population growth. There are several valuable texts available for those interested in further study of demography.
Two excellent introductory texts are:
Peter R. Cox: *Demography.* 4th Ed. Cambridge, England, Cambridge University Press, 1970. 469 pp.
Donald J. Bogue: *Principles of Demography.* New York, John Wiley & Sons, 1969, 899 pp.

More mathematical introductions are:
Mortimer Spiegelman: *Introduction to Demography.* Rev. Ed. Cambridge, Mass., Harvard University Press, 1968. 514 pp. (Spiegelman includes an extremely large bibliography covering a wide range of topics related to population size, control, measurement, and so forth.)
Nathan Keyfitz: *Introduction to the Mathemat-* *ics of Population.* Reading, Mass., Addison-Wesley Publishing Co., 1968. 450 pp. (This highly technical and mathematical text is especially careful in its presentation of interrelationships among various measures of population composition and vital rates.)

A very complete source is the two-volume work:
Henry S. Shryock and Jacob S. Siegel: *The Methods and Materials of Demography.* Washington, D.C., Bureau of the Census, 1973, 888 pp.

The most useful and complete source of world population data is the
United Nations Demographic Yearbook, *published annually since 1948. For many nations and areas of the world the Yearbook includes the most recent available information on*

population sizes, vital rates, and many more specialized demographic statistics.

For up-to-date trends in population studies, refer to

Studies in Family Planning, *a monthly bulletin published by* The Population Council, *New York City.*

A recent book discussing the background and the causes of overpopulation is:

Thomas McKeown: *The Modern Rise of Population.* New York, Academic Press, 1976. 168 pp.

Overpopulation often leads to poverty. Young boys in a slum district of New Delhi, India, are working in a grinding shop to help support their families.

6

POPULATION CHANGES AND THE HUMAN CONDITION

6.1
INTRODUCTION

Some people live in harsh, hot deserts and tend small flocks of goats and sheep. Others live in the Arctic and, even today, hunt whales from small open boats. Some are students and live in well-heated dormitories. In many parts of the world young children work in their parents' fields. Others wander city streets, selling small items to provide incomes for their families. Some city dwellers ride to plush offices in comfortable cars. In the depressed areas of some cities, the poor and homeless sleep outside at night and search during the day for some way to support themselves.

A study of human populations includes a study of all such human conditions. The situation is extremely complex. Any change in world conditions affects different people differently. If the price of fuel doubles, some families can afford the increase without particular concern. Others cannot. For example, a rise in fuel prices means that the cost of fertilizers will also rise. If some struggling farmers cannot afford to buy the fertilizer they need, their families may starve. As another example, a shift in global wind patterns may change climates in many regions of the world. These changes may bring a welcome increase of rainfall to some areas, cause

The Sahara desert. (Courtesy of Sygma)

A snow house in the Arctic. (Courtesy of the American Museum of Natural History)

105

A cabin in the mountains of Colorado.

An urban place is a city where many people live close together. Rural regions are places where there are few people and large areas of uninhabited land.

destructive floods in others, and leave some unfortunate groups of people in a condition of severe drought.

When a baby is born, the family is most immediately affected. There is one more mouth to feed, one more person to clothe, care for, and educate. The birth of one new infant doesn't affect the world or even the local community very much. But large population changes do. Global resources are finite. There is a limited supply of food, energy, water, space, and materials. These must be divided among the entire population of the world.

6.2
LAND USE, URBANIZATION, AND RURALIZATION

Most of the Earth is not heavily populated. Few people live on high mountains, in tropical rain forests, in deserts, or on tundra. Antarctica, Greenland, Canada, the U.S.S.R., Australia, and most of the countries of Africa and South America have fewer than 15 persons per square kilometer. By contrast, the Netherlands has over 1000 persons per square kilometer. You may have read that the Earth is overpopulated. But the problem is not one of lack of space. The problems are pollution, a lack of resources, an inefficient use of land, and a psychological sensation of being "crowded." Later chapters will deal with resources such as energy and food, and with pollution caused by human activities. This section discusses the settlement patterns of people.

People feel crowded in the world today because many want to live in the most hospitable environments possible. People want to live where the climate is pleasant and where they will have easy access to food, water, and a livelihood of some kind. In the United States, 90 percent of the population lives on 10 percent of the land. In New York City, hundreds of people live in a single apartment house, and many poor families live in a single room. Yet you can drive for miles across the prairies of Montana and seldom see a house.

Urban Settlements

In the twentieth century, millions of people have moved from the country to large cities. This process is called **urbanization**. Urbanization has been caused by and, in turn, has produced profound social and economic changes.

The first cities rose along the Tigris and Euphrates rivers roughly 6000 years ago. Since then, great cities have grown and fallen in many areas of the world. Most people, however, have traditionally lived in the country. In 1600, only 1.6 percent of the population of Europe lived in cities of over 100,000. By 1800, only 2.2 percent of the people lived in large cities. In fact, before 1800

A modern apartment complex in Boulder, Colorado.

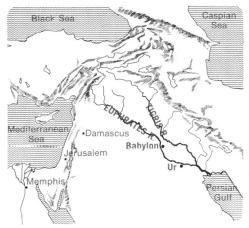

The Fertile Crescent, home of many early civilizations.

An apartment complex in Srinagar, India.

no country was predominantly urban. Between the time of the fall of the Roman Empire (about the 5th century A.D.) and the beginning of the nineteenth century, no European city had one million inhabitants. Thus, on the eve of the Industrial Revolution, Europe was essentially an agricultural continent. Outside of Europe, settlement patterns were even more rural.*

In the hundred years from 1800 to 1900, cities grew rapidly. In 1900, at least twelve cities had populations of over one million. By 1975, nearly 40 percent of the inhabitants of the world lived in urban areas.† Worldwide, about 140 cities had populations of over one million. By the year 2000, over 50 percent of the world's population will probably live in urban places. There will probably be more than 250 cities of over one million inhabitants.

If population in the twentieth century has grown rapidly, urban growth has exploded. World population tripled in the period from 1800 to 1960. In the same time, the population living in urban centers increased more than 40 times!

The extraordinary growth of cities

---

*An exception may have been East Asia. Archeological evidence suggests that the first city to exceed one million persons may have been the capital of the Khmer Republic (Cambodia). Tokyo and Shanghai may have had populations of over one million by 1800. Moreover, by 1800, many parts of East Asia boasted very large central cities surrounded by highly productive agricultural areas.

†*United Nations Demographic Yearbook*, 1975.

A single family dwelling in Telluride, Colorado.

I. Dispersed agricultural settlement (e.g., midwestern United States and Canada)

II. Agricultural villages (e.g., European farming villages)

III. Coastal settlements (e.g., fishing villages)

IV. Settlements along a strip (e.g., along banks of a river, tracks of a railroad, or by a highway)

V. Centralized urban places (e.g., Mexico City)

VI. Suburbanized urban places (e.g., Boston)

VII. Urban clusters (e.g., megalopolis of the northeast corridor of the United States: Boston-NYC-Philadelphia-Wilmington-Baltimore and Washington, D.C.)

FIGURE 6–1. Schematic diagram of patterns of human settlement.

Street person sleeps on the sidewalk in New York City. (Photo by Daniel S. Brody and Stock, Boston)

during the twentieth century has been primarily due to migration from agricultural areas to urban places. The rapid movements of migrants lead to serious and, in many cases, unsolved problems. Cities are often unable to provide clean water, adequate housing, transportation, education and other services to newcomers.

Modern cities have grown haphazardly. Most are places of vast contrast. In certain districts wealthy people live in plush apartments with easy access to fine stores, theaters, and restaurants. In other places masses of unemployed people live in central slums. Most North American cities have followed a characteristic growth pattern. Several decades ago many factories and warehouses were located in the central areas. These industries employed large numbers of unskilled or semiskilled workers. But gradually the cost of land, taxes, and services in the cities increased. Manufacturers began to move their facilities to the outskirts of the

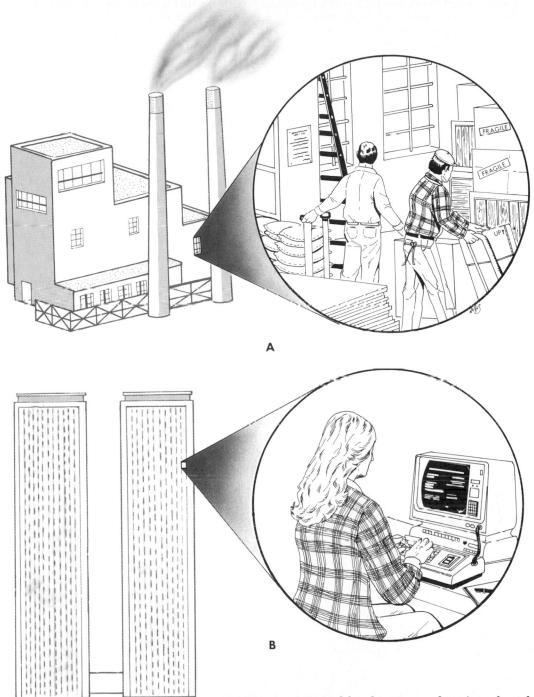

FIGURE 6–2. *A,* Several decades ago many factories and warehouses were located in the central cities, supplying jobs for unskilled or semiskilled workers. *B,* Today, large managerial headquarters have moved into the cities. There are many jobs for skilled workers, but fewer jobs for the unskilled.

cities. Roughly at the same time, many middle class families also moved away from the central regions to suburban areas. Stores and shops moved outward to satisfy their customers. As a result, fewer jobs were available for semi-skilled workers in the cities. Thousands of people became unemployed. Many of these people did not have enough money to move to the suburbs where jobs were more available.

As factories have moved away, large

In 1977, the city of New York verged on the edge of bankruptcy. The city universities were shut down because the government could not afford salaries for the teachers. Other vital services were imperiled. A series of loans were required to restore normalcy, but the threat of bankruptcy still exists.

managerial headquarters have moved into the cities. Although these operations offer many jobs for skilled workers such as executives, managers, and secretaries, they provide too few jobs for the unskilled. As a result, many residents of urban areas in the United States are without work. Thus, there are many dismal slums in the central cities. The cost of welfare and housing has become so great that urban governments are facing serious economic problems.

Transportation

Until the twentieth century, the size of a city was limited by modes of trans-

A deserted building in New York City. An example of urban decay.

portation. A worker had to live close to his or her place of work. Ancient Rome, for example, was a city of slightly over a million people. It had narrow streets and apartment houses of six to eight stories. People usually worked close to their homes. Modern methods of transportation permit people to travel long distances daily. As a result, there has been no need to cluster housing, stores, and factories. Shopping centers, businesses, and homes are scattered about without any particular order. If a kindly magician could rearrange the people, houses, businesses, schools, hospitals, and power plants of Los Angeles, vast improvements could be made. Homes could be moved closer to places of work. Business districts could be concentrated, leaving space for large parks and recreation areas. There would be less air pollution because cars would be used for shorter and fewer trips. Less energy would be consumed. More leisure time would be available. All in all, the city would be a much more pleasant place in which to live.

No one, however, has the power to level Los Angeles, or any other city, and start anew. What, then, can be done to ease the problems caused by urbanization? Solutions are frustratingly difficult. It is expensive to move homes and roadways. In the developed nations, many older cities are undergoing a process of "urban development." Old sectors are being remodeled or destroyed and replaced. Some of the new areas are well planned so that living conditions are improved. Others are poorly planned. People are forced out of their old neighborhoods and into crowded slums. Many new buildings are architecturally sterile and unpleasant and expensive to live in.

The typical city in the developed world, then, has a densely populated inner core with businesses and with both luxury and slum housing. Surrounding the central city is a suburban region that is expanding rapidly. Road building encourages new suburban growth. In turn, this growth creates demand for new roads. The city expands. The combined effects of these patterns can lead to **urban sprawl.**

As an example of the changing patterns of land use in this century, consider the United States. In 1920, 40,000 square kilometers of land had been ur-

Suburban region in Boulder, Colorado.

banized. By 1975, over 170,000 square kilometers had been urbanized. Most of this increased area was farmland in 1920.

In the developing world, cities also expand. However, in these regions, people cannot generally afford to live in the suburbs and commute to the city. So, the central regions of the city have become very crowded. It is difficult and expensive to provide adequate services for all of these people. As a result, sanitation is poor. Disease and social problems are common. The cities in the developing countries are often characterized by dismal living conditions.

Even though the living conditions of many twentieth century cities are horrible people migrate there voluntarily. Why? The reason must be that conditions in rural areas are even worse. Migration to the city has been called a push-pull phenomenon. People are pushed out of rural areas by hopeless poverty and pulled by the lure of the city. India is characterized by what is termed a push-back phenomenon. Conditions in the city are even worse than in many rural areas, so that many migrants return to their homes.

Few developing nations have the money needed to plan urban develop-

Many people live in small tents on the sidewalks in the slums of New Delhi, India.

A farmer's house in the United States. A single family lives isolated from others. (Photo courtesy of Ken Heyman)

ment. Masses flock to the cities because of dismal rural poverty. When they arrive, they find dismal urban poverty. The urbanization process is much too rapid to permit orderly development.

Rural Regions

There are many different types of settlement patterns in rural areas. In the United States and Canada single families live isolated from others. The single greatest problem with this pattern is that people have difficulty receiving emergency medical care. In many rural regions people live in small clusters of houses surrounded by farms. This pattern is found in Europe, Asia, Africa, and South America. Often a larger town serves as a central business district for several small farming villages. Many of the problems of rural land use are associated with problems in agriculture, and these will be discussed in Chapter 10. In many areas of the world, young people are moving away from farms and into the cities. Sometimes, so few people of working age remain that there is not enough labor to tend the fields.

6.3
CONSEQUENCES OF POPULATION DENSITY

As the population density in a given area increases, each person's share of available supplies of food, land, water, fuel, and other resources must necessarily decrease. In the past, people in many parts of the world have raised their standards of living despite the rising population. They have been able to do this by using the available resources with greater efficiency. For example, people have been able to raise more and more food on a given area of land. However, the population growth in other regions has been so rapid that widespread starvation is common. Since we cannot predict the future advances of technology accurately, we cannot predict the maximum possible human population. The total world supplies of food, energy, and resources are discussed in Chapters 7, 8, 9, and 10.

Before humans are eliminated by death through starvation and thirst, it is certain that the quality of life on Earth will change. Many forests and wild places will disappear and be replaced by cities and indoor environments. Some individuals will welcome the change; others will be harmed. But what will happen to society as a whole as population continues to increase?

Some people claim that high population density promotes unhappiness, stress, violence, and political upheaval. In a series of experiments with mice and rats, John Calhoun studied the effects of extreme crowding.* He built

* See John B. Calhoun in *Scientific American,* Vol. 206, p. 139, February, 1962, and "Environment and Society in Transition," *Annals of the New York Academy of Sciences,* June, 1971.

cages supplied with enough food and water for many more rats than the space would normally hold. A few animals were placed in each cage and were allowed to breed. The population and the density grew quickly. Then the animals began to act strangely. Females lost their ability to build proper nests or care for their infants. Some of the males became sexually aggressive. Most simply retreated and did not communicate with others. In short, the normal processes of socialization were destroyed.

However, humans act differently from animals. In human societies, high population densities do not always lead to serious social problems. For example, the Netherlands is one of the most densely populated countries in the world. Yet the standard of living is high and crime rates are low. Hong Kong has the highest population density of any city in the world. Nearly 40 percent of residents of Hong Kong share their dwelling unit with nonrelatives. Almost 30 percent sleep three or more to a bed. Indeed, 13 percent sleep four or more to a bed. Most people live in a single room and share that room with at least eight other people.* Even

*R. E. Mitchell: "Some Social Implications of High Density Housing." *American Sociological Review,* Vol. 36, pp. 18–29, 1971.

under these conditions of extreme crowding, there is little or no evidence of antisocial behavior caused by the crowding itself.

On the other hand, in some societies crowding and antisocial behavior appear to be related. In the United States, there are 35 times more robberies in proportion to the population in cities than in the country. A person living in the city is 2 to 4 times more likely to be murdered, raped, or beaten than a person in the country.*

Human beings respond to many different types of influences. New York City is a crowded place that has a high crime rate. Yet Hong Kong is even more crowded and has a lower crime rate. Obviously, complex social issues are important to explain human behavior. Crime and antisocial behavior may be linked partially to crowding. Perhaps more important are other factors — the strength of the family unit, a person's sense of self-worth, and the availability of jobs.

*Lawrence E. Hinkle and William C. Loring: *The Effect of the Man-Made Environment on Health and Behavior.* U. S. Public Health Service, 1975.

A view of Hong Kong, the most densely populated city in the world. (Photo courtesy of Ken Heyman)

6.4
THE URBAN ENVIRONMENT AND HEALTH

A carcinogen (kär sin′ə jen) is a material that causes cancer in humans or in animals.

The Medieval European cities must have been unhealthy, unsanitary places to live in. In most of them the water was polluted, sewage collected in local cesspools, and garbage fermented in open dumps. Many people, especially children, must have died from infectious diseases. Occasionally large epidemics took their toll. The Black Plague is a disease that is spread by fleas that live on rats. In the 14th century this plague swept through Europe, killing one third of the total population of several nations.

In modern times, infectious disease still takes a large toll in many cities of the developing world. In the developed world, however, epidemics of typhoid, influenza, tuberculosis, and plague are no longer common. Medical attention is generally available, water is free of disease bacteria, and sanitation is usually adequate. There has been, however, an increasing concern that the urban environment still poses dangers to human health. We know that the air in many cities is polluted, and in some cities the water quality, too, is poor. Tens of thousands of different types of chemicals are manufactured yearly. Hundreds of thousands of tons of these chemicals are dispersed into the envi-

The Environmental Protection Agency estimates that 63,000 different chemicals are in common use in the United States today (Thomas H. Maugh, *Science,* Vol. 199, p. 162.)

ronment. People ingest small concentrations of pollutants whenever they breathe, drink, and eat. How do these materials affect human health?

Scientists have shown that some pollutants cause disease. For example, cigarette smokers have a much higher risk of contracting lung cancer, other lung diseases, and heart disease as compared with nonsmokers. Workers in asbestos factories inhale small asbestos fibers. These fibers collect in the lungs and eventually cause cancer and other diseases. Manville, New Jersey, is the site of the world's largest asbestos processing plant. By one estimate, over one fourth of all the people in the town are suffering from environmental illnesses. Many have already died from these diseases, and many more will probably succumb in the future.

A great many other chemicals are suspected of causing cancer. Some of these are present in polluted air and water. Others are found in liquor and processed foods. People have asked, "We're eating, drinking, and breathing potential carcinogens every day. Is it possible that these environmental pollutants are leading to an increase in cancer?" It is hard to answer this question because it is very difficult to *prove* that pollution causes cancer.

For example, scientists have learned that if air pollutants are concentrated and rubbed onto the skins of rats, many of the animals will contract skin cancer. This makes us think that urban pollutants may be carcinogenic. Yet there is no absolute proof. Perhaps the concentrated pollutants cause cancer, but normal urban air does not. Perhaps pollution causes cancer in rats but not in humans.

In another study, scientists mapped the relative probability of dying of intestinal cancer in different parts of the United States (see Fig. 6–3). As you see, a person living in the Northeast has a relatively high risk of dying of intestinal cancer. Before you conclude that the environment in the Northeast causes cancer, however, several other factors must be considered. Perhaps people in the Northeast smoke more cigarettes or get less exercise than people in other parts of the country. Perhaps they eat a different diet. The racial distribution in the Northeast may be unique, and different races have been

Medieval cities must have been unsanitary, unhealthy places to live in. (Photo of the French city of Carcassonne, courtesy of the French Tourist Association)

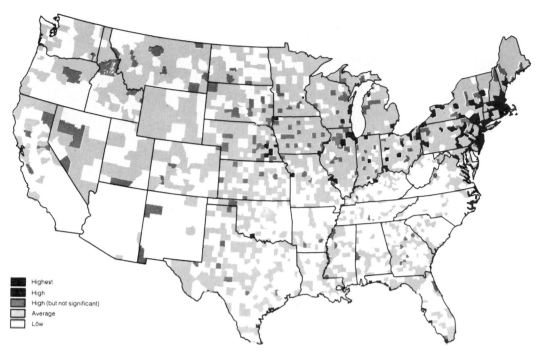

Highest
High
High (but not significant)
Average
Low

FIGURE 6–3. Deaths from cancer of the large intestine in the United States. (Courtesy of Dr. William Blot and *The Sciences*)

shown to have different susceptibilities to cancers.

Cancer is difficult to trace because there are no apparent immediate causes and effects. A person doesn't suddenly gasp, choke, and collapse after smoking a pack of cigarettes. Years or decades may pass before an acute illness is apparent. Similarly, environmental carcinogens are slow to act.

When a fifty-year-old person contracts cancer, no one can positively trace the cause of the disease. There-fore, the geographic distribution shown in Figure 6–3 does not prove that pollution causes this disease. It only raises suspicions.

6.5
CURRENT WORLDWIDE POPULATION TRENDS

Predictions of the future are unreliable. Dramatic upsets such as major

---

WHAT IS CANCER?

Normal cells in our bodies are being replaced all the time. When you get a sunburn, the burned skin peels and falls off, and a new layer takes its place.

Sometimes cells start to reproduce in an unorganized, uncontrolled way. They spread beyond their normal limits and invade other areas of the body. This abnormal growth is called **cancer,** and a clump of such cells is called a **tumor.** There are many different kinds of cancer, and there may be different causes of this disease as well. The topic will be discussed further in appropriate sections of the book.

The elderly are often not economically independent. The sign this Mexican beggar wears says, "I am 80 years old. Therefore, no one will give me work. To live I beg your charity. Thank you." (Photo by Anton Turok)

wars and shifts of climate are unfore-seeable. Prediction, then, is based on the assumption that at least some social order and control is maintained in the world. If the world is relatively peace-ful in the years ahead, the population will eventually stabilize when birth rates decline. Two questions arise. When will the population stop grow-ing? How large will the population of the world become?

Questions are easier to ask than to answer. Perhaps the population will grow quickly in some areas on Earth and decline in others. Perhaps growth will be slow for many years. Perhaps growth will be rapid for a century, then stop, and start again later.

The world population in 1978 was about 4.5 billion. The United Nations predicts that world population will reach 12 billion by 2075. If fertility is higher than projected, the total popula-tion may be nearly 16 billion. If the fertility declines very rapidly, the total population may be slightly less than 10 billion.

Two striking projections emerge. First, the population of the world is likely to triple in the next century. Sec-ond, much of the growth is likely to occur in the countries of the Southern Hemisphere — Africa, Latin America, and South Asia. In 1975, for instance, about half the world's population lived in the south. In a century, roughly three fourths will live there. The difference will be due to a very rapid birth rate in these regions.

Rapid population growth is an espe-cially difficult problem in the develop-ing countries. At the present time, many people do not have enough to eat. Many more live mainly on rice, wheat, or potatoes and do not get enough pro-tein. In most nations agricultural pro-ductivity is increasing. But because the population is also increasing rapidly, people still starve. In poor societies where resources are limited, housing, medical attention, clean water, sanita-tion, and education are difficult to pro-vide. Pollution control is expensive and is often ignored.

As mentioned earlier, poverty is par-ticularly severe in cities in the develop-ing nations. Many poor farmers whose crops have failed move to the cities. They have few skills, and unemploy-ment is high. It is difficult to feed ev-

A beggar in India.

A woman holds her starving child in the Sahel region of North Africa. (Courtesy of the Agency for International Development)

eryone. (After all, the farmers left their farms because they couldn't feed themselves.) Many are weakened by hunger and die of infectious diseases.

In the developed countries birth rates have been declining rapidly in recent years. In the United States, Canada, and most of the nations in Europe, an average of less than two babies are born to each woman. If current trends continue, the populations of these nations will start to decline sometime after the year 2000 (see Section 5.4). In the meantime, there is generally adequate food, housing, and clothing for almost all of the people in developed nations.

Even in the developed societies increases in population cause problems. There is a growing shortage of land for housing, roads, farming, and recreation. Reserves of energy and metal are being seriously depleted. Pollution is becoming an extremely serious issue.

There are pockets of poverty in all developed nations. This poverty exists because food and other essentials are not distributed evenly — not because there is not enough.

Sketch drawn by Major Powell showing the original descent of the Colorado River.

6.6
CASE HISTORY: POPULATION CHANGES AND RECREATION — THE GRAND CANYON

Imagine yourself floating downriver through the Grand Canyon. For the moment the current is smooth and placid, but around the bend you hear the roar of white water. Soon you are committed to the rapids and dwarfed by standing waves that tower over your head. You pull hard at the oars to avoid that "hole" that could flip your boat or trap

it while it fills with water. Somewhere in the back of your head you hold the guidebook description of the rapid: moderate difficulty, duration ¼ mile. You are assured that the waters ahead will not become more turbulent, and that there are no Niagara-sized falls around the next bend. A few more hard pulls on the oars and you are home free.

When Major Powell and his men first ran the river in 1869 there were, of course, no guidebooks. The men never knew when the river would be calm or fast. There were times when they felt that one more rapid or an overpowering falls would crush them. Three of the soldiers finally gave in to their fears and decided to hike across the desert toward civilization. They didn't know that they had already negotiated the worst of the river, and that during their overland journey they would be ambushed by Indians and killed.

Today, many thousands of whitewater enthusiasts run the river every year. Some ride forty-foot motorized rafts while others seek the challenge in tiny kayaks that can barely be seen amid the waves. In many places the rocks of the canyon walls drop precipitously to the water's edge. In others,

The Grand Canyon of the Colorado.

A large commercial outfitter preparing rafts for the journey downstream. (Courtesy of Hunt Worth)

An independent kayaker in the rapids of the Colorado. (Courtesy of Hunt Worth)

Part of Major Powell's expedition in Canyon of Desolation, Green River. Photo taken August 19, 1871. (Courtesy of U.S. Geological Society)

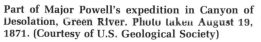

Here is an account of Major Powell's experience of the Colorado River. "We step into our boat, push off, and away we go, first on smooth but swift water, then we strike a glassy wave and ride to its top, down again into the trough, up again on a higher wave, and down and up on waves higher and still higher until we strike one just as it curls back, and a breaker rolls over our little boat, ... the open compartment of the "Emma Dean" is filled with water and every breaker rolls over us. Hurled back from a rock, now on this side, now on that, we are carried into an eddy, in which we struggle for a few minutes, and are then out again, the breakers still rolling over us. Our boat is unmanageable, but she cannot sink, and we drift down another one hundred yards through breakers, how, we scarcely know. We find the other boats have turned into an eddy at the foot of the falls and are waiting to catch us as we come, for the men have seen that our boat is swamped.... Our boat is bailed and on we go again."*

*J. W. Powell, *The Exploration of the Colorado River and Its Canyons.* (See Bibliography.)

One user day is one person being on the river for one day. If five people spend ten days on the river, they spend a total of fifty user days.

smooth sandy beaches offer comfortable camping sites. By the early 1970's it became obvious that most of the prime beaches were becoming polluted. Blackened fire sites and collections of human waste were despoiling the canyon. The National Park Service realized that something had to be done. There were simply too many people on the river. The Park Service placed severe restrictions on the handling of wastes and fires. They limited the traffic through the canyon to 89,000 user days per year. They also stipulated that 92 percent of the traffic should be carried by commercial guide services, and 8 percent of the permits could be used by private groups.

When the National Park System was first established in 1872, very few tourists were adventurous enough to brave the dangers and hardships of the parks. Therefore, the government sold rights (concessions) to private companies to build transportation systems, hotels, and restaurants to serve the parks and encourage tourists. The present boating outfitters grew out of this original concession system. In the Grand Canyon, people pay $50 per day or more to run the rapids. The most elite tours serve shrimp cocktails and ice cream for evening meals. Expensive advertisements for raft trips appear in many travel and adventure magazines. The river has become big business.

The problem is that private river runners feel cheated. Noncommercial groups are allowed only 8 percent of the user days, but there are many more people who would like to float the canyon on their own. A lottery is held every year. In 1977, there were 515 applications for private permits. Only 30 were approved.

Private boaters claim that the present system discriminates in favor of the rich. Just about anyone who cares to spend the money and travel with a guide can make the journey. But enthusiasts who wish to row or paddle their own boats and travel cheaply usually fail to obtain permits. Some environmentalists fear that a trend may develop. In the future, is it possible that hiking, backpacking, canoeing, scuba diving, rock climbing, and horseback riding will be similarly regulated? The problem is that some type of rules and regulations must be enforced when population pressures threaten an ecosystem. Yet it is hard to write laws that are universally just. Moreover, political lobbyists and commercial interests often sway legislative decisions. The challenge is to preserve natural ecosystems without discriminating unfairly against anyone.

CHAPTER SUMMARY

6.1 INTRODUCTION

There are many different types of human societies. Everyone must share global resources. There is a limited supply of food, energy, water, space, and materials.

6.2 LAND USE, URBANIZATION, AND RURALIZATION

Urban settlements. In the twentieth century there has been a widespread migration of people from the country to the city. In the developed nations, many factories, warehouses, and stores have moved from the central cities to suburban regions. As a result, millions of unskilled workers have become unemployed. Suburban development has been haphazard, and as a result space and transportation facilities are used inefficiently.

In the developing nations, poverty in the central cities is even more dismal, and there is little money for urban development or welfare.

Rural regions. Rural regions have been depopulated in recent years by migration to the cities.

6.3 CONSEQUENCES OF POPULATION DENSITY

The eventual limits of population growth are set by world resources. Experiments with animals suggest that crowding produces antisocial behavior. No one knows how crowding affects behavior in people. Crime rates are high in some crowded regions and low in others.

6.4 THE URBAN ENVIRONMENT AND HEALTH

The probability of dying of some forms of cancer is higher in many urban areas than in rural ones, but we are not sure why.

6.5 CURRENT WORLDWIDE POPULATION TRENDS

(a) The population of the world is likely to triple in the next century. (b) Much of this growth is likely to occur in the developing nations. Shortages of food, energy, and materials will become even more acute.

KEY WORDS

Urbanization

Urban sprawl

TAKE-HOME EXPERIMENT

1. Using whatever reference sources are available, try to build a demographic history of your community for the past fifty years. Briefly outline the history of population changes in the community. What ethnic group or groups have migrated into or out of the area? How have population movements been related to patterns of employment? Discuss.

PROBLEMS

1. **Urbanization.** Outline a very brief history of the growth of cities from ancient times to the present.

2. **Urbanization.** There were more jobs in the central cities in the United States in 1977 than in 1967. Explain why this growth has not helped the unemployed who live in the ghettos.

3. **Urbanization.** Many blacks in the United States moved from southern farms to the northern cities in the early 1900's. Jobs were relatively plentiful. Explain why many of the grandchildren of these early migrants cannot find work at present.

4. **Urban regions.** Discuss the growth of urban regions. What is urban sprawl? How has it developed?

5. **Suburban regions.** Discuss the statement: "Modern transportation systems have helped develop suburbia. In turn, suburbia has helped develop modern transportation systems."

6. **Urban regions.** Explain why urban sprawl leads to pollution and inefficient use of energy, time, and space.

7. **Urban regions.** Explain why it is difficult to correct the problems caused by urban sprawl.

8. **Urbanization.** Briefly contrast the process of urbanization in the developed and the developing worlds.

9. **Population density.** Describe some unpleasant effects of high population densities.

10. **Population density.** Describe some unpleasant effects of very low population densities.

11. **Settlement patterns.** Describe the settlement pattern of the town or city in which you live. Does the pattern in your locale resemble any of the patterns of Figure 6–1? Discuss.

12. **Population density.** Discuss the effects of crowding on rodents. Do people in crowded conditions necessarily act in the same manner? Discuss.

13. **Population density.** Population density is usually defined in terms of number of people per unit of land area. Does a high-rise luxury apartment have a high population density or a low one? From the population density alone, could you predict whether crime rate will be high or low? Discuss.

14. **Population density.** The Netherlands has been used as an example of a nation that has a high population density and a low crime rate. Although its agriculture is very efficient, there is little farmland, and people must import food to survive. What would happen if the worldwide population density were as high as the density in the Netherlands? Discuss.

15. **Health.** Discuss the difference between the health problems of people who live in cities today with those of a century ago.

16. **Urban health.** It has been shown that people living in the Northeastern part of the United States have a higher probability of dying from intestinal cancer than the people in other parts of the country. What do these data prove? What do they suggest? Discuss.

17. **Worldwide population.** What is the worldwide population today? What is it predicted to be in one hundred years? Discuss some problems of rapid population growth.

18. **Population growth.** Compare population growth in the developed and the developing nations. Compare the problems related to population growth in these two regions.

QUESTIONS FOR CLASS DISCUSSION

1. Describe the environment that you live in as urban, rural, or suburban. Obtain a map of your region and use different colored pencils to shade areas occupied by: (a) factories and warehouses, (b) retail businesses, (c) low-density housing, (d) high-density housing, (e) agricultural areas, and (f) open spaces. Try to design a more efficient system of land use. How would you alter the present system to make your region more efficient or more pleasant? Discuss.

2. Obtain a copy of your local zoning ordinance. Do the zoning laws promote efficient or inefficient land use? Would it be possible to improve the efficiency of land use in your neighborhood? Discuss with government officials any proposals you might have and report on their reaction.

BIBLIOGRAPHY

An introduction to population distribution and settlement patterns is found in:
John C. Bollens and Henry J. Schmandt: *The Metropolis — Its People, Politics, and Economic Life,* Third Edition. New York, Harper and Row Co., 1975. 401 pp.
Melville C. Branch: *Planning Urban Environments.* Stroudsburg, Pennsylvania, Dowden, Hutchinson, and Ross, Inc., 1974. 254 pp.
Robert W. Burchell and David Listokin (eds.): *Future Land Use.* New Brunswick, New Jersey, Rutgers University Press, 1975. 369 pp.
Maurice Yates and Barry Garner: *The North American City,* Second Edition. New York, Harper and Row Co., 1976. 513 pp.

A concise summary of the effects of changes in population size and distribution is presented in:
Lennart Levi and Lars Andersson: *Psychosocial Stress: Population, Environment, and Quality of Life* New York, Spectrum Publications, Inc., 1975, 142 pp.
Lawrence E. Hinkle and William C. Loring (eds.): *The Effect of the Man-Made Environment on Health and Behavior.* U.S. Department of Health, Education and Welfare, Publication Number (CDC) 77–8318. 315 pp.

Several books sound an alarm for our crowded planet. One of the most popular of these is:
Paul R. Ehrlich: *The Population Bomb.* New York, Ballantine Books, 1968. 201 pp. (Ehrlich includes a bibliography of similar discussions.)

On the other hand, there is an important argument for encouraging moderate population growth expressed in a very provocative work:
Alfred Sauvy: *General Theory of Population.* (Translated by Christophe Compos.) New York, Basic Books, 1969. 550 pp.

A very interesting group of papers on population growth is presented in the two-volume work:

National Academy of Sciences: *Rapid Population Growth.* Baltimore, Johns Hopkins Press, 1971. (Vol. 1, 105 pp.; vol. 2, 690 pp.)

The United Nations publishes several volumes on population. Among the most useful are:

The Population Debate: Dimensions and Perspectives. New York, The United Nations, 1975. (Vol. 1, 676 pp.; vol.2, 726 pp.)

The Determinants and Consequences of Population Trends. New York, The United Nations, 1973. 661 pp.

For a discussion of population policies see:

World Bank Staff Report: *Population Policies and Economic Development.* Baltimore, The Johns Hopkins University Press, 1974. 214 pp.

There are several books that discuss the relationship between environmental pollution and cancer. Two short and concise references are:

William M. Boland: *Cancer and the Worker.* New York, The New York Academy of Sciences, 1977. 77 pp.

Robert J. Harris: *Cancer — The Nature of the Problem,* Third Edition. Baltimore, Maryland, Penguin Books, 1976. 175 pp.

An excellent book which recounts the adventures of the first exploration of the Grand Canyon is:

Major J. Powell: *The Exploration of the Colorado River and Its Canyons.* New York, Dover Publications, 1961. (First published in 1895 under the title *Canyons of the Colorado.*) 400 pp.

7

WORLD RESOURCES — ENERGY AND MATERIALS

7.1

INTRODUCTION

All natural ecosystems receive their energy from the Sun. Sunlight is trapped by green plants and is used to produce energy-rich chemicals. Plant matter is then eaten by animals. As animals eat, move about, and in turn are eaten, the stored energy in the plant matter is gradually converted to heat. Eventually, this heat is lost to space. The lost heat is replaced by a daily inflow of sunlight, so the system never "runs out of energy."

Primitive people lived within natural systems. They collected fruits and vegetables and hunted game. Before the use of fire, each person needed approximately 2000 kcal of food energy per day. In later times, primitive farmers domesticated animals, raised grains and vegetables, and used fuels for cooking and heating. Energy requirements per person (per capita) rose to about 12,000 kcal per day.

Since wood was the primary fuel, these farmers still were using renewable energy sources. It is clear that people no longer live within this ancient pattern of energy flow. By 1860, small amounts of coal were being mined. Steam engines had been invented, and a resident of London used about 70,000 kcal per day. At that time, people in western Europe were using fuels faster than they were being replaced. This was possible because large supplies of fossil fuels represented a reservoir of stored energy. But such a situation cannot continue forever. If consumption is higher than production, then the supply must run out someday. In the United States in 1975, per capita energy consumption was approximately 235,000 kcal per day. Most of this energy was derived from the use of fossil fuels. This enormous rate is unique in the history of the world. At no other time and in no other place have people used energy faster than North Americans do today.

Rapid energy consumption is not confined to the United States. All over the world people are burning large quantities of fossil fuels. We know for a fact that this consumption cannot continue for long. Supplies of coal, oil, and gas will soon be depleted.

Many types of materials are used in industrial societies. Fertilizers are needed to grow large quantities of food.

Remember, the calorie is a unit of energy. 1 kcal = 1 kilocalorie = 1000 calories.

125

1. Man without fire
(2000 kcal/day)

2. Primitive agriculture
(12,000 kcal/day)

3. ca. 1860
(70,000 kcal/day)

4. ca. 1980
(230,000 kcal/day)

Energy consumption by people.

Transportation networks, houses, clothing, and appliances are made out of materials such as metal, plastic, glass, cloth, and wood. Recall from Chapter 2 that energy and matter are fundamentally different. Matter can be recycled. If we manufacture an engine out of iron and other metals, the engine may break down and rust, but the metal never disappears. It can be remelted and recast into a new engine. The iron atoms never get "tired" or "used up." Therefore, in theory, the world's supply of metals need never be depleted. They can be used again and again forever.

Energy, on the other hand, is not a material. Energy is defined as *the ca-*

pacity to perform work or transfer heat. Where does the energy go after it has been used? If coal is used in a locomotive, the fuel burns, releasing gases, smoke, and soot. But what becomes of its energy? This nonsubstance, called "the capacity to perform work" is hard to keep track of. If we can find used iron and reuse it, why can't we find used energy and reuse it? Or better yet, if energy isn't really matter, could we find some for nothing? These two questions plagued scientists for a long time, and the search for answers led to the study of heat-motion, or **thermodynamics.**

7.2
THE FIRST LAW OF THERMODYNAMICS — (OR YOU CAN'T WIN)

Since the beginning of civilization, people have been faced with the problem of how to lift heavy loads. Build-

Drawing by Pieter Pourbus, the Elder, of a man-powered crane, in use in the sixteenth century. The development of clever machines led some early inventors to believe that it would be possible to build a perpetual motion machine. However, we are now certain that it is impossible to build a perpetual motion machine.

A person who can lift 50 kg unaided

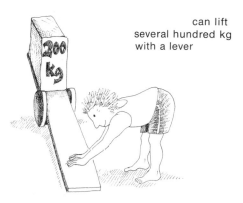

can lift several hundred kg with a lever

ings have been traditionally built of wood or stone, and engineers have had to move these heavy items from the ground onto the walls and roofs. In the early days, we can imagine that people worked together to lift stones and timbers. Later, clever inventors developed simple machines such as the lever, the wheel, and the inclined plane. A person who can lift 50 kilograms unaided can lift hundreds of kilograms with a lever and even more with a well designed system of gears or pulleys. In the early days, engineers believed that machines actually reduced the amount of work required to lift an object. Scientists now understand that such devices do not actually reduce the amount of work. Instead, they spread it out over a longer period of time and smooth out the effort. However, this difference can easily be overlooked, and a device such as a lever can be mistakenly thought of as a "work-saver." Since many "work-savers" had been invented, people reasoned that if

The search for perpetual motion was at its height from about 1650 to 1750. By the year 1775, the Paris Academy of Sciences refused to accept schemes for perpetual motion. Some inventors continued the search into the 1800's, but by that time most scientists had realized that the search was futile.

you were clever enough you could build a machine that would do all the work for free. They called this imaginary device a **perpetual motion machine.** If you owned a perpetual motion machine you could "turn it on" and it would lift all the boulders you wanted while you sat and watched. It may be difficult for the modern reader to appreciate that the search for the perpetual motion machine seemed entirely reasonable. Many very clever people looked for a solution. However, all attempts failed. The failures have been so consistent that we are now convinced that the effort is hopeless. We believe that it is a fundamental law of nature that it is impossible to build a perpetual motion machine. This is one statement of the **First Law of Thermodynamics.**

This law can also be expressed in terms of conservation of energy: *Energy cannot be created or destroyed.*

Don't ask for a proof of the First Law. There is none. The First Law is simply a statement about human experience with energy. If it is impossible to create energy, then it is hopeless to try to invent a perpetual motion machine, and we may as well turn to some other method of doing work.

7.3
THE SECOND LAW OF THERMODYNAMICS — (OR YOU CAN'T BREAK EVEN)

Throughout most of history people have used their muscles or the muscles of animals to perform work. The modern industrial age began in 1769 when James Watt developed the steam engine. A steam engine is a device that converts heat from burning fuel to work. Heat and work are both forms of energy. Thus, it is possible to design an engine that burns wood or coal and *converts* the energy from the burning fuel into work. The work output can pump water, lift rocks, grind grain, or drive a locomotive up a hill.

Early steam engines operated at about 5 percent efficiency. That means that only 5 percent of the potential energy of a fuel was converted to work, and 95 percent was "lost" to the environment as heat. It was soon found that

Steam locomotive.

higher efficiencies could be obtained by improving the design of the engines. Clever inventors therefore worked hard at trying to convert *all* the energy in fuels to work. They all failed. This failure has been expressed as another law of nature, called the **Second Law of Thermodynamics.** The Second Law states that *it is impossible to convert all the energy in a fuel into work.* Thus, in any heat engine, some energy is "lost." The energy is "lost" only in the sense that it is no longer available to do work. Instead, this energy warms the environment.

Suppose a lump of coal is burned in a steam locomotive. Some of the heat is converted to useful work, and the engine travels up a hill. Now imagine that the fire burns out and the engine rolls back down the hill. There is friction in the wheel bearings, between the wheels and the track, and between the locomotive and the air. As a result, a large mass of air and metal is heated slightly. According to the Second Law, it is impossible to collect the heat from the warm air and metal and reuse it to drive the locomotive back up the hill. Thus, the energy from the coal cannot be recycled. This observation is a general one, and explains why energy, once used, cannot be reused. *Materials can be recycled but energy cannot.*

7.4
OUR FOSSIL FUEL SUPPLY

In 1978, approximately 90 percent of the total energy used in the United States was derived from fossil fuels — coal, oil, and natural gas; 5 percent was from nuclear fission. Another 5 percent was from renewable sources — mostly power from falling water. Small amounts of energy came from solar, wood, and geothermal sources.

The Second Law assures us that someday our fossil fuel reserves will be depleted. How much time do we have? Most authorities believe that natural gas is our least abundant fossil fuel. The peak consumption rate will probably be reached about 1985, or before most of the readers of this book have reached middle age. Already, contractors in many areas are unable to supply gas lines for heating new houses because the industry claims it will not have enough fuel for new customers. The scarcity of natural gas is environmentally unfortunate, since it is less polluting than any other widely used fuel.

Geothermal energy is energy derived from the Earth's heat. See page 000.

Oil

Petroleum is perhaps the most versatile fossil fuel. Crude oil, as it is pumped from the ground, is a heavy, gooey, viscous, dark liquid. The oil is refined to produce many different materials such as propane, gasoline, jet fuel, heating oil, motor oil, and road tar. Some of the chemicals in the oil are extracted and used for the manufacture of plastics, medicines, and many other products. It is difficult to imagine what would happen to our civilization if the supply of liquid fuels ran out. Automo-

---

A large fraction of the public behaves as if there were no energy problem. This summer gasoline consumption has been setting records. Sales of automobiles during August were at a peak for the month. Use of electricity has been at an all-time high. Consumption so far this year is more than 7 percent above that of a year ago. The present behavior is consonant with polls which indicate that a majority of citizens are uninformed about energy problems. Only 48 percent of the people know that we must import oil to meet needs.

Our energy reserves and producing capacity continue to deteriorate. Four years ago the rate of domestic production of crude oil was 9.4 million barrels per day (mbd) and imports were about 5.7 mbd. Today, even with Alaskan oil, domestic production is 8.4 mbd while imports of crude oil and products are about 8.5 mbd. During the first 6 months of this year, imports were 31 percent above those of the corresponding period of last year. In part the spurt is related to decreasing availability of natural gas for industry. Low-priority consumers of natural gas are switching to oil products. Thus, when the next curtailment of foreign oil imports comes, the damage could be serious.

Editorial by Philip H. Abelson, *Science,* Vol. 197, September 30, 1977.

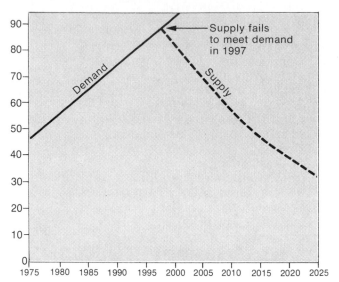

FIGURE 7–1. One projected estimate of the supply and demand for petroleum through the year 2025.

1975 to 2025. According to this graph, oil production will continue to increase until about 1997. At that time, many of the richest fields will be depleted and production will slow down. Yet the need for oil will continue to increase. People will want more oil than is available. Therefore, a real and permanent shortage will result. As shown by the graph, oil will still be available, but there won't be enough to supply people's wants.

How accurate is Figure 7–1? Of course, no one is sure. Perhaps there is more fuel in the ground than we expect; perhaps there is less. Perhaps demand for fuel will increase more than expected; perhaps it will decrease. Using different estimates of supply and demand, it is possible to draw different graphs. Some of these are shown in Figure 7–2. But even the most optimistic of the four estimates tells us that oil will be scarce by the year 2004.*

What will happen when the petroleum reserves are depleted? Will people all learn to drive teams of horses to town and rediscover the use of wood as a fuel? The answer is almost certainly no. Several stop-gap measures are available, and a few more long range solutions will probably be found.

*Andrew R. Flower, "World Oil Production." *Scientific American*, March, 1978, p. 42.

biles, airplanes, most home furnaces, and many appliances could not operate. Many industries would have to redesign their factories. Yet, reliable estimates indicate that before the year 2000 there will not be enough petroleum to meet worldwide demand. Figure 7–1 shows one projected graph of oil production and demand for the years

A CLOSER LOOK AT THE ASSUMPTIONS USED TO CALCULATE FUTURE OIL SUPPLY

(a) How much fuel will be needed in the near future?

For the past twenty-five years worldwide fossil fuel consumption has been increasing at a rate of about 5 percent per year. Recently there has been much talk about conservation, but actual consumption has increased, not declined. Most observers feel that as long as fuels are available, energy consumption will continue to grow at current rates. Perhaps energy use will diminish somewhat, but it is hard to foresee drastic reduction during the next two decades.

(b) How much oil is actually available?

There is wide disagreement about how much oil is actually available on Earth. Some people feel that many potential oil fields have been overrated. If this is true, oil shortages will be upon us before 1997. In recent years new oil fields have been discovered in northern Alaska, under the North Sea in Europe, offshore on the East Coast of the United States, and in other locations. Could it be possible that there is really much more oil than we expect? Perhaps. But worldwide fuel consumption is so great that even if there is 1½ times

as much undiscovered oil as even optimists predict, oil shortages will exist by the year 2010.

(c) Political factors

Approximately 55 percent of the known oil reserves lie in the Middle East. The nations that control this oil may decide to limit production at any time. If they do, oil shortages could start sooner than predicted. In 1973, Arab nations reduced their oil exports. As a result, shortages were felt even though there was enough oil in the ground to meet demands.

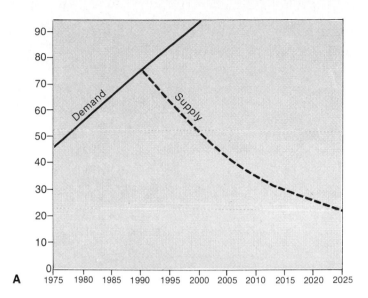

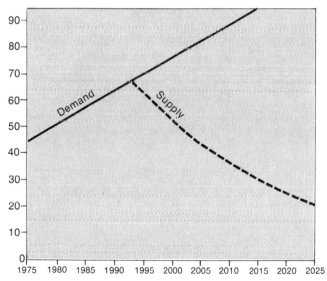

FIGURE 7–2. Other projections for petroleum supply and demand through the year 2025.

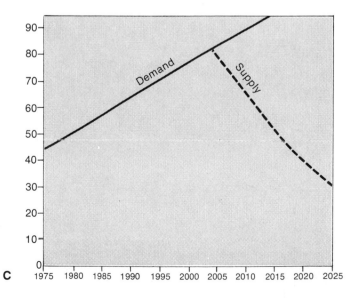

Will people be driving horses when the oil runs out?

automobiles, in most home furnaces, or in many industries. One solution to the problem is to convert coal to liquid or gaseous fuels. The theory of converting solid to liquid fuels is well understood. It has been practiced both in the laboratory and in industry. In fact, conversion of coal was common in the 1920's. During World War II the Germans converted coal to gasoline for military use. By 1978, there were over fifty plants in various parts of the world that produced gaseous or liquid fuels from coal. Most of these were small industrial operations. Large scale conversion of coal is still in the developmental stage. The major problem is one of economics. In the United States in 1978, gas from coal cost about 1¼ times as much as the most expensive imported gas. Undoubtedly, gas from coal will become relatively cheaper than other gas in future years. In fact, conversion of solid to liquid and gaseous fuel will probably be common before the end of the century.

Coal

"Coal is cheap, hated, abundant, filthy, needed." Title of an article by Jane Stein, Smithsonian Magazine, February, 1973.

Large reserves of coal exist in many parts of the world. As shown in Figure 7–3, we can expect widespread availability of this fuel at least until the year 2200. However, problems arise. When coal is mined, large areas of land surface are disturbed. More air pollutants are released from burning coal than from burning oil or gas. These issues will be discussed further in Chapters 8 and 12.

Shale is a type of rock.

Another difficulty arises because coal cannot be used directly in conventional

Oil Shale and Tar Sands

Until now, most of our petroleum has been extracted from underground wells. However, large quantities of oil also exist within the pores of certain rocks. Huge deposits of oil-bearing shale exist in the United States in Colorado, Wyoming, and Utah. Sand fields, laden with oil, have been discovered in Africa, the United States, and Canada. We don't know how much oil can be obtained from oil shales and tar sands. A Canadian firm is currently extracting oil from tar sands. The richest deposits of oil shale are believed to add another 25 percent to the world's usable supply of liquid petroleum. Today it is usually cheaper to import oil than to mine shale. But the situation may change in the near future. As imported oil becomes more costly, and as the technology for mining shale becomes more highly developed, shale deposits may be used more. Another problem arises because large quantities of water are needed to extract oil from shale. In the semiarid regions where shale is found, the water is also needed to irrigate farmlands. Thus it is impossible to predict how much shale will actually be mined.

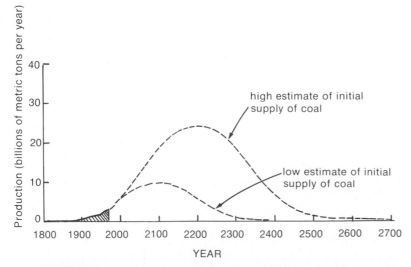

FIGURE 7–3. Past and predicted world coal production based on two different estimates of initial supply (Adapted from M. King Hubbert: The energy resources of the earth. *In: Energy and Power, A Scientific American Book.* San Francisco, W. H. Freeman & Co., 1971.)

A pilot plant that produces synthetic gasoline from coal. (Courtesy of U.S. Energy Research and Development Administration)

Oil shale country in western Colorado. (Courtesy of Atlantic Richfield Co.)

Mining uranium ore undergound near Grants, New Mexico. (Courtesy of Ranchers Exploration and Development Corporation)

7.5
NUCLEAR FUELS

Oil shale, tar sands, and chemical conversion of coal will undoubtedly extend the life of the fossil fuel age. But such stop-gaps cannot be the basis for a long range survival of our technological society. Instead we must turn either to renewable resources or to nuclear energy.

Orders for New Nuclear Power Plants

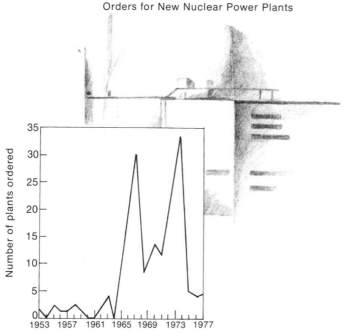

The use of nuclear energy has sparked one of the most significant debates of our times. As will be discussed in more detail in Chapter 9, there are some people who believe that nuclear reactors are absolutely essential to our society. Others think that these reactors should be banned.

In 1978, only 5 percent of the energy used in the United States was provided by nuclear fuels. By the year 2000, we can probably expect this figure to rise to somewhere between 10 percent and 20 percent. Several problems arise, however. Nuclear reactors are expensive to build. Many citizen groups oppose construction of the "nukes," and the cost of uranium fuels is expected to rise in the near future. In 1978, uranium cost about $18 per kilogram. But there is little doubt that accessible, inexpensive, high grade ores will soon be used up. Then the price will rise. By the end of the century, uranium may cost over $100 per kilogram. At this time our uranium supply will not be exhausted, but the price of electricity generated in this manner will be high. Therefore, we cannot expect conventional nuclear fission to be a source of inexpensive energy.

In recent years **breeder reactors** have been developed (see Chapter 10). Breeders can produce many times more energy per kilogram of uranium than conventional fission reactors can. Some people predict that important environmental problems will be solved and that breeders will therefore extend the life of our nuclear fuel supply for many thousands of years. However, not everyone agrees. In fact, the question is a matter of considerable controversy.

Another possibility can be considered. Scientists may someday harness the hydrogen **fusion** reaction.* The technological problems yet to be solved are so difficult that some authorities expect fusion never to be functional. At the other extreme some optimists predict fusion to be feasible within ten years. However, most members of the scientific community feel that the development of such a power plant is highly unlikely before the year 2000. If fusion is found to be practical, the energy available to us will be enormous.

---

*The development of fission and fusion reactors will be discussed further in Chapter 9.

TABLE 7-1. ESTIMATED URANIUM RESERVES (AS URANIUM OXIDE, U_3O_8)

Cost of Uranium Oxide, $/kg*	Thousands of Metric Tons Available	Year of Exhaustion
18	600	1986
22	700	1988
33	900	1990
65	1000	1992
110	4000	?
220	7000	?
450–1100	more than 5,000,000	?

*Dollar value in 1977.

7.6
RENEWABLE ENERGY SOURCES

In the early nineteenth century, wood was the world's most widely used source of energy. If a forest is managed properly and loggers cut trees only as fast as they regrow, fuel can be harvested indefinitely. Our present rate of energy consumption is so great that the world's forests cannot begin to meet the demand. However, several other renewable sources are available. En-

Wood is still used as a heat source in many rural regions.

ergy can be obtained from the Sun, the power of falling water (hydropower), the winds, and the tides. Large quantities of heat are also available from the internal layers of the Earth (although strictly speaking this energy supply is not renewable). If all the renewable sources of energy were exploited efficiently, they could provide enough power for our civilization and could replace fossil and fission fuels. Many of the technological problems needed to harness these sources have already been solved. However, the development of renewable energy sources has been quite slow, and their full potential is still very far from being realized. Some of the social, economic, and technical problems involved in harnessing renewable energy sources will be discussed in the next chapter.

7.7
ORES AND FERTILIZERS

The Second Law of Thermodynamics assures us that since energy cannot be recycled, our fossil fuel deposits will someday be depleted. There are no such restrictions on metals or ores. As mentioned earlier, these materials are never used up. They can be used over and over again indefinitely.

What happens, then, to "used" iron in our modern world? Some is recycled and recast. The rest is thrown into garbage dumps and dispersed through land and water. This far-flung material could be recovered, but it would be very expensive to do so. Even if labor costs are not considered, large quantities of energy are needed to collect and concentrate materials that have been scattered about the countryside. Since energy supplies are also limited, many discarded materials may never be reusable.

To date, people have not exhausted the supply of any essential mineral. Furthermore, most forecasters predict adequate supplies at least until the end of the century. But beyond that time it is difficult to know what will happen. Several factors are important in predicting future availability of minerals.

Ore Deposits

Prospectors or geologists seldom find pockets or veins of pure metal. Rather they find valuable minerals mixed with other types of rock. An **ore** is considered to be a rock mixture that contains enough valuable minerals to be mined profitably with currently available technology. A rare and valuable mineral such as uranium may be mined commercially if it exists in concentrations as low as 0.1 percent by weight in the rock. On the other hand, an iron deposit is considered to be an ore only if the rock contains at least 30 percent metal. The **mineral reserves** of a region are defined as the estimated supply of ore in the ground. Reserves are depleted when they are dug up. But our reserve supply may be increased by either of two processes. First, new deposits may be discovered. In addition, there are many known deposits that are now un-

A small-time mining operation.

economical to mine. For example, a deposit containing 20 percent iron is not considered an ore reserve because it is too expensive to extract the metal from the rock. If technology improves so that the materials can be refined cheaply, or if the market price of iron goes up, the deposit will suddenly become an ore reserve.

Many of the very high-grade concentrated ores are being used up rapidly. These mines are essentially nonrenewable. Once they are gone, our civilization will have suffered an irreplaceable loss. But our technological life will not end with the exhaustion of these rich reserves. There are many poorer reserves that can be mined. However, mining of low-grade ores presents many problems. Some of these are discussed in the following paragraphs.

Availability of Energy

To extract metal from ore, the dirt and rock must be first dug up and crushed. Then the ore itself must be separated and chemically converted to the metal. Finally the metal must be refined and purified. Each step, especially the chemical conversion, requires energy. Moreover, low-grade ores require much more energy to process than do high-grade ores. If energy

A large open pit coal mine near Farmington, New Mexico. This was level mesa land before the mining operation.

Large amounts of energy are required for the production of steel. (Courtesy of Bethlehem Steel)

supplies are exhausted, it may be impossible to mine and refine many ores.

Pollution

Most mining processes cause much pollution of land, water, and air. For example, sulfur is found in large quantities in many ore deposits. When sulfur reacts with oxygen and then with water, sulfuric acid is produced. This and other acidic substances often run off from mining operations onto the land. Such pollution, known as **acid mine drainage,** finds its way into streams and rivers where it kills fish and disrupts normal life cycles. Some of the sulfur is converted to gaseous air pollutants such as hydrogen sulfide and sulfur dioxide. Sulfur, of course, is not the only polluting chemical from

mining operations. Many other byproducts of mining cause serious environmental disruption.

Just as more energy is required to handle low-grade ores than high-grade ores, more pollution generally results from processing these impure materials. This pollution is becoming an increasingly serious problem.

Land Use

The world is running short of food, energy, and recreational areas as well as of high-grade mineral deposits. What should our policy be if a valuable ore or fuel lies under fertile farmland or a beautiful mountain? Which resource takes precedence? Vast coal seams lie under the fertile wheat fields in North America. If large areas of low-grade

Today concrete is often used to reduce steel consumption in many construction projects.

Approximately 25,000 liters of water are used to refine a ton of petroleum, and nearly 50,000 liters are needed to produce a ton of steel.

metal ores must be exploited, the problem will extend to these reserves as well. These difficult questions will be discussed in more detail in the next chapter.

Future Demands for Metals

In order to determine how long current mineral resources will last, we must estimate how much will be used in the future. But how can we predict future levels of demand? No one knows how much industrialization will occur in developing nations. Will most of the people living in India own automobiles and televisions someday? If so, it will not be easy to satisfy future demands for minerals.

In recent years many substitutes for metals have been developed. Reinforced concrete, fibrous glass, and plastics are now used in many applications that once called for steel or other metals. If metals can be replaced by non-metals for many uses, current reserves will last longer.

Rates of Recycling

Valuable metals and other materials are often discarded rather than recycled. More efficient recycling would certainly extend the life of our concentrated mineral reserves. Chapter 14 discusses the complex scientific, economic, social, and political aspects of recycling.

7.8
WATER RESOURCES

A rock climber scaling a dry, vertical, granite wall can live on about 1½ liters of water per day. This water is used for drinking only. By contrast, water consumption per person in the United States averages about 2000 liters (approximately 500 gallons) per day. Only a very tiny portion of that is used for drinking. Large quantities are used in the home for washing, cooking, bathing, and flushing toilets. Even larger quantities are used in industry. Factories, refineries, and electrical generating stations use water as a coolant, as a raw material, and to flush wastes. Even more water is used to irrigate grain and vegetable crops. In the United States, farmers consume six times as much water as all other users combined.

Water is truly a renewable resource. Every year large quantities are removed from the world's rivers and every year the rivers are refilled by rainfall. Therefore, in an exact sense, we never "use up" our water supply. But the quality or location of the water may change so that it is less available or less useful to people.

For example, the Colorado River system is a major source of water for much of the Southwestern United States. Some of this water is withdrawn for home use. After having passed through sinks, bathtubs, and toilets, the water is generally piped through sewers to treatment centers. Here it is partially purified and discharged back into the river system. But some of the water is not returned to the river system. Some of the "lost" water is used to irrigate lawns. Some is used for drinking. Much is lost through evaporation both in the home and during purification. Of course the water isn't really lost, but it is transported into the air, not back into the river. Some of the river water is used by industries. Much is used for irrigation. Some of the irrigation water filters through the soil and returns to the river, but most of it evaporates and falls as rain somewhere else. As the river nears the California border, large quantities are piped to major cities in Nevada and Southern California. So much water is pumped away that in some years the river dries up completely near its mouth. The Colorado is a river that never quite makes it to the sea. All its water is taken away before it gets there.

Sometimes water isn't consumed, but it is polluted so badly that its usefulness becomes limited. Many industries along the upper Mississippi River use large quantities of water for washing, cooling, or rinsing. This water is usually partially purified and returned to the river. Little is actually consumed. But the water that is discharged is not as pure as it was before it entered the factory. Pollution continues all along the entire Mississippi basin. As a result, the river in New Orleans is dirty, smelly, and oily. There is water in the river but it is unpleasant to swim in, expensive to purify, and poisonous to many kinds of fish.

Specific problems of water use and purification will be discussed in more detail in Chapter 13.

7.9
CASE HISTORY: WATER AND COAL IN THE SOUTHWEST UNITED STATES

During the late summer of 1977, I was driving through northeastern Arizona. This region is dry and semiarid and small herds of cattle and sheep graze amid sparse grasses and sage. Just outside of Kayenta my car broke down, and I was forced to hitchhike to town to get help. An old, 1950 vintage flatbed truck slowed down, but did not stop. The driver shouted, "Jump in, I don't have any first gear and can't stop!" I ran along and hopped onto the moving vehicle. The engine groaned, the bearings rattled, but we slowly accelerated onto the highway. The driver, a Navajo rancher, was hauling a load of hay

The Colorado River.

south from Green River, Utah. In other areas with wetter climates, late summer is the time to gather in the hay. But there is too little rainfall in this region to grow any hay. Therefore, the Navajo and Hopi farmers must drive north to buy feed for their animals. The man who was driving the truck explained to me that it is very difficult to make a living raising cattle if you cannot grow your own hay. Every fall he sells many of his animals. Most of the money is used to buy hay to feed the rest over the winter. Calves and lambs are born in the springtime and the animals are raised on grass until fall. None of the farmers of the region make much money. He was hoping that his animals would bring a high price at the auction this year so he could afford a rebuilt transmission for his truck.

Later I spoke with Keith Smith, the Navajo Chapter President in Kayenta,

A typical Navajo Ranch near Kayenta, Arizona.

Aerial view of open pit coal mine near Kayenta, Arizona. (Photo by Sam Bingham, Chinle, Arizona)

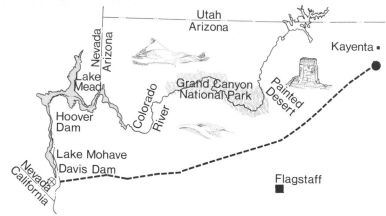

Dotted line shows the route of the coal slurry pipeline.

Keith Smith, chapter president of the Navajo tribal council, Kayenta, Arizona.

Animals are allowed to graze during most of the year, but most farmers in the region must buy hay for the winter months.

about the water problems in the region. He told me, "The last few years the water table has been going down. During June and July, people's wells have gone dry. Ranchers have to drive to town to haul water in pick-up trucks so the cattle won't die." After a silence he continued, "A rancher can't hardly even break even if he has to haul water in a pick-up."

There is a large coal mine a few miles south of Kayenta. Coal from this mine is sold to the Mohave Power Project nearly 480 kilometers (300 miles) away. There the fuel is burned to produce electricity for Las Vegas and Southern Claifornia. The coal is shipped from Kayenta to Mohave via a coal-slurry pipeline. Pipeline shipment is cheaper than shipment by rail. The problem is

"A man can ride under the power lines and feel the energy traveling through them. But the electricity is going to Phoenix, Tucson, Las Vegas, and Los Angeles. We can't tap into that power and many farms have no electricity. I have to drive eight miles in the pickup truck to get water because I don't have a pump and the water table is going down." Ernest Yellowhorse, Cove, Arizona.

Many farming families in northeast Arizona live in hogans made of mud and sticks.

Coal contains mainly pure carbon. Liquid fuels contain carbon and hydrogen. Therefore, hydrogen must be added to convert coal to liquid fuel. Water, H_2O, is the cheapest source of hydrogen and must be used in the conversion.

that it uses water. The coal is first ground into a fine powder. Then it is mixed with water. The whole mixture, called a **slurry,** is pumped through a large pipe. Water for the project is taken from several large wells near Kayenta. The pipeline uses approximately 7500 liters (1900 gallons) of water *per minute* and operates 24 hours per day, 7 days a week.

If the coal were shipped by rail instead of by pipeline, the price of fuel would rise slightly. However, the ranchers in the region would have enough water to maintain their stock.

As the energy crisis worsens, the coal companies will need even more water than they use now. Recall that one solution to the oil shortage is to convert coal to liquid fuel. This conversion requires large quantities of water. The water is an absolutely essential chemical ingredient in the conversion. Where will the power companies find the water? Even now, the farmers near Kayenta cannot get what they need. Experts feel that even if other consumers were rationed more severely than they already are, there still might not be enough water to manufacture the fuel that we need. In some regions, water, not energy, is becoming the limiting resource.

CHAPTER SUMMARY

7.1 INTRODUCTION

Modern society is dependent on fossil fuels and it is certain that the fossil fuel supply will be exhausted in the future.

Energy is the capacity to perform work or transfer heat. Materials can, in theory, be used over and over again. Energy cannot.

7.2 THE FIRST LAW OF THERMODYNAMICS — (OR YOU CAN'T WIN)

The **First Law of Thermodynamics** states that energy cannot be created or destroyed: It is impossible to build a **perpetual motion machine.**

7.3 THE SECOND LAW OF THERMODYNAMICS — (OR YOU CAN'T BREAK EVEN)

The potential energy of a fuel is *never* completely converted into work. Some is always lost to the surroundings. Energy cannot be recycled.

7.4 OUR FOSSIL FUEL SUPPLY

In 1978, approximately 90 percent of the total energy used in the United States was derived from fossil fuels. Natural gas is the least abundant fuel. Worldwide oil production is not expected to meet demands by the year 2000. Supplies may be increased by extracting oil from shale or tar sands or by converting coal into liquid fuel. Coal is relatively abundant, and supplies are expected to last for approximately 200 to 300 years.

7.5 NUCLEAR FUELS

Low-priced uranium is scarce, and nuclear fuels will become expensive as the richer ores are depleted.

7.6 RENEWABLE ENERGY SOURCES

Energy can be obtained from wood, the Sun, falling water, the winds, and the tides. These sources are renewable and, if used properly, will never be depleted.

An ore is said to be depleted when concentrated deposits are dispersed so widely that large quantities of energy are needed to mine them. The future supply of metals and fertilizers depends not only on the size and purity of mineral deposits but also on (a) the availability of energy; (b) the ability of engineers to solve pollution problems; (c) decisions as to land use, especially with regard to the question, "Should mining preempt farming or are they compatible?"; (d) future demands for metal; and (e) future rates of recycling.

7.8 WATER RESOURCES

Water resources are renewed continually. But some river systems are extensively exploited and others are seriously polluted.

KEY WORDS

Energy
Perpetual motion machine
First Law of Thermodynamics
Second Law of Thermodynamics

Ore
Mineral reserve
Acid mine drainage
Slurry

TAKE-HOME EXPERIMENTS

1. **Work and heat.** Take a bowl of cold water and record the temperature. Beat the water vigorously with a fork or, if available, an electric egg beater. Record the temperature again. What do you observe? Explain.

2. **Soluble minerals in soil.** Mix 25 to 30 g (about an ounce) of soil with enough warm water so that you can pour the mixture into a filter. A coffee filter will do if laboratory filter paper is not available. The filtered liquid should be clear. If it is not, simply pour it back through the same filter and let it run through again. Now evaporate all the water by warming it on the *lowest* setting of a hot plate, or by leaving it to stand overnight on a warm object such as a radiator. You will notice a small residue in the bottom of your container. This residue is a mixture of various minerals that dissolve in water. For example, if the material is yellow-brown, iron may be present.

To make a further test, dissolve some washing soda (not bicarbonate of soda or baking soda; it must be washing soda, which is a form of sodium carbonate) in a

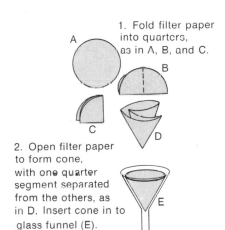

1. Fold filter paper into quarters, as in A, B, and C.

2. Open filter paper to form cone, with one quarter segment separated from the others, as in D. Insert cone in to glass funnel (E).

How to Make a Filter

little water. Dissolve your soil minerals in 2 or 3 drops of water and add a few drops of the carbonate solution. If the mixture becomes cloudy, you can be fairly certain that your sample contains some metals. Would it be possible to build a mine and extract the metals from the soil? Would it be profitable? Discuss.

PROBLEMS

1. **Energy.** Discuss the difference between renewable and nonrenewable resources. What are some advantages and disadvantages of using each type of resource? Discuss.

2. **Energy.** What is energy? Which of the following sources of energy are renewable and which are nonrenewable: coal, sunlight, falling water, uranium, wood?

3. **First Law of Thermodynamics.** Is a lever a simplified type of perpetual motion machine? Discuss.

4. **First Law of Thermodynamics.** Write two statements of the First Law of Thermodynamics.

5. **Second Law of Thermodynamics.** Write two statements of the Second Law of Thermodynamics.

6. **Second Law of Thermodynamics.** What happens to the energy of a gallon of gasoline when it is burned in the engine of an automobile? What happens to the matter in the gasoline?

7. **Second Law of Thermodynamics.** An electric generator converts heat from fuels into work. Can generators be built to convert all the heat into work? Discuss.

*8. **The Second Law of Thermodynamics.** If someone placed a refrigerator in the center of a room and left the door open, would the room be heated, cooled, or would the temperature remain unchanged. Explain.

9. **Fossil fuel supply.** Predict some problems of energy production, supply, and use in the years 2000, 2050, 2200. On what do you base your predictions?

10. **Fossil fuel supply.** True or False: "The world will run out of petroleum around the year 2000." Defend your statement.

11. **Fossil fuel consumption.** Name ten devices that can operate only if liquid fuels are available.

12. **Fossil fuel consumption.** Discuss three alternative energy sources that could be used when petroleum supplies are exhausted. Name some benefits and problems associated with each.

*13. **Oil shales.** Predict some environmental side effects of mining oil shales.

14. **Uranium reserves.** Two geologists, estimating the availability of uranium for fission reactors, arrive at different answers. Explain how such differences can occur.

15. **Ores.** What is an ore? Explain how an ore deposit can be depleted. Explain how ore reserves can increase.

16. **Metal reserves.** There are approximately 10 billion tons of iron dissolved in the world's oceans. Do you think that it might be practical to mine sea water for its mineral content? Discuss.

---

*Indicates more difficult problem.

17. **Land use.** Discuss why land use problems are likely to become more severe if low-grade ores are mined.

18. **Ore deposits.** Briefly discuss five factors that govern the future supplies of metals.

*19. **Ore deposits.** Large quantities of many minerals exist under the sea. Refer to the five factors listed in Problem 18 above and use these guidelines to discuss some of the problems of mining undersea deposits.

*20. **Mineral reserves.** Petroleum is generally burned as a fuel, but in many applications the chemical compounds in the oil are used for the manufacture of plastics and other materials. Explain why petroleum cannot be economically recycled after it is burned, but if it is used for the synthesis of plastics it can be recycled many times.

*21. **Depletion of mineral resources.** A noted environmental scientist reported that the world tin reserves may be depleted in 1990. In making this prediction, he assumed that (a) mining technology and world economic activity will remain constant, (b) consumption levels and population will remain constant, and (c) no new deposits will be discovered. From this information alone, do you feel that you can agree with his conclusion? Defend your answer.

22. **Recycling.** Does recycling of automobiles really conserve iron, or does it only conserve energy? Defend your answer.

23. **Recycling.** Sand and bauxite, which are the raw materials for glass and aluminum, respectively, are plentiful in the Earth's crust. If we are in no danger of depleting these resources in the near future, why should we concern ourselves with recycling glass bottles and aluminum cans?

24. **Water.** If water resources are renewable, why are there water shortages?

25. **Water.** A certain factory uses water for rinsing and then returns almost all of it to the river. Explain how the company might affect water resources even though it does not directly consume water.

QUESTIONS FOR CLASS DISCUSSION

1. Large deposits of valuable metal ores may exist on the Moon and on Mars. Discuss some advantages and disadvantages of mining these ores.

2. Recent reports indicate that many carcinogenic (cancer-producing) compounds are produced when liquid fuels are synthesized from coal. In addition, over one third of the fuel content of the coal is consumed during the conversion. Using this information and the material in the text, list the categories of direct costs and external environmental costs required to produce liquid fuel from coal.

3. Large quantities of water flow unused through the great river systems of the far north such as the Yukon River in Alaska and the McKenzie River in northern Canada. Engineers have suggested that water could be pumped from these basins to cities and farms farther south. Discuss some advantages and disadvantages of this suggestion.

4. There was a shortage of cement in the United States during the spring and summer of 1978. Many construction projects were slowed down, and the price of building rose. (The cost of an average home increased $150.00.) At least part of the shortage occurred because pollution control laws forced some cement processing operations to shut down. Using this example, discuss how pollution control laws affect our daily lives. How would our lives be affected if there were no pollution control laws?

BIBLIOGRAPHY

Several books that discuss energy resources include:

Philip H. Abelson (ed.): *Energy: Use, Conservation and Supply.* Washington, D.C. American Association for the Advancement of Science, 1974. 154 pp.

Ford Foundation Report: *A Time to Choose, America's Energy Future.* Cambridge, Mass., Ballinger Publishing Co., 1974. 511 pp.

S. S. Penner and L. Icerman: *Energy — Demands, Resources, Impact, Technology, and Policy.* Reading, Mass., Addison-Wesley, 1974. 373 pp.

Marion Shepard, Jack Chaddock, Franklin H. Cocks, and Charles M. Harmon: *Introduction to Energy Technology.* Ann Arbor, Mich., Ann Arbor Science Publishers, 1976. 300 pp.

A few books dealing with the mineral resources of the world include:

David N. Cargo and Bob F. Mallory. *Man and His Geologic Environment.* Reading, Mass., Addison-Wesley, 1974. 548 pp.

Eugene N. Cameron (ed.): *The Mineral Position of the United States, 1975–2000.* Madison, The University of Wisconsin Press, 1973. 159 pp.

Gary D. McKenzie and Russell O. Utgart (eds.): *Man and His Physical Environment.* Minneapolis, Minn., Burgess Publishing Company, 1975. 387 pp.

Ronald Tank (ed.): *Focus on Environmental Geology.* New York, Oxford University Press, 1973. 474 pp.

National Academy of Sciences Report: *Mineral Resources and the Environment.* Washington, D. C., National Academy of Sciences, 1975. 348 pp.

8

ENERGY: POLLUTION, CONSUMPTION, AND CONSERVATION

8.1
INTRODUCTION

People first started using fire over 500,000 years ago. For millenia the primary fuel was wood. In the early 1900's, coal was used extensively in industrial societies. Today, petroleum and natural gas are the most widely used fuels. The Age of Oil is expected to exist for only a short while. Roughly speaking, we can say that petroleum started to become widely used in about 1920. By the year 2020, only a hundred years later, petroleum production will probably not be able to meet demands. People will have to search for other fuels (Table 8–1).

One of the most crucial debates of our time is centered on the question: What will the primary energy sources be when the oil runs out early in the 21st century? One school of thought argues that: (a) People must learn to conserve energy. *Conservation* will extend the life of our present fossil fuel reserves and provide the time needed to develop new energy systems. (b) The primary fuels of the early 21st

century will be *coal* and *uranium*. These two fuels will be used to power many new centralized electric generating plants. The electricity will be used to heat homes, drive electric cars and trains, and perform many of the functions now requiring gas and oil.

A second school of thought agrees that conservation is vitally important. But these people believe that large-scale centralized generation of electricity is not the answer. Future energy needs should be met by massive use of solar and wind energy. In addition, there should be return to the use of wood and other plant products as fuel.

In this chapter we will explore some of the issues involved in this debate.

8.2
ENERGY PRODUCTION AND POLLUTION

If fossil and nuclear fuels are to be used in the future, several serious environmental problems are certain to arise. Land surfaces are disturbed

147

TABLE 8–1. COMPARISON OF VARIOUS ENERGY SOURCES

Energy Source	Current Use	Future Availability of Energy Supply	Current Cost	Pollution Problems
Oil	41%	Shortages by year 2000	Cheap	Considerable
Gas	27%	Shortages by year 1985	Cheap	Cleanest fossil fuel
Coal	19%	Two to three hundred year supply	Cheap	Severe problems
Nuclear	7%	Unknown	More expensive than fossil fuel	Severe problems
Solar	Less than 1%	Excellent	Slightly more expensive than oil	Negligible
Hydroelectric	4%	Small expansion	Cheap	Some problems
Geothermal	Less than 1%	Small expansion	Cheap	Some problems
Waves and tides	Negligible	Excellent	Slightly more expensive than oil	Considerable
Wind	Negligible	Excellent	Slightly more expensive than fossil fuels	Negligible
Wood, plant matter & garbage	1%	Small but continuous	Cheap	Some problems
Hydrogen fusion	0	Excellent (if reactors can be built)	Unknown	Serious thermal pollution problems

Air pollution from burning of fossil fuels.

whenever fuels are mined. The consumption of fuels causes serious air pollution problems. The waste heat from electric generators disrupts ecosystems and may even alter world climate. Health and safety problems arise whenever nuclear or fossil fuels are used.

Oil. Oil drilling on open prairies or deserts does little environmental damage. However, in recent years, oil companies have begun to search for petroleum in more hostile environments, such as in the arctic or under the ocean floors. Large oil spills have occurred off the coasts of southern California and Louisiana. In Alaska, major problems arose in transporting the oil from a region far above the Arctic circle to large cities farther south. After a lengthy court battle, oil companies were given permission to build a pipeline across Alaska. The completed pipe crosses previously untracked wilderness, delicate icy tundra, mountains, rivers, and active earthquake zones. In winter, outside temperatures drop to −50° (−58° F). In the summer they soar to +35° C (+95° F). Environmentalists have argued that remote wilderness areas have been altered. Tundra and caribou migration routes have been disturbed. If the pipe should break, rich salmon fisheries and vast waterfowl areas will be destroyed.

In Europe, valuable oil fields have been found under the North Sea. To obtain this oil, engineers must first build fixed platforms in the ocean. Drilling rigs are then mounted on these steel islands. Despite great care, accidents occur in all types of drilling operations. Broken pipes, excess pressure, or difficulty in capping a new well have repeatedly led to blowouts, spills, and oil fires. When these accidents occur on shore, small areas of farmland are destroyed. But entire ecosystems are not affected. When accidents occur at sea, millions of barrels of

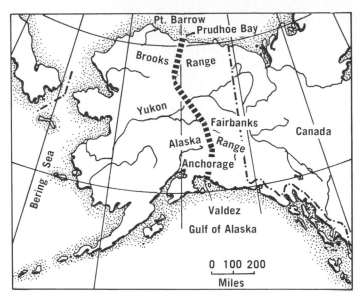

Route of Alaska pipeline.

Pipeling zigzags over the tundra so that the pipe will resist fracture if the earth moves. (Wide World Photos)

The search for oil under the sea.

Aerial view of a strip mine in southeastern Montana. To the left, piles of rubble are being reclaimed and blended into the terrain. (Courtesy of Bureau of Reclamation, photo by Lyle Axthelm)

Decker Coal Mine in Montana. (Courtesy of Bureau of Reclamation, photo by Lyle Axthelm)

oil can be dispersed throughout the waters. Fish and other marine life are poisoned. Ocean ecosystems are disrupted.

Another problem arises when oil is transported over the ocean. Large supertankers carry oil around the globe. Many accidents have occurred in recent years, and ecological damage has been significant. This issue will be discussed in more detail in Chapter 13.

Coal. Coal can be mined either in underground tunnels or in exposed open pits. **Tunnel mines** do not directly disturb the surface of the land, but the earth and rock dug out to make the tunnel must go somewhere. This material, called **mine spoil**, is often seen as ugly surface pollution of land near mining operations. Sometimes the land above the tunnel collapses, leaving gaping holes on the surface. Perhaps most significantly, underground mining is dangerous and unhealthy for the workers. Fires, explosions, and cave-ins claim many lives. In addition, fine coal dust suspended in the air enters the miner's lungs. This dust gives rise to a series of serious and often fatal ailments known as **black lung disease.**

Tunnel mines alter the flow of ground water and pollute underground streams. They are also often more expensive and less efficient than open pit mines. For these reasons, mining companies have been switching many of their operations to open pits, also called **strip mines.** To dig a strip mine, the surface layers of topsoil and rock are first scooped off (stripped) by huge power shovels. This exposes the underlying coal seam. The coal is then removed, and a new cut is started. If the land is not reclaimed, strip mines leave behind vast holes and huge piles of rubble.

Strip mining is the cheapest way to mine many coal fields—that is, cheapest for the mining company. But this cost analysis does not include economic externalities. Neither arguments nor figures are needed to convince anyone that open, rootless dirt piles are uglier and less useful than a natural forest or prairie. The uglification is a loss to all of us. The dirt piles erode. When hillsides erode, the dirt clogs streams and kills fish. Furthermore, sulfur deposits are often associated with coal seams. This sulfur reacts with water in the presence of air to produce sulfuric acid that pollutes the streams below.

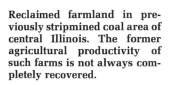

Reclaimed farmland in previously stripmined coal area of central Illinois. The former agricultural productivity of such farms is not always completely recovered.

Perhaps even more crucial is the fact that millions of hectares of coal lie under farmland in many parts of the world. For example, large deposits exist under the fertile wheat fields of the Great Plains in North America. Is the land more valuable as a wheat field or as a coal mine? At the present time, a mining company can buy a farm that lies over a coal seam, mine the coal, destroy the land, and still realize a large profit. But many people feel that this monetary structure ignores economic externalities. A wheat field can provide needed food for many years to come. Coal is used only once and then it is gone.

Perhaps people can have both wheat and coal if old mines are reclaimed. After the coal is removed the holes can be refilled with subsoil and covered with topsoil. If the land is then fertilized, it is possible to plant wheat and corn over the mines after about five years. However, the soil generally is not as rich as it was originally.

The price of coal in 1977 was about $20 to $25 per metric ton. If strip mined lands were restored adequately, the price of coal would rise by about $1 to $1.50 per metric ton. Are users of coal willing to pay this price? If it is not paid directly, the cost is ultimately paid in the form of rising farm prices.

The United States has roughly 16 million hectares (40 million acres) of coal deposits that can be strip mined in the near future. Such vast areas magnify the problems outlined above. If coal companies were charged the true environmental cost of ruined streams and farmland, they would find it profitable to reclaim the land.

8.3
ELECTRICITY

One of the major issues in environmental planning today concerns the problems of generating electric power. People who believe in a future based on coal and nuclear energy feel that a great many new electric generating facilities must be built in the next twenty years. Others disagree. They argue that the production of electricity leads to depletion of resources, water pollution, air

> A hectare is the metric unit of land surface. One hectare equals 10,000 square meters, or a little less than 2½ acres.

> One metric ton = 1000 kilograms = 2205 pounds = 1.1 short tons.

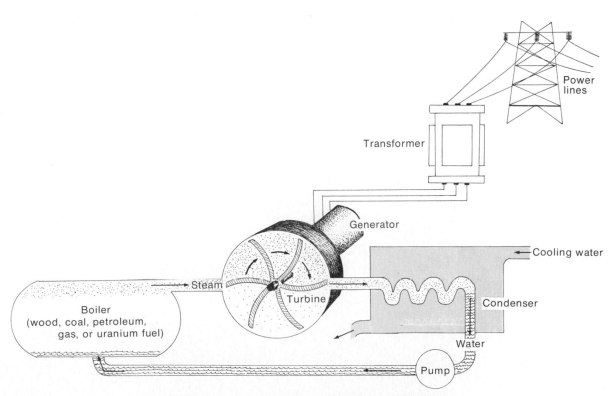

FIGURE 8–1. Schematic view of an electric generator.

pollution, solid wastes, radioactive wastes, land uglification, and thermal pollution. If people would learn to conserve electricity, they could benefit from the use of this convenient energy supply without building more power plants and adding an extra burden on the environment.

Most of our electricity is now produced in **steam turbines.** The operating principle here is uncomplicated. Some power source, such as coal, gas, oil, or nuclear fuel, heats water in a boiler to produce hot, high-pressure steam. This steam expands against the blades of a turbine. A turbine is a device that spins when air or water is forced against it. You can think of it as a kind of enclosed windmill. The hot, expanding steam forces the turbine to spin. The spinning turbine then operates a generator which produces electricity. After the steam has passed through the turbine it is cooled, liquefied, and returned to the boiler to be reused. Normally the steam is cooled with river, lake, or ocean water.

The *cooling cycle* is essential to the entire process (see also Section 8.4). Recall that the Second Law of Thermodynamics states, "It is impossible to convert all the energey of a fuel into work." But the efficiency can be improved if the exhaust gases are cooled. Even the best electric generators operate at only about 40 percent efficiency. This means that for every 100 units of potential energy in the form of fuel, only 40 units of electrical energy can be produced. The other 60 units of heat energy are lost to the environment.

Although a modern generator is only 40 percent efficient, it is more efficient than other common heat engines. A large diesel is 38 percent efficient, an automobile is 25 percent efficient, and a steam locomotive is only 9 percent efficient. Therefore, electricity is an efficient way to perform work. Less fuel is needed to operate an electric lawn mower or car than gas-powered machines.

Electric Heat. In modern society, electricity is often used to provide heat. Advertisements advise people to "live better electrically," and buy stoves, toasters, space heaters, and water heaters. But electric heaters are thermodynamically inefficient. Sixty percent of the heat is discarded at the power plant. On the other hand, a gas-fired stove can be nearly 100 percent efficient, and home furnaces are 60 to 80 percent efficient. Electricity is essential for many functions, but it is wasteful when used to produce heat. Much energy could be conserved if electricity were used only where it is needed and if fuel were used directly when heat is needed.

8.4
THERMAL POLLUTION

The amount of heat that must be removed from an electrical generating facility is quite large. A one-million-kilowatt power plant running at 40 percent efficiency heats 10 million liters of water by 35° C (63° F) every *hour.* It is

More energy is lost when the electricity is transmitted through "power lines." If we count this loss, the overall efficiency of generating electric power and delivering it to your home is only about 33 percent.

"It not the humidity — it's the thermal pollution." (*American Scientist,* Sept.-Oct., 1971)

FIGURE 8–2. This large cooling tower dwarfs the nuclear reactor at a reactor site in southern Washington.

not surprising that such large quantities of heat, added to aquatic systems, cause ecological disruptions. The term **thermal pollution** has been used to describe these heat effects.

What happens when the outflow from a large generating station raises the water temperature of a river or lake? Fish are cold-blooded animals. This means that their body temperature increases or decreases with the temperature of the water. In natural systems, water temperatures are relatively constant. When the temperature is raised, all the body processes of a fish (its metabolism) speed up. As a result, the animal needs more oxygen, just as you need to breathe harder when you speed up your metabolism by running. But hot water holds less dissolved oxygen than cold water. Therefore, cold-water fish may suffocate in warm water. In addition, warm water can cause outright death through failure of the nervous system. In general, not only fish but also the entire aquatic ecosystems are rather sensitively affected by temperature changes. For example, many animals lay their eggs and plants dis-

perse seeds in the springtime when the water naturally becomes warm. If a power plant heats the water in midwinter, some organisms may start reproducing. But if the eggs are hatched at this time, the young may not find the food needed to survive.

Not all power plants discharge waste heat directly into natural environments. Many use special lakes called **cooling ponds.** In other places, large **cooling towers** (see Fig. 8–2) have been built. These alternatives cost money. Once again, the price people are willing to pay is a measure of the environmental quality they want.

We have seen that the mining and refining of fossil fuels and the production of electicity cause widespread environmental problems. What are the alternatives?

8.5
SOLAR ENERGY—ACTIVE SYSTEMS

Every 17 minutes the solar energy that falls on our planet is equivalent to human energy needs for a year at the

1977 consumption level. In the *least* sunny portions of the United States (excluding Alaska) an area of only 80 square meters (a square approximately 9 m, or 29 ft, on a side) receives enough sunlight to supply the total energy demands of the average American family. In addition the technology is available *today* to trap and use much of this energy. Power can be produced without creating a pollution problem. Yet solar energy is not in common use.

ADVANTAGES OF USING SOLAR ENERGY

Limitless supply.
Produces no air pollution.
Produces no water pollution.
Produces no noise.
Produces no thermal pollution.
No possibility of a large scale explosion or disaster.
Conserves the Earth's resources.
Technologically available for widespread use immediately.

Production of Hot Water — Solar Collectors

A few years ago, I lived in the mountains in a cabin without running water. I set a black plastic pipe in a spring on the hillside above the cabin, and laid the pipe along the sunny surface of the hillside toward the cabin. At the bottom of the pipe I installed a simple faucet. The water would fill the pipe in the morning and then sit in the sun all day. The sun would heat it, so that by evening there was always a plentiful supply of hot water for bathing and washing. You can see, therefore, that a solar hot water heater is not necessarily a complex device. Even a bucket of water sitting in the sun all day will become warm. A **solar collector** improves the efficiency of the design. One type of collector consists of a coil of copper pipe welded to a blackened metal base. The whole assembly is covered by a transparent layer of glass or plastic. The operating principle is uncomplicated. Sunlight travels through the glass and is absorbed by the blackened surface. Metal conducts heat readily, so the water in the pipe gets hot. The glass traps the heat within the collector so it does not easily escape back into the atmosphere (Fig. 8–3).

A solar collector of this type can be connected to a home heating system. The hot water produced in the collector is stored in an insulated tank. It can be used directly for washing or bathing. Or it can be pumped through radiators to heat the house. Of course, sunshine is not available at night or on cloudy days. Hot water can conveniently be stored to heat the house overnight. But it is expensive to build a system large enough to heat and store enough water

to supply heat during several days of cloudy weather. Therefore, most solar systems are installed together with a conventional furnace. The solar collector is used on sunny days and the furnace is used when it is cloudy. Naturally, such a dual system is initially more expensive than a simple furnace. But then large amounts of fuel are saved every year. If one calculates the total cost of a solar collector, furnace, and fuel over a fifteen-year-period, one finds that in all cities in the United States solar heating is cheaper than electric heating. Solar heating can be slightly more expensive or slightly less expensive than oil heating depending on the location of the house (Table 8–2). Since the price of fuel is almost certain to rise in the near future, the economic advantage of solar heating will increase. In addition, solar heat is clean, nonpolluting, and completely renewable. Why, then, aren't solar heaters widely used? That question is hard to answer.

Perhaps the main reason is that even in the most favorable locations solar heat isn't that much cheaper than oil heat. Therefore, the high initial cost of solar collectors discourages many people. It is sometimes hard to borrow the extra money. Many builders find that they can realize the highest profits if they build inexpensive houses with conventional heating systems. The builder is not concerned if the homeowner or tenant is burdened with high fuel bills for years or decades to come.

The student must remember that current cost estimates do not include economic externalities. If people were

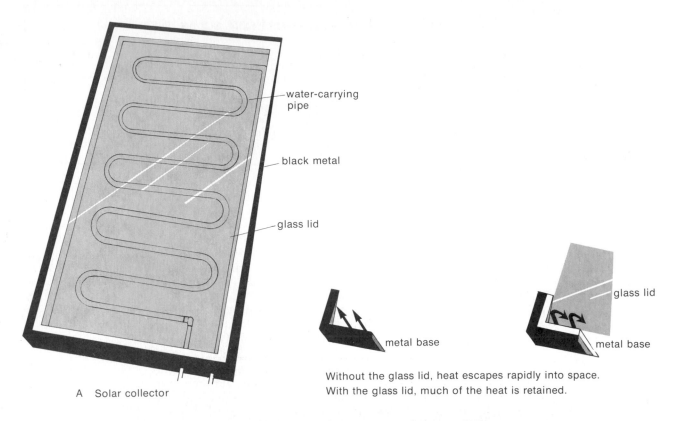

water-carrying
pipe

black metal

glass lid

glass lid

metal base

metal base

A   Solar collector

Without the glass lid, heat escapes rapidly into space.
With the glass lid, much of the heat is retained.

B

**FIGURE 8–3.   *A*, Solar collector. *B*, Solar collectors on a residential home.
(Courtesy of Energy Systems Division of the Grumman Corporation,
manufacturers of Sunstream Solar Collectors)**

A magnifying glass focusing sunlight.

charged for depletion of oil and gas reserves and for environment pollution, solar heating would be considerably cheapter than other systems.

Production of High-Temperature Steam

If you take an ordinary magnifying glass and focus sunlight onto a piece of paper, you can easily burn a hole in the paper. The lens concentrates the solar energy from a large area to a small one so that high temperatures can be realized. Sunlight can also be concentrated through the use of specially designed

TABLE 8–2.   COMPARATIVE COSTS OF OPERATING VARIOUS HEATING SYSTEMS IN THE UNITED STATES*

| Region | Typical Yearly Energy Bill for Various Systems | | |
	100% Oil	100% Electric	70% Solar with 30% Oil
Northeast	$550.00	$1650.00	$440.00
Middle Atlantic	410.00	1320.00	400.00
South Atlantic	300.00	600.00	300.00
West North Central	400.00	930.00	460.00
Mountain	420.00	930.00	410.00
Pacific	160.00	320.00	160.00

*Source: *Congressional Record*, February 27, 1975.

This table was published before a recent tax amendment was passed in the United States allowing special deductions for solar heating systems. These tax advantages serve to reduce the overall cost of solar heat. At the present time, a system of 70 percent solar and 30 percent oil is the cheapest heating system in most parts of the country.

There are wide differences of opinion on the immediate feasibility of solar energy systems. Three quotations, below, illustrate the debate.

"The prospects for solar are brighter than most imagine. If it tries hard enough, the U.S. can meet up to one quarter of its energy needs through solar technology by the year 2000. And for the year 2020 and beyond, it is now possible to speak hopefully and unblushingly of the U.S. becoming a solar society." President's Council on Environmental Quality as reported in Chemical and Engineering News, April, 1978.

"The big utility companies desperately want to develop energy that will utilize their existing grid systems and permit them to go on selling electricity to the populace. A more revolutionary form of energy such as solar power, might put the utility industry out of business." Jack Anderson, column, Oct. 10, 1977.

"I am very pessimistic about solar energy. Heating water seems all right at present. Space heating is marginal. . . . Solar heating may possibly improve. . . but the engineering has not yet been done and there is no reliable cost estimate." Hans Bethe as reported in Federation of American Scientists Newsletter, March 1978.

Forest of mirrors focus sunlight on the 200-foot-high "power tower" at the Department of Energy's Solar Thermal Test Facility at Sandia Laboratories. Each of the 222 heliostat arrays at the world's largest solar facility contain 25 four-foot-square mirrors.

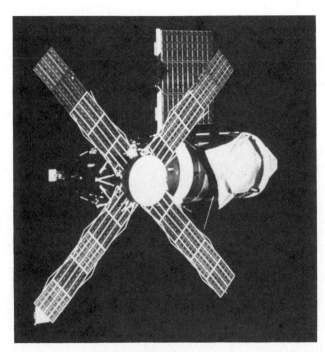

The Skylab space station, powered by a group of photoelectric cells mounted on "windmill" arms. (Courtesy of NASA)

mirrors. A large solar furnace of this type, completed in France in 1970, reaches temperatures up to 3500° C, hot enough to melt any metal. Such a device can be used to make hot steam to drive a turbine and produce electricity. In the United States two "solar power towers" have been planned. One in Albuquerque, New Mexico, was put into operation in the fall of 1978.

Solar Generation of Electricity. In the late 1900's, scientists discovered that light can knock electrons off the surface of various metals. Thus light energy can be converted directly to electrical energy. A device that produces electricity from sunlight is called a **solar cell.** Solar cells are commonly used today to convert sunlight to electricity in spacecraft. They are quiet and trouble-free. They emit no pollution and appear to have a long life expectancy. Today, it would be entirely possible to build an electric generating station on Earth using solar cells. A 1000-

megawatt plant (equivalent to a large fuel-burning facility) would occupy only 10 square kilometers if it were built in the southwestern American desert. The major problem at the present time is that solar cells are so expensive that a solar plant is uneconomical even though the fuel is free. But the situation may be changing. New processes are being developed to manufacture solar cells more cheaply. Many experts hope that economical units could be in production by the mid-1980's.

Of course, solar generating stations alone could not entirely replace fossil and nuclear fuels, for the sun does not shine all the time.

There are several ways to store solar energy. Perhaps the most practical is to use solar electricity to produce hydrogen fuel according to the equation

$$\text{Water} + \begin{array}{c}\text{Electrical}\\\text{Energy}\end{array} \rightarrow \text{Hydrogen} + \text{Oxygen}$$

Hydrogen is a versatile and useful fuel that can burn in air and be used as a replacement for gasoline and other liquid fuels.

8.6
OTHER RENEWABLE ENERGY SOURCES

Hydroelectric Energy

Many early settlers in North America used the power of falling water to drive their mills and factories. Today many large rivers are dammed. The energy of water dropping downward through the dam is used to produce electricity. Energy produced in this way is called **hydroelectric energy.**

Only 3½ percent of the total power in the United States is derived from hydroelectric sources. In all probability, this energy source will not become more important in the future. There are simply not enough good locations for building large dams. Moreover, other environmental problems arise. It would be a great aesthetic and recreational loss if places like the Grand Canyon or the Snake River Gorge were dammed. Some people have argued that future development of large dams and centralized hydroelectric power stations may not be the best way to harness the energy in our streams and rivers. Many

A large hydroelectric dam. (Courtesy of Bureau of Reclamation; photo by E. E. Hertzog)

small dams and mills were built along the Eastern seaboard during the 1800s. Most of these are now abandoned. If small electric turbines were placed in these existing dams, significant quantities of electricity could be generated.

Geothermal Energy

Energy derived from the heat of the Earth's crust is called **geothermal energy.** In various places on the globe, such as in the hot springs and geysers of Yellowstone National Park in Wyoming, hot water is produced near the surface. Although no one suggests harnessing Old Faithful for generating electricity, there are several places where hot underground steam is available. The power company must simply dig a well and pipe the free steam into a turbine. The Pacific Gas and Electric Company is producing some electricity from a generator connected to wells in central California. This project is new in the United States, but an Italian facility at Larderello has been in operation since 1913.

Even optimistic supporters, however, do not expect a large portion of our power to be supplied in this manner. There are not enough hot springs at or near the surface of the earth. Continuous exploitation for more than a century or two is expected to exhaust the water or heat content of these wet wells. Geothermal energy is not always free from pollution. Underground steam or hot water is often contaminated with sulfur compounds, which must be removed before they are discharged to the air or to a lake or river.

Energy from Waves and Tides

People have always marveled at the power of ocean waves and tides. In recent years engineers have designed systems to harness some of this energy. A commercial tidal generating station is now operating along the coast of Brittany in France.

Tidal dams present serious ecological problems. Recall that bays and inlets provide shelter, nurseries, and

Section of the geothermal steam field where Pacific Gas and Electric Company generates 396,000 kilowatts of electricity from underground steam.

habitats for many saltwater fish. Industrial operations in these areas destroy the estuary systems.

Energy from the Wind

The power of wind has been used since antiquity to drive ships, pump water, and grind grain. What is its potential in modern society? There is more than enough wind energy avail-

Wind generator.

able to supply the world's energy needs. At present it is technically feasible to build windmills capable of producing electricity. These can be built either as small, home-sized units or as large central generators. It is slightly more expensive to harness wind power than solar power. But if windmills were mass produced, then energy from the wind could be cheaper than energy from nuclear power plants. Therefore, if people would use the best available technology, this energy source could play a significant role in replacing fossil fuels.

Energy from Wood, Plant Matter, and Garbage

We live in a wasteful society. Half of the household trash in the United States and Canada is paper. Huge piles of bark, wood scraps, and logging wastes rot slowly near many sawmills. Mounds of animal manure lie unused near large feedlots. If people would collect these wastes and use them as fuel, considerable quantities of energy could be salvaged. In France, there are twenty generators that burn garbage for use as domestic energy sources. A large facility in Paris produces electric power for 130,000 people, and 20 percent of the total steam used for heating in the entire city. In the United States some organic wastes are recycled but most are not. If these discarded materials were used as fuel, the total energy output would be greater than the 1975 production of power from nuclear power plants.*

Once again, our failure to use a valuable resource is a measure of social attitude. Individual and industrial patterns are slow to change, even when such change is beneficial to all.

Summary

Clearly, the burning of garbage is not a final answer to the energy problem. But if garbage, solar, geothermal, tidal, and wind energies were all exploited, the total contribution would be very substantial. In fact these energy sources

*C. C. Burwell, *Science*, Vol. 199, March, 1978, p. 1041.

A century ago most ships were powered by the wind. Today there is some renewed interest in building ocean-going, commercial sailing ships. This space-age "dynaship"—still in the planning stage—would use electric motors to manage sails so that a single person on the bridge could control the entire vessel. (Drawing by Marion Mackay)

all together could eliminate the need for fossil fuel.

So far, this chapter has explored the various types of energy production. Recently, people have been asking, "Why do we need so much energy in the first place? Would it be possible to reduce consumption without altering our life styles appreciably? The answer is yes. The remaining sections will consider energy consumption and conservation.

8.7
ENERGY CONSUMPTION AND CONSERVATION FOR TRANSPORTATION

The two *least* efficient modes of transportation, the automobile and the

Noonday traffic in Boulder, Colorado.

The bicycle is an energy-efficient means of transportation. (From D. Plowder: *Farewell to Steam.* Brattleboro, Vermont, The Stephen Greene Press, 1966)

BART, a modern mass transit system in San Francisco.

airplane, are the two major transportation industries in the United States.

The automobile, in particular, has modified our style of living. As mentioned in Chapter 6, many cities are experiencing urban sprawl. Houses are far from shopping centers and places of work. The shopping centers themselves are dispersed. For example, in Boulder, Colorado, there are no food stores in the central part of the city. A person shopping in the downtown mall area must travel nearly 1.5 km to buy groceries. Many people live in the suburbs and must commute large distances daily. What can be done? Solutions are available, but difficult. No one can reorganize the city overnight. The system is set in concrete.

People can't easily move existing buildings, but they can change transportation patterns. For example, a heavy, luxury, 1975 model automobile in poor mechanical condition with power steering, automatic transmission, and air conditioning travels 8 miles or less on a gallon of gasoline. Many 1979 model subcompacts drive at 35 miles per gallon or more on the freeway. It isn't hard to understand that if people use carpools and small cars, much less fuel is needed than if drivers travel alone in luxury automobiles. Consider another factor. In the United States, most automobile trips are for distances of less than five miles. People

TABLE 8-4. HOW YOU CAN SAVE FUEL IN TRANSPORTATION

Method	Fuel Savings as Compared to Wasteful Practices
Walk or use a bicycle, stay at home	Saves 100%
Use mass transit	Saves 50% (or more)
Carpool	Saves 50% (or more)
Keep car well tuned	Saves 20%
Drive smoothly (no jerks, fast starts and stops)	Saves 15%
Purchase car without an air conditioner	Saves 10%
Keep tires inflated to proper pressure	Saves 5%

drive to the corner market to pick up a newspaper and a quart of milk or to mail a letter. These trips are costly. Large amounts of gasoline are needed to start a cold automobile. As a rough approximation, a car that is capable of operating 25 miles per gallon on long trips operates at 5 miles per gallon on short trips. If people reduced the number of short trips, walked, or rode bicycles, large quantities of fuel would be saved (Tables 8–3 and 8–4).

Mass Transit. Mass transit — buses, trains, and trolleys — can provide clean, efficient transportation. At the present time, only 3 percent of urban transport in North America is carried by public transportation systems, and 97 percent is carried by the automobile. In some cities bus and trolley fares are being reduced in an effort to encourage use. Yet most drivers still prefer their private cars.

As a result, a dangerous situation has evolved. If fuel shortages were to develop quickly, the existing mass transit system could not possibly handle urban transportation needs. Millions of workers would be stranded — unable to get to work. Cities would be paralyzed. Therefore, it seems vitally important to build public transportation systems now to avert future disaster.

Many modern cities are becoming so crowded that the automobile is no longer an efficient means of transportation. An automobile is designed to provide direct, quick, comfortable, and private transportation. There is no need to walk to a bus stop, wait for the bus, and then mix with other commuters and

TABLE 8-3. ENERGY REQUIREMENTS FOR TRANSPORTING PEOPLE AND FREIGHT

A. People*

Bicycle	425 pkm/l
Walking	280 pkm/l
Intercity bus	35 pkm/l
Train	20 to 85 pkm/l, ranging from poorly patronized intercity routes to commuter trains
Automobile	4n to 16n pkm/l, depending on model and condition of car, where n = number of passengers
Jumbo jet	8 to 10 pkm/l, for estimated average passenger loads
SST	5.7 pkm/l, for estimated passenger loads

B. Freight**

Pipeline	93 km/l
Railroad	63 km/l
Waterway	62 km/l
Truck	11 km/l
Airplane	1 km/l

*In passenger-kilometers per liter of fuel (pkm/l).
**The value of 1 km/l for air freight is used as a standard for reference.

Rush hour traffic in Denver, Colorado.

shoppers. However, in recent years, expressways have become so crowded that commuting by automobile has lost much of its advantage. During rush hours, highways are choked with automotive smog. Traffic is often stalled for long periods of time, so travel is slow. Accidents are common. Moreover, parking spaces are often hard to find, so people cannot drive directly to their destination anyway. If quiet, efficient, comfortable mass transit systems were built in most cities, and if roadbuilding projects were curtailed, mass transit might become more popular.

8.8
ENERGY CONSUMPTION AND CONSERVATION FOR INDUSTRY

North American industries are generally more wasteful of fuel than European industries are. One prime example of this waste is the inefficient use of steam. Recall from Section 8.3 that steam is used to drive turbines to pro-

duce electricity. After the steam has passed through the turbines it is no longer useful for producing work. But it can be used to provide heat for homes, industry, or a variety of other purposes. In North America, most of this steam is "thrown away." That is to say that the steam is cooled by discharging its heat into the atmosphere or into rivers, lakes, or oceans. If this steam were used for other purposes, large quantities of energy would be saved. Many European electric companies sell their excess steam, thereby saving money and energy.

Another way to reduce industrial energy consumption is to increase recycling of materials. In general, production of manufactured goods from recycled wastes consumes less energy than the production from raw materials. However, total manufacturing *costs* reflect the price of labor, transportation, and capital investment as well as fuel bills. Unfortunately, when economic externalities are neglected, recycling operations often seem more expensive than primary production. Unless in-

"Chemical industry reaches energy goal. Chemical companies saved 14.2% of fuel used per unit of output in 1977, compared to a base year of 1972." The companies report that process modifications and new processes designed to conserve higher-priced energy account for the most savings.

Chemical and Engineering News, May 8, 1978

dustries are offered economic incentives or threatened with penalities, we can expect little significant increase in industrial recycling.

It is also possible to reduce excess industrial energy by consuming less. For example, if appliances were built to last longer, and if people used less paper for packaging, less energy would be needed for production.

8.9
ENERGY CONSUMPTION AND CONSERVATION FOR HOUSEHOLD AND COMMERCIAL USE

Most of the residential and commercial uses of power involve the burning of fossil fuel to heat air and water. Of course, solar energy could be easily used for many of these applications.

Space Heating. Fire for heating is one of the most ancient uses of fuel. Without space heating people could not live in many areas. Most modern homes are heated by oil- or gas-fired furnaces. Furnaces operate at 60 to 80 percent efficiency. But there is an increasing trend to "clean, efficient" electric heat. Electric heaters are, in fact, clean and 100 percent efficient in the home. But, recall that the electric generators that supply the power for your 100 percent efficient heater operate at 40 percent efficiency and are not at all clean. Therefore, whenever electricity is used for heating, large quantities of energy are wasted. Electric space heaters, hot water heaters, stoves, ovens, and clothes dryers are all wasteful. All could be replaced by natural gas, oil, or propane-powered appliances.

Fuel consumption for home heating can be reduced by half if adequate insulation is installed. (Photo by Nancy Craft)

Insulation. Many older houses in North America were built with no insulation at all. If an uninsulated house in the midwestern United States is fitted with adequate insulation, heat consumption will be reduced by half. The added cost of the insulation would be paid back in one or two winters. Even today, when fuel prices are rising and energy supplies are diminishing, many poorly insulated homes are being built.

Passive Solar. Section 8.5 discusses systems for heating homes with solar energy. It is also possible to use solar energy without special collectors. A

WHAT IN THE WORLD ARE YOU DOING?

I'M CONSERVING ENERGY...

HOW?

BY TURNING OFF THE LIGHT WHEN THE CAVE IS NOT IN USE.

(B. C. by permission of Johnny Hart and Field Enterprises, Inc.)

TABLE 8–5. HOW YOU CAN SAVE FUEL IN HOME HEATING

Method	Heat Savings Compared to Energy Inefficient House
Have your furnace maintained, cleaned, and tuned properly	Saves 10%–20%
Add extra insulation in ceilings and walls	Saves 30%–50%
Add storm windows	Saves 10%
Calk leaky windows and doors	Saves 5%–15%
Cover windows with drapes and shades at night	Saves 5%–15%
Close off and do not heat unused rooms.	Saves (variable)
Turn thermostat down	Saves 3% per °F

Draperies can also be used to reduce air conditioning bills. The direct sunlight entering south-facing windows heats the house when you want it to be cool. Pull the drapes shut and you can save 15 percent of your air conditioning bill.

simple window uses the lighting value of the sun. A house without windows would be dull and dreary indeed! Windows may also allow solar heat to enter the house. When the sun shines directly on a window facing south, the radiant energy enters the room and warms it. But heat also escapes back out through the glass. During the nighttime, heat escapes and no sunlight enters. Most homes built today are poorly planned. More heat is lost through the windows than is gained. But this trend can be reversed through proper planning (Table 8–5). A few suggestions follow: (a) A house should be built with large windows facing south and only small windows on the north side. Since the sun does not strike the north side directly, these windows are sources of heat loss. But south-facing windows are a source of heat gain on sunny days, even in winter. (b) Still air is an excellent insulator. Heat loss can be reduced considerably by building windows with two layers of glass separated by an air space. Windows built in this manner are commercially available. Alternatively, storm windows can be used to trap air and reduce heat loss. (c) During the nighttime heat is lost through windows and, of course, nothing is gained. If heavy draperies or shutters are used to close the windows when it is dark, large savings are possible. (d) Five to fifteen percent of a normal heat bill can be saved if leaks near the edges of doors and windows are sealed with caulking and weather stripping.

Some homes built in sunny climates, such as in New Mexico or Arizona, have been designed so that the windows alone provide nearly all the heat that is needed in the winter. On the other hand, conventional houses are heated by a furnace and then lose 20 percent of their heat through the windows.

In North America most homes are heated to 22° C (72° F) during the winter. People feel that this is the expected comfort range. But the 22° C temperature is merely a matter of habit. In 1932, the preferred room temperature for most people was 19° C (66° F). The comfort range rose slowly to 19.5° C in 1941, and to 20° C by 1945. Today most homes are heated to 22° C during the winter. Thirty years ago, people normally wore long underwear and sweaters during the winter and used less fuel. Today, fashions operate in disregard of fuel consumption. Men wear jackets, ties, and long pants in business offices during the summer and then turn on the air conditioner. Similarly, women wear dresses in winter and turn up the heat to warm their legs.

Use of Appliances. Modern appliances add comfort and convenience to

Passive solar home near Telluride, Colorado. Sunlight entering through the large, south-facing windows provides a significant heat source. This house is located in the mountains of Colorado, yet in its first year of use, no auxiliary heat was needed until mid-November.

our daily lives. For example, a sewing machine is truly a work saver. In a minute or less a person using a sewing machine can sew a seam that would take an hour or more by hand. What's more, the machine uses little energy. But much energy is wasted by improper use of appliances. As mentioned earlier, an electric range uses 2½ times as much energy as a gas range to do the same job. An automatic dishwasher uses the same amount of energy whether it is full or half empty. And many people run their washers with half a load. Table 8–6 lists some ways that you can save energy used by your home appliances.

From fuel mining to fuel transportation to fuel consumption, the environment always loses when energy is consumed. Early Americans thought that since you could neither damage nor deplete the environment, you may as well take the most convenient route to your goal. Today we know that we can damage and deplete our planet. Old attitudes must, therefore, be reevaluated.

8.10
ENERGY CONSERVATION AND THE ECONOMY

Many people in business, labor, and government have argued against conservation. They claim that efforts to save energy will cause unemployment. United States Secretary of Energy, James S. Schlesinger, said, "Restraining energy growth means restraining

TABLE 8–6. HOW YOU CAN SAVE FUEL FOR HOME APPLIANCES

Method	Savings (Dollars/Year)*
Place extra insulation around hot water heater	Saves $19/year
Place insulation around hot water pipes	Saves $20/year
Hang clothes on line in summer rather than use dryer	Saves $25/year
Disconnect drying cycle from dishwasher	Saves $25/year
Put covers on pans when cooking	Saves $5/year
Use wool or down blankets rather than electric blankets	Saves $8/year
Use proper size light bulbs, turn lights off when not in use	Saves (variable)
Turn off pilot lights on gas stove	Saves $10/year

*Energy savings here have been calculated using a value $.05 per kilowatt hour.

the growth of jobs. It means unemployment.*"

Certainly a sudden fuel shortage would cause widespread unemployment. Factories would be forced to shut down, gas station attendants and automobile mechanics would be out of work, shippers and truckers would become unemployed. But many economists argue that a gradual conservation program would *increase* jobs. Suppose, for example, that people in the United States gradually shifted toward increased use of mass transit. Automobile factories would slow production. But

---

*Address to AFL-CIO convention, Los Angeles, December 9, 1977.

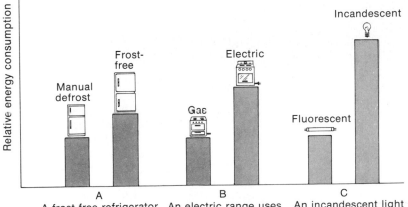

Fuel consumption in the home could be reduced without any loss of comfort if more efficient appliances were used.

A frost-free refrigerator uses 1½ times as much energy as a conventional one

An electric range uses twice as much energy as a gas range

An incandescent light uses three times as much energy as a fluorescent light

A

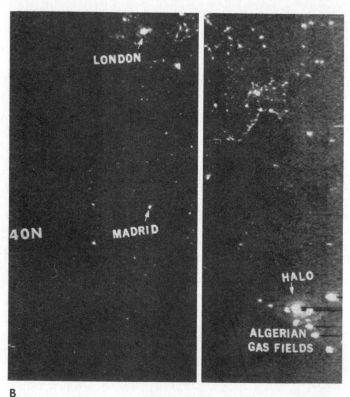

LONDON

40N   MADRID

HALO

ALGERIAN
GAS FIELDS

B

A dramatic example of the differences in energy consumption between the United States and Europe can be seen in these two photographs. At midnight, when most people are fast asleep, vast regions of the United States are lighted up, as can be seen from this satellite photograph (*A*). On the other hand, most of Europe is dark, and much less electricity is needed. (*B*).

(*A*, U.S. Air Force photo; *B*, U.S. Defense Meteorological Satellite Program photo)

the automobile workers would not necessarily lose their jobs. They could be employed to manufacture buses and trains. Workers who now specialize in road construction could build and repair rail systems. If cars and appliances were maintained longer and thrown away less frequently, more repair specialists would be needed. Carpenters and plumbers could be hired to install solar collectors and to add extra insulation to homes.

Conservationists argue that if energy consumption continues to grow until, suddenly, we run out of oil or coal, the economy will collapse disastrously. If people start now to build a low-energy society, changes can be made gradually. There is no need for large-scale unemployment. As is discussed in the next section, people in Sweden use much less energy than Americans do, and employment is high.

Energy use is also linked with world trade. Most nations of North America, Europe, and Asia import some or most of their oil. These imports cost money. A country has an equal **balance of trade** when the total value of all imports equals the total value of all exports. In the United States, so much money is spent on oil that there is a negative balance of trade. More money is spent overseas than is received. This inequality threatens the national economy. If less imported oil were needed, the international economic position of the United States would certainly improve.

8.11
COMPARATIVE CASE HISTORY: ENERGY CONSUMPTION IN THE UNITED STATES AND SWEDEN

Both the United States and Sweden have well-developed economic sys-tems. In both countries people are well fed. Premium medical care is available. Infant deaths are low. Educational levels are high. Most families have a car, telephone, television, refrigerator, and vacuum cleaner. Yet the average person in Sweden uses less than two thirds the energy that an American does. North Americans may well look to people of other nations to see how consumption can be reduced.

Transportation. Overall, Swedes use approximately one fourth the energy for transportation as compared to Americans. They save fuel in many ways: (a) People frequently walk or use bicycles for short trips. (b) Mass transit is used more frequently. (c) On the average, Swedish cars consume half as much fuel per mile as American cars do.

Space Heating. Winters are considerably longer and more severe in Sweden than they are in the United States. Yet energy consumption for heating per person is less. Swedish houses are generally better insulated and more ef-

> Energy consumption in Canada is nearly equal to that in the United States.

A view of Stockholm, Sweden. (Courtesy of Swedish Information Service)

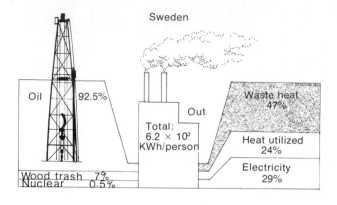

Sweden

Oil 92.5%

Total: 6.2 × 10³ KWh/person

In / Out

Waste heat 47%

Heat utilized 24%

Electricity 29%

Wood trash 7%
Nuclear 0.5%

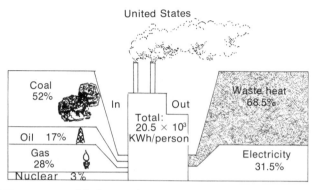

United States

Coal 52%

In / Out

Total: 20.5 × 10³ KWh/person

Waste heat 68.5%

Electricity 31.5%

Oil 17%

Gas 28%

Nuclear 3%

FIGURE 8–4. Use of fuel to produce electricity in Sweden and the United States. (kWh is an abbreviation for kilowatt hours, a unit of energy.) (From *Science,* Vol. 194, Dec. 1976, p. 1001)

ficient than their American counterparts.

Industrial Consumption. Recall that excess heat is generated whenever fuels are burned to produce electricity. In the United States most of this heat is thrown away. In Sweden waste heat is used extensively to heat homes. The result is shown in Figure 8–4. In addition, Swedish industries are generally highly efficient because newer, innovative technology is in use. Thus, less energy is used to produce and refine a kilogram of steel, oil, paper, cement, or chemicals in Sweden than in the United States.

There are several reasons why Swedish consumption is so low. Heavy government taxes have raised fuel prices so that, for example, gasoline costs twice as much as it does in North America. But price alone is not the only factor. The government in Sweden has taken vigorous steps to promote energy conservation. Some of these are listed in the box below.

The Swedish example teaches us that it is possible to use considerably less energy without lowering the quality of life.

United States

Building codes permit construction of leaky, poorly insulated homes.

Low priority to energy conservation in housing loans.

Mass transit marginal to poor. Bus service unavailable to many suburbs of United States cities. Intercity rail system drastically reduced in recent years.

Sweden

It is illegal to build without adequate insulation.

High priority to energy conservation in housing loans.

Mass transit highly efficient. Bus service four minutes apart during peak hours in major cities. Intercity rail system provides direct service to most regions.

CHAPTER SUMMARY

8.1 INTRODUCTION

Some people believe that by the year 2000, society must depend on coal and nuclear fuel, while others argue that massive use of solar, wind, and agricultural energy sources will provide the answer.

8.2 ENERGY PRODUCTION AND POLLUTION

Oil drilling and transportation in hostile environments has led to serious environmental disturbances.

If coal is mined in **tunnel mines,** the

surface contours of the land are left un-touched. But the mining operations are dangerous, expensive, and polluting. **Strip mines** are often cheaper to operate, but the surface of the land is disturbed. In many cases, valuable farmland is destroyed. Strip mines can be partly reclaimed if we are willing to pay the price.

8.3 ELECTRICITY

To produce electricity some power source, such as coal, gas, oil, or nuclear fuel, heats water in a boiler to produce hot, high-pressure steam. The steam expands against the blades of a turbine and the spinning turbine operates an electric generator. The steam, its useful energy now spent, flows into a condenser where it is cooled and liquefied. Generally, river or lake water is used to cool the condenser.

For every 100 units of potential energy in the form of fuel, 40 units of electric energy are produced. Sixty units of energy are lost to the environment.

8.4 THERMAL POLLUTION

Power plants discharge large quantities of heat into waterways. This effect, called thermal pollution, causes widespread ecological disruption.

8.5 THE USE OF SOLAR ENERGY

(a) Flat plate solar collectors are inexpensive and relatively easy to build. The hot water produced can be used economically for space or water heating.

(b) Large solar furnaces have been built from specially oriented mirrors. These devices concentrate the light so that high temperatures can be obtained.

(c) Solar cells can convert sunlight directly to electricity.

8.6 OTHER RENEWABLE ENERGY
SOURCES

(a) Hydroelectric energy (power from falling water) is limited because there are not enough good locations for building large dams.

(b) Geothermal energy (heat derived from the Earth's crust) may become more promising as more well sites are explored.

(c) Energy can also be obtained from tidal action, from the wind, and from burning garbage.

8.7 ENERGY CONSUMPTION AND
CONSERVATION FOR
TRANSPORTATION

The two least efficient modes of transportation, the automobile and the airplane, are the two major industries in the transportation field. Significant effort should be made now toward building mass transit systems so that in twenty-five years, when fuel shortages become even more acute, people's habits and the structures of cities will be geared toward less waste.

8.8 ENERGY CONSUMPTION AND
CONSERVATION FOR INDUSTRY

Energy is required in all manufacturing processes. Industries in the United States are generally more wasteful than European industries. Thus, reevaluation of planned obsolesence, the reduction of the amount of paper used in packaging, and generally reduced consumption would greatly help alleviate environmental stresses caused by manufacturing and by waste disposal.

8.9 ENERGY CONSUMPTION AND
CONSERVATION FOR HOUSEHOLD
AND COMMERCIAL USE

Domestic energy consumption could be curtailed by construction of tight, well-insulated buildings, curtailing the use of resistance electric heat, using appliances efficiently, initiating changes in people's habits and fashions so that heating and cooling requirements would be reduced.

8.10 ENERGY CONSERVATION AND
THE ECONOMY

Many economists feel that high energy consumption is necessary for a healthy economy. Others disagree strongly.

KEY WORDS

Open pit mine (strip mine)
Tunnel mine
Black lung disease
Steam turbine
Thermal pollution
Cooling pond
Cooling tower

Solar collector
Solar cell
Passive solar heating
Hydroelectric energy
Geothermal energy
Mass transit
Balance of trade

TAKE-HOME EXPERI-MENTS

1. **Efficiency of electrical consumption.** Look at your last month's electric bill and record the number of kilowatt hours of electricity used at that time. Now initiate an energy conservation program in your home. Report on the conservation measures used. Look at the electric bill one month later, and record the amount of electricity saved. (Caution — if your house is heated or cooled electrically, there may be a large monthly variation in electric bills due to change of seasons. For example, less heat is needed in March than in February. If the records are available, compare electric consumption for a month this year with the consumption for the same month last year.) Did your program save money? Discuss any inconveniences raised by the conservation practices.

2. **Solar collectors.** Remove the inner dividers from each of four identical ice cube trays. Fill each tray with the *same* amount of water, until it is about 2/3 full. (Between 1½ and 2 cups will do for the average-size tray. Use a measuring cup.) Set all four trays in the freezer compartment of a refrigerator until the water is frozen. Cover two of the trays with clear plastic wrap, sealed around the edges with tape or a large rubber band. Replace the trays in the freezer and let them stay overnight. Next day, about midmorning, prop the four trays on a small box on a waterproof surface outdoors. Let two of them (a covered one and an open one) slope toward the north and the other two slope south. Let them sit in the sun and note the rate at which they melt. Which design was

the most efficient collector of solar energy? Explain your results.

3. **Efficiency of transportation.** Obtain a map of your local community. Mark the locations of your residence, your school, the grocery store, post office, bank, and the five other stores that you visit most frequently. Measure the distance from your home to each of these locations. What type of transportation do you usually use to travel from one to another? Discuss the efficiency in terms of fuel consumption, time, and cost of the transportation system you use. Would it be difficult to improve the fuel efficiency of this system? Discuss.

4. **Electric meter.** Find the electric meter in your house or apartment. If you live in a private house, the meter is probably located on the outside wall, whereas if you live in an apartment, it is probably in the basement. As shown in the photograph, a meter has several small dials and a horizontal disc. The disc spins when electric energy is being used. The rate of spinning is proportional to the amount of electricity consumed. Thus, if many appliances are plugged in, the disc will spin rapidly, and if little electricity is being used in the house, the disc will barely move. Turn off all the electrical appliances in your house, and unplug the refrigerator and freezer. The disc should now be stationary. Plug in one 100-watt

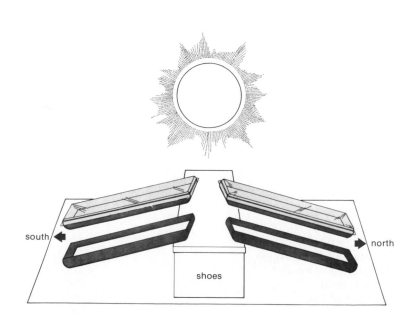

Household electric meter.

light bulb. How many seconds are required for the disc to make one complete revolution? Plug in a 50-watt light bulb and measure the time required for one revolution. Now plug in an appliance whose power requirement you do not know. Record the time required for one revolution of the disc. How much power is consumed by the appliance? (Don't forget to plug in your refrigerator and freezer when you finish this experiment.)

5. **Fuel for cooking.** Turn on your stove and wait until the burner reaches constant heat output. Pour one cup of cold water in a pan, cover it, and measure the time needed for the water to boil. Cool the pan, pour another cup of cold water in it, and measure the time needed for the water to boil with no cover on it. Compare your results.

PROBLEMS

1. **Energy strategy.** Discuss two different energy strategies proposed for the United States. What are the similarities and differences of the two approaches?

2. **Oil.** Explain why an oil spill in the North Sea is more harmful to the environment than an oil spill in the plains of Texas.

3. **Oil.** Explain why an oil pipeline across Alaska is more harmful to the environment than an oil pipeline across Texas.

4. **Coal mining.** Discuss the relative advantages and disadvantages of strip and tunnel mining.

*5. **Coal mining.** Explain how unrestrained strip mining of coal could cause the price of meat and bread to rise. Use this example to discuss the nature of economic externalities, introduced in Chapter 1.

6. **Electric power generation.** Explain the function of the cooling system in a steam turbine. Is a cooling system needed at a hydroelectric facility?

7. **Efficiency of generation of electricity.** Compare the efficiency of each of the following methods of warming the air in a room: (a) Oil is burned in a home furnace. (b) Oil is burned in a generating plant to make electricity. The electricity is then transmitted to the house to operate an electric heater.

8 **Thermal pollution.** Define thermal pollution. What are the sources of thermal pollution?

*9. **Thermal pollution.** Since marine life is abundant in warm tropical waters, why should the warming of waters in temperate zones pose any threat to the environment?

*10. **Thermal pollution.** Warm water carries less oxygen than cold water. This fact is responsible for a series of disturbances harmful to aquatic organisms. Explain.

11. **Waste heat and climate.** What types of energy sources do not add any additional heat to the environment?

12. **Solar energy.** Describe the construction and function of a flat plate collector.

*13. **Solar heating.** Discuss the advantages and disadvantages of using solar energy for home heating. Explain what factors you would evaluate in deciding whether or not to install solar heat if you were building a new home.

14. **Solar generation of electricity.** Discuss the potential for using solar energy to generate electricity. Explain why this potential is not now being exploited.

15. **Solar generation of electricity.** Discuss the environmental impact of placing large solar generators in desert regions. Compare this impact with the problems of coal-fired generators.

16. **Hydrogen economy.** Explain how an economic system based on hydrogen fuel would function differently from a system based on petroleum.

17. **Alternative energy sources.** Discuss briefly the prospects for solar, geothermal, tidal, wind, and garbage energies. What problems do these methods entail?

*18. **Hydroelectric energy.** Some people have suggested that the large northern rivers such as the Yukon in Alaska or the McKenzie in Canada should be dammed for the production of hydroelectric power. The electricity would then be transmitted south to large North American cities.

*Indicates more difficult problem.

Compare and contrast the environmental problems of transporting oil with the problems of transporting electricity across the Northern wilderness.

19. **Transportation.** Give some reasons why it is often difficult to alter patterns of surface transportation.

20. **Transportation.** It takes a few hours to fly across the United States. A comparable journey by train takes several days. Yet the cost to the passenger is about the same. Discuss the impact of this economic structure on world fuel reserves.

21. **Transportation.** Discuss present transportation patterns in the United States. How could they be made more efficient?

*22. **Energy of Transportation.** Referring to Table 8–2, compare the relative fuel consumption of: (a) a single person driving in a heavy, luxury car; (b) four people riding in a heavy, luxury car; (c) one person driving in a small sub-compact car; (d) four people riding in a small sub-

_____

*Indicates more difficult problem.

compact car; (e) travel by bus; (f) travel by bicycle.

*23. **Transportation.** In 1975, 97 percent of urban traffic was carried by the automobile, and only 3 percent was carried by mass transit systems. Thus, even if mass transit systems were to double the number of passengers they carry, there would be relatively little impact on overall fuel consumption rates. Should these figures be used as an argument against building new mass transit systems? Defend your answer.

24. **Energy consumption for industry.** Compare the use of waste steam in North America with its use in Europe.

25. **Energy conservation for industry.** Discuss some factors that might make recycling economically attractive.

26. **Home heating.** Why is a furnace more efficient than an electric heater?

27. **Home heating.** Explain how proper placements of glass can conserve home heating costs.

28. **Home appliances.** List three electric appliances that are efficient and three that are inefficient. Discuss.

QUESTIONS FOR CLASS DISCUSSION

1. Prepare a class debate. On one side argue in favor of a coal–nuclear energy future; on the other side argue in favor of a nonnuclear, nonfossil fuel future.

2. Many people suggest that the government should provide low cost loans for solar collectors. Others disagree. They argue that solar energy would only benefit those living in suburbs or in the country. City dwellers would not benefit. They argue further that if solar energy were such a good idea, it would be profitable enough to work without government support. Discuss these arguments.

3. Gasoline taxes have traditionally been used for the construction of new roads. This practice has been considered fair because the roads are paid for by those who use them most. Increasingly, economists and social philosophers feel that many of our traditional concepts of fairness must be reevaluated in the light of environmental problems. Do you feel that there should be a reevaluation of road tax use? If so, how would you allocate funds? If not, explain.

4. List five ways to improve the heating efficiency in your home or apartment. If you live in a private home, estimate the cost of implementing these changes. If you live in an apartment house, do you think that such changes would be a worthwhile investment for the owner?

5. Some businesses or manufacturers can operate only if large quantities of energy are readily available. Examples are automobile sales and maintenance and manufacture of glass bottles. Others could operate efficiently in an energy-conservative society. Two examples are computer manufacturing and sales, and movie theaters. Divide the class into several groups. Have each group analyze a portion of the yellow pages of the local telephone directory. List the businesses that depend on high energy consumption and those that could operate in a conservative society. Discuss how a slow change to a more conservative society could be realized without large-scale unemployment in your city.

BIBLIOGRAPHY

The debate on future energy use has been published in several journals as quoted in the text. Two books that discuss opposing views on the energy questions are:

Fred Hoyle: *Energy or Extinction? The Case for Nuclear Energy.* Salem, New Hampshire, Heinemann Educational Books, Inc., 1977. 81 pp.

Amory B. Lovins: *Soft Energy Paths.* Cambridge, Massachusetts, Ballinger Publishing Co., 1977. 231 pp.

A three-volume work that reviews many aspects of the relationship between energy use and human welfare is:

Barry Commoner, Howard Boksenbaum, and Michael Corr (eds.): *Energy and Human Welfare.* New York, Macmillan, 1975. Vol. 1, 217 pp: Vol. 2, 213 pp.: Vol. 3, 185 pp.

Several other recent and comprehensive books on energy and the environment are:

Philip H. Abelson (ed.): *Energy: Use, Conservation, and Supply.* Washington, D. C., American Association for the Advancement of Science, 1974, 154 pp.

Barry Commoner. *The Poverty of Power.* New York, Alfred A. Knopf, 1976, 314 pp.

Ford Foundation Report: *A Time to Choose — America's Energy Future* Cambridge, Mass, Ballinger Publishing Co., 1974. 511 pp.

Allen L. Hammond, William D. Metz., and Thomas H. Maugh, II: *Energy and the Future.* Washington, D. C., American Association for the Advancement of Science, 1972. 184 pp.

S. S. Penner and L. Icerman: *Energy.* Reading, Mass., Addison-Wesley, 1974, 373 pp.

Marion L. Shephard, Jack B. Chaddock, Franklin H. Cocks, and Charles M. Harmon: *Introduction to Energy Technology.* Ann Arbor, Mich., Ann Arbor Science Publishers Inc., 1976. 300 pp.

H. Stephen Stoker, Spencer, L. Seager, and Robert L. Capener: *Energy from Source to Use.* Glenview, Ill., Scott, Foresman and Company, 1975, 337 pp.

Two books on special topics are:

J. Richard Williams: *Solar Energy Technology and Applications.* Ann Arbor, Mich., Ann Arbor Science Publishers, 1974, 120 pp.

Marvin M. Yarosh (ed.): *Waste Heat Utilization.* Springfield, Va., National Technical Information Service, 1971. 348 pp.

An excellent book on mass transit is:

Tabor R. Stone: *Beyond the Automobile.* Englewood Cliffs, N. J., Prentice-Hall, 1971. 148 pp.

A nontechnical book that pleads the case against the automobile is:

John Burby: *The Great American Motion Sickness.* Boston, Little, Brown and Co., 1971, 408 pp. (Also available from Consumers Union, Mt Vernon, N.Y.)

Three how-to-do-it books on conserving energy in your own environment are:

Bruce Anderson: *Solar Home Book.* Harriville, New Hampshire, Cheshire Books, 1976. 298 pp.

George S. Springer and Gene E. Smith: *The Energy-Saving Guidebook.* Westport, Connecticut, Technomic Publishing Co., 1974. 103 pp.

Carol H. Stoner: *Producing Your Own Power.* Emmaus, Pennsylvania, Rodale Press, 1974. 322 pp.

9

NUCLEAR ENERGY AND THE ENVIRONMENT

9.1
THE NUCLEAR CONTROVERSY

Early in 1976, two articles about nuclear power appeared at the same time in popular journals. One was written by Barry Commoner in *The New Yorker*. The other was written by Hans Bethe in *Scientific American*. Both authors are well known. Commoner is a biologist and environmentalist. Bethe is a physicist who received the Nobel Prize in 1967 for his discovery of the nuclear reactions in stars. They could hardly have disagreed with each other more. Commoner concluded that "the entire nuclear program is headed for extinction. It will leave us with a monument, which people will need to care for with vigilance, if not affection, for thousands of years — huge stores of radioactive wastes, and the powerless, radioactive hulks of the reactors that produced them." Bethe, on the other hand, stated that the United States must have sources of energy other than from fossil fuel. He reasoned that the only important source between now and the end of the century is nuclear fission. Bethe also minimized the various objections that have been raised against nuclear power. He pointed out that the risks involved (other than from nuclear weapons) "are statistically small compared with other risks that our society accepts."

The opposing sides in the controversy are by no means limited to two people. For example, under the heading "No Alternative To Nuclear Power," 32 scientists (one third of whom are Nobel Prize winners) supported Bethe's position.[*] On the other hand, another group of scientists, also including many Nobelists, submitted a declaration against nuclear power.[†]

How can knowledgeable people differ so sharply? Part of the answer, at least, is apparent. They differ in the statements they believe to be true. They differ in the assumptions they make about the probability of accidents. They differ in the judgments they make about the value of the expected benefits and the human cost of the expected damages. But perhaps the most significant gap that separates the two sides lies in their views of the human condition. Those who favor a nuclear future make the following assumptions: (1) We must have the energy sources needed for continued growth. (2) Alternative energy sources, such as solar en-

---

[*]*Bulletin of the Atomic Scientists*, March, 1975.
[†]Prepared under the auspices of the Union of Concerned Scientists, Cambridge, Massachusetts.

177

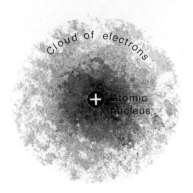

Cloud of electrons

Atomic nucleus

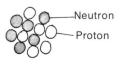

Neutron

Proton

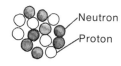

Neutron

Proton

FIGURE 9–1. **The atom. (***Left:* **carbon-12.** *Right:* **carbon-14.)**

ergy, are interesting but are not likely to contribute significantly in this century. (3) Coal, which is available now, is environmentally a more damaging choice than nuclear power.

The opponents of nuclear power argue that we do not need to use so much energy. As we saw in the Case History about Sweden in Chapter 8, other societies use less energy and provide good conditions of life. They argue further that alternatives such as solar energy could be used now. The anti-nuclear argument also holds that the more conservative path will ultimately be the better choice for preserving and improving the human condition.

9.2
RADIOACTIVITY

During the 19th century scientists learned that ordinary matter is a collection of fundamental units called **atoms.**

An atom consists of a small, dense, positively charged center called a **nucleus.** The nucleus is surrounded by a diffuse cloud of negatively charged **electrons.** The nucleus is composed of positively charged **protons** and neutral particles called **neutrons.**

Greek philosophers suggested that matter was made up of atoms, but they had no experimental proof of their theory.

Chemical elements are considered to be the stuff of which all other substances are composed. About 104 elements are known. Most of these are found in nature, but some have been made synthetically. Some well-known elements are hydrogen, carbon, nitrogen, oxygen, and uranium. All the atoms of a given chemical element have the same number of protons in the nucleus. Thus all hydrogen atoms contain 1 proton, all carbon atoms have 6, nitrogen 7, iron 26, and uranium 92. The neutrons add mass to the nucleus but have relatively minor effects on the chemical behavior of the element.

As stated above, all carbon atoms have 6 protons in the nucleus. Most of them have 6 neutrons as well. An atom with 6 protons and 6 neutrons is called carbon-12 (6 protons + 6 neutrons = 12). A carbon atom that has 8 neutrons is called carbon-14 (6 protons+8 neutrons = 14). Elements whose atoms have the same number of protons but different numbers of neutrons are called **isotopes.**

Most isotopes of most atoms are stable. Suppose we take a container full of copper-63 atoms, set it on a table, and wait. If we wait one, two, ten thousand, or ten trillion years, the copper atoms will all remain as copper atoms. But suppose we had, instead, a container filled with radium-226. If we set this container down and waited, many things would happen. Radium nuclei are unstable. Even if we do nothing to the radium atoms, the nuclei will spontaneously decompose, releasing energy and particles. This process is called **radioactivity.**

Let us examine our container of radium-226 more closely. If we could see the individual atoms, we would note that here and there a particle flies out of the radium nucleus at high speed. What remains behind is no longer a radium nucleus, but rather the nucleus of a new atom (radon). If we watched these interesting events for 1600 years, we would find that, after this time, *half* of the radium atoms would have decomposed, and the other half would have remained unchanged. The time required for half of the atoms to decompose is called the **half-life.** Some half-lives are longer than 1600 years. The half-life of uranium-238, for example, is 4.5 billion years. Other half-lives are

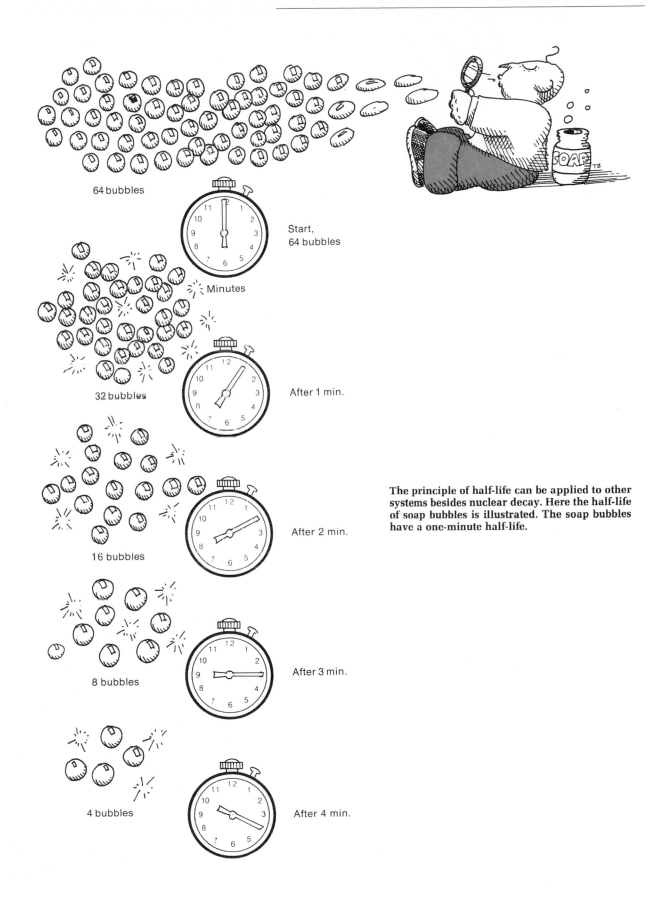

64 bubbles

Start,
64 bubbles

Minutes

32 bubbles

After 1 min.

16 bubbles

After 2 min.

The principle of half-life can be applied to other systems besides nuclear decay. Here the half-life of soap bubbles is illustrated. The soap bubbles have a one-minute half-life.

8 bubbles

After 3 min.

4 bubbles

After 4 min.

shorter. Carbon-14 has a half-life of 5770 years. Iodine-137 has a half-life of 24 seconds.

Uranium-238 decomposes slowly. As a result, energy is released at a low rate. Iodine-137, with a half-life of 24 seconds, decomposes much more rapidly, and the release of energy is much faster. As an analogy, think of two logs, one rotting slowly and the other burning rapidly. The rotting log decomposes over a long period of time but always remains cool. On the other hand, the burning one is hot during its short lifetime. Similarly, a radioactive element with a long half-life emits low levels of radiation. But isotopes that have short half-lives are analogous to the burning log, for they emit dangerously high levels of radiation.

9.3
BIOLOGICAL EFFECTS OF RADIATION

On several occasions during the past 75 years, groups of people have been exposed to large doses of radiation. Much information about what radiation does to the body has been provided by the effects of the atomic bombs dropped on Hiroshima and Nagasaki. In addition there have been various nuclear accidents and some chronic exposures to radioactivity. If humans are exposed to a dose of between 100 and 250 rads, they will develop fatigue, nausea, vomiting, diarrhea, and some loss of hair. However, most of them recover completely from the immediate illness. For doses around 400 to 500 rads, however, the outlook is not so rosy. During the first few days, the illness is similar to that of the previous group. The symptoms may then go away almost completely for a time. But beginning about three weeks after the exposure they will return again. The radiation impairs bone marrow function. Bone marrow is essential in producing materials needed by the body to fight infection and stop bleeding. Fifty percent of those exposed in this dose range will die. Of these most will die of either infection or bleeding. If, instead, the dose administered is about 2000 rads, the first week of the illness will again

A rad is a measure of radiation dosage. It is expressed in terms of energy received per unit of body weight. One rad = 100 ergs/g of body tissue. See Appendix A for discussion of units of energy.

be the same as in the previous groups. But these people become very ill in the second week rather than in the third week, with severe diarrhea, dehydration, and infection leading to death.

Radiation and Cancer

There are many delayed effects of radiation that occur months or years after exposure. None of these is better studied or of more concern than cancer. There is overwhelming evidence that exposure to radiation increases the incidence of cancer. Before the dangers of radiation were appreciated, early workers were careless in their handling of radioactive materials. A great many later suffered from various cancers. The famous case of the radium dial workers in the 1920's deserves mention. Many women were responsible for painting the dials of watches with a phosphorescent radium paint in use at that time. Radium is radioactive. They routinely dipped the end of the brush in their mouths before applying the paint to the dial face. In later years this group experienced a very high incidence of bone tumors. Within a decade after the atomic attacks on Hiroshima and Nagasaki, many of the survivors died of leukemia. In more recent times, a large number of these survivors have died of other forms of cancer. In 1954, the United States government conducted a series of nuclear bomb tests in the Pacific. Many of the island people living in the region have since died of leukemia and other forms of cancer.

"I hope you weren't planning on a big family, Miss Whipple."

(From *Industrial Research,* September 1973)

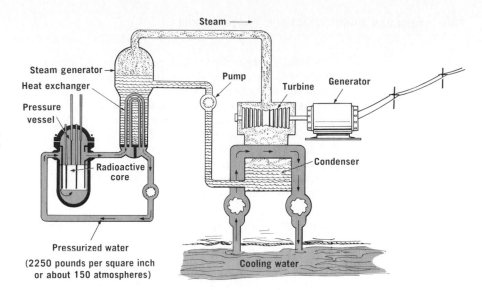

FIGURE 9–2. Schematic illustration of a nuclear plant powered by a pressurized water reactor.

Radiation not only may affect individuals alive at present, but may endanger future generations as well. If the genetic materials in the sperm cells of a male or the egg cells of a female are altered, infants may be born deformed. Alterations of this type are called **mutations.** In every animal species studied in the laboratory, radiation has been shown to produce mutations. Though it is unethical (and impractical) to perform genetic experiments on humans, we have every reason to believe that radiation causes mutations in people as well.

9.4
NUCLEAR FISSION REACTORS

Recall that the nuclei of radioactive isotopes spontaneously emit small particles. In 1939, scientists discovered a nuclear reaction that was much different from this natural radioactivity. In this new reaction, atomic nuclei did not merely decompose by expelling a small particle. Instead the nuclei actually split apart into two pieces of more or less equal size. This phenomenon is called **nuclear fission.** Considerable energy is released. But there are two conditions that must be satisfied for fission to take place.

(a) Only certain nuclei of unstable atoms are fissionable. For example, uranium-238 is not fissionable, but uranium-235 is.

(b) For fission to occur, the nucleus must first absorb a neutron. Stray neutrons may fly about here and there in the natural environment, but they are rare. Therefore, if a mass of uranium-235 rests in the earth, as for example in a uranium mine, nothing much happens. However, if these atoms are given a small start, the reaction will continue by itself. Study the schematic reaction in Figure 9–3.

You see that if one neutron is absorbed by one uranium-235 atom, two fission products, energy, and two or three neutrons are released. These neutrons can then strike other uranium atoms and trigger a continued fission reaction. Obviously, if every neutron released strikes another uranium-235 atom, the reaction will accelerate very rapidly. Such a reaction is called a **chain reaction.** An explosion will occur. The result is an atomic bomb.

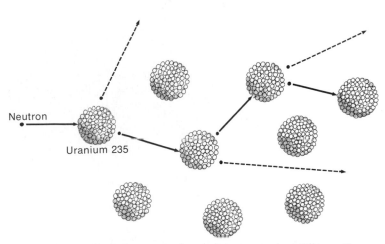

FIGURE 9–3. The fuel for a nuclear reactor contains only a small concentration of uranium-235 atoms. Therefore many neutrons escape without hitting a uranium-235 nucleus. The reaction continues in a controlled fashion and cannot blow up like a bomb.

A **nuclear reactor** is designed so that some but not all of the released neutrons strike other uranium-235 atoms. In this way the reaction can continue at an even, controlled rate. Some energy is released, but not too much. Therefore it is possible to obtain useful power without causing an explosion.

In a nuclear reactor the heat energy from the fission is used to boil water and create steam. The hot steam is used to drive a turbine to produce electricity.

9.5
SAFETY IN NUCLEAR FISSION REACTORS

Can a nuclear reactor blow up like a bomb? The design and operation of a

FIGURE 9–4. A typical fuel assembly, consisting of many fuel rods, being lowered into place in a reactor. (From D. R. Inglis: *Nuclear Energy: Its Physics and its Social Challenge.* Reading, Mass., Addison-Wesley, 1973, p. 91.)

nuclear reactor are largely centered on the control of neutrons. The nuclear fuel that is used contains only a low concentration of uranium-235. Therefore, most of the neutrons released during fission do not hit other fissionable nuclei. The reaction continues in a slow, controlled fashion. The fuel elements cannot blow up like a bomb.

Other kinds of serious accidents. The fuel elements in a nuclear reactor are long rods of uranium fuel surrounded by a metal casing (see Fig. 9–4). In some reactors, these fuel rods are immersed in water. The water serves to cool the fuel rods and also to slow down fast neutrons. Rods made of cobalt or other types of materials are spaced throughout the radioactive core. These rods absorb neutrons. If the rods are raised, more neutrons are available and the fission reaction speeds up. If the rods are lowered, neutrons are absorbed and the reaction slows down.

What would happen if a pipe carrying cooling water to the reactor burst? Such an accident is called a **loss of cooling accident.** Heat would build up very rapidly. In about 45 seconds tem-

FIGURE 9–5. Control-rod driving mechanism of a pressurized water reactor (PWR), partially disassembled. (From Inglis, *ibid.*)

cooling water. If everything works as planned, the emergency system will prevent the reactor core from heating up and escaping out of control.

What happens if all cooling systems fail? In about half an hour, the temperature will rise until the fuel rods melt. The concentrated mass of melted fuel will grow even hotter. Within a few hours the concrete shell surrounding the reactor core will break and tons of white hot radioactive material will melt its way into the ground. (This series of events is sometimes jokingly called the China Syndrome, meaning that the molten mass is moving toward the other side of the globe.) Large quantities of radioactive matter may then be released from the ground to the air or to the underground waterways.

What is the probability of such a disaster? Let us return to the first step, the bursting of a water pipe. This is no ordinary household pipe that readily springs a leak during a cold snap when the occupants happen to be on vacation. Nonetheless, the pipe was manufactured by people in some factory and it could be faulty. Even if the pipe does burst, there is always the emergency cooling system. For a true disaster to occur, *both* systems would have to fail.

The United States government has made an estimate of nuclear safety. The key conclusions of this estimate were as follows:

"Of course it's perfectly safe. Any accident would be in complete violation of the guidelines established by the Nuclear Regulatory Commission."

(© 1980, Sidney Harris)

Event	Chance of Occurrence
Complete fuel meltdown of any one reactor in any one year.	1 in 17,000
An accident killing as many as 1000 people by acute radiation sickness in any one year.	1 in a million
The "worst" accident occurring to any one reactor in any one year, which would cause 3300 deaths, 45,000 "early illnesses," over 1500 latent fatal cancers, and an approximately equal number of genetic defects.	1 in a billion

It is very easy to show that, in Bethe's words, these risks "are statistically small compared with other risks that our society accepts." However, opponents of nuclear power reject these predictions on many levels. Here are some of their objections:

peratures in the reactor core would rise to 1480°C. At this temperature the water left in the reactor core would react with materials in the fuel rods to produce hydrogen gas. When hydrogen gas disperses in air, the mixture can be violently explosive.

Emergency cooling systems have been designed to prevent such a catastrophe. As soon as the main water supply breaks, the emergency system takes over and provides a backup supply of

- Various tests of small-scale emergency cooling systems have all resulted in failures. These findings are very different from the predictions listed above.
- The government study examines the risks to the people *within 25 miles* of nuclear reactors. It therefore neglects the possible effects of an accident on the entire United States population, let alone on the population of the world.
- Private insurance companies have not been willing to insure nuclear power plants for full liabilities. Some people say that this means that the insurance companies do not believe the government predictions.
- The entire analysis is based on assumptions (or guesses) about the chances of various accidents. Some scientists disagree with the predictions. Some engineers have resigned from their jobs at nuclear power plants in protest against what they believe are unacceptably great risks. Perhaps the most severe criticism of all is the fact that the report failed to include accidents *that had already occurred when the report was issued.* Many malfunctions (but no catastrophic accidents up to the time of this writing) have plagued nuclear plants throughout the lifetime of the program. One such incident is described in the Case History at the end of this chapter.

"Routine" Radioactive Emissions. Even with the best of designs and with accident-free operation, some radioactivity is routinely released from all nuclear reactors. For example, the fuel cartridges are encased in metal shells. These shells are thin and sometimes develop small leaks. Radioactive materials flow through the holes into the water. Even if there were no leaks, neutrons pass into the water and make some of its impurities radioactive. This, too, is a source of "routine" emission.

Since the beginning of time, all life on Earth has been subject to some exposure to radioactivity. There are naturally radioactive materials in the Earth's crust, and more radiation reaches us from outer space. Some people argue that compared with these natural sources the extra radiation from power plants is trivial. Others disagree. Scientists have pointed out that natural radioactivity is suspected to cause 10 percent of all cancer deaths. Any extra exposure would cause even more harm.

9.6
DISPOSAL OF RADIOACTIVE WASTES

Recall that when uranium nuclei split in a nuclear reactor, they break into two fission products (Equation 9.1). The fission products are also radioactive. These radioactive fission products are responsible for a major and as yet unsolved environmental problem. Let us look at the situation more closely.

Since one uranium atom splits into two radioactive fission products, there is an immediate doubling of the number of radioactive atoms on Earth. But this is only a small part of the problem. Uranium-235 has a half-life of over 700 million years. It was compared earlier to a slow, rotting log. But many of

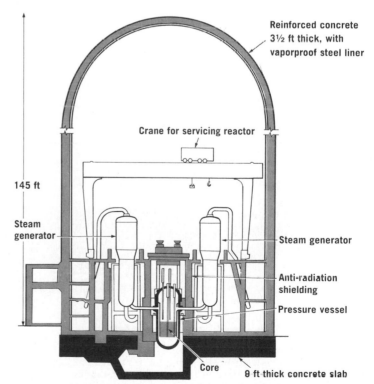

FIGURE 9–6.   Containment structure for a nuclear power plant. (© 1971 by The New York Times Company; reprinted by permission)

the fission products have half-lives measured in centuries, years, months, or seconds. Since these products decompose much faster than uranium does, they emit much higher levels of radiation. They can be compared to a rapidly burning log. There is no way to speed up or slow down the decay of these products. The wastes of a nuclear reactor emit harmful radiation for many thousands of years. Since it is impossible to destroy them, they must be stored somewhere where they do the least harm.

As an example of the nuclear waste problem, consider plutonium-239. Plutonium doesn't occur naturally anywhere on Earth. All the plutonium on our planet has been manufactured either in nuclear reactors or in weapons development programs. The half-life of plutonium-239 is 24,000 years. If there is one gram of plutonium today, there will be one-half gram after 24,000 years. After another 24,000 years, half of the remaining quantity will be decomposed. Therefore, after 48,000 years there will be one quarter of a

"A thousandth of a gram of plutonium taken into the lungs as invisible speck of dust will kill anyone . . . Even a millionth of a gram is likely, eventually, to cause lung or bone cancer." John McPhee, *The Curve of Binding Energy,* Ballantine, 1975.

gram left. The plutonium produced today will have to be cared for by thousands of future generations.

So far, no satisfactory "permanent" disposal for nuclear wastes has been found. After most of the uranium in the fuel rods has been used up, the entire rod is dissolved in acid. The unused uranium is recovered. The hot, corrosive, radioactive liquid waste is pumped into holding tanks. A large government nuclear facility in Hanford, Washington, has been storing its wastes in large underground steel and concrete tanks. So far over 50 million gallons of waste have been stored. But eventually tank linings corrode and leak. All in all, a total of approximately 450,000 gallons of radioactive wastes have leaked out of these tanks into the environment.*

The people at Hanford have been building new tanks and pumping wastes from old leaky ones into the new ones. This game of "musical tanks" is expected to continue until a more satisfactory solution is found.

The nuclear waste problem is perhaps the single most controversial issue of the nuclear power debate. What do you do with highly poisonous "garbage" that will remain active and dangerous for many thousands of years? Governments cannot be held responsible — they cannot be expected to survive for these time periods. An ideal storage place must be far from sites of geological changes such as earthquakes. It must be safe from enemy attacks, explosion, and sabotage. The site should be far enough away from natural ecosystems so that no radiation leaks into the environment.

According to some authorities there are several safe locations for nuclear wastes. Many salt mines and rock formations have been stable for millions of years and are expected to be stable for millions more in the future. These may provide ideal locations for waste disposal sites. But other people disagree. In fact, the disagreement has been so strong that all proposed sites have so far been rejected. People are afraid of nuclear wastes in their environment. One United States congressman was asked recently where he thought nu-

Three one-million-gallon capacity, double-shell tanks (shown during construction) were completed in the spring of 1977 and tied into Hanford waste management operations. After completion the tanks were covered with a minimum of six to seven feet of soil. The tanks employ the latest monitoring and leak detection equipment and are linked by computer to the integrated tank surveillance system.

Chemical and Engineering News, April 10, 1978, p. 17.

Control room of a nuclear plant.

clear wastes should be stored. He answered, "Somewhere other than in my congressional district."* Some government officials have announced that if a satisfactory disposal site is not found, future construction of nuclear power plants may be halted.

9.7
BREEDER REACTORS

Recall that a conventional fission reactor uses uranium-235 as a fuel. Only 0.7 percent of the uranium found in nature is uranium-235. The remaining 99.3 percent is uranium-238. If the abundant uranium-238 could be used as fuel, the energy content of uranium reserves would be increased over 100 times. It is possible to convert uranium-238 to plutonium-239 and then use the plutonium as a fission fuel.

uranium-238 + 1 neutron → plutonium-239 + 2 electrons
plutonium-239 + 1 neutron → fission products + energy + 2 or 3 neutrons

A reactor that produces plutonium and then uses it as a fission fuel is known as a **breeder.** Breeders increase the life of the world's uranium resources. But they also increase the danger from nuclear reactors because

(1) The fuel must be more concentrated in a breeder than in a conventional reactor. The situation is inherently more dangerous because energy is released in a more concentrated form.

(2) Breeders manufacture plutonium-239 and then use it as a fuel. Plutonium is a dangerous, radioactive element. It remains in our environment for hundreds of thousands of years (see page 186).

*New York Times, April 16, 1978.

Nuclear fusion reactor (experimental).

9.8
NUCLEAR FUSION

Most of the energy derived from middle-aged stars such as our sun comes from hydrogen fusion reactions.

Fission reactions occur when heavy nuclei split apart. **Fusion** reactions occur when nuclei of light elements are joined together. Isotopes of hydrogen will fuse if they are fired at each other at very high speeds. If a large mass of hydrogen isotopes fuses in a very short time, the reaction cannot be contained and it goes out of control. This is the explosion of the "hydrogen bomb." On the other hand, useful energy could be extracted from fusion if it were possible to devise a controlled reaction.

To initiate fusion, the temperature must be raised to a minimum of about 40 million degrees Celsius. Major problems exist. It is impossible to build a container that will withstand these temperatures. Some scientists have been developing a sort of "magnetic bottle." This isn't a physical bottle at all. Rather it is a magnetic field that contains the hydrogen. Other scientists have been working with high intensity laser beams to control the reaction. But the problems haven't yet been solved. No one has developed a controlled fusion reactor. But if the problems are solved, what kind of world would be created?

First, there would be abundant energy. It is estimated that if everyone used as much energy as the average North American does, readily abundant fuels could provide energy for about a billion years. This statement, by itself, does not describe the kind of world that such an abundance would create. But it does imply the likelihood of drastic social and environmental change.

Perhaps the foremost difficulty would be the thermal pollution that such an abundance could create. In fact, some scientists believe that this problem sets an upper limit on the rate at which any new energy sources could be developed.

Could the fusion reactor get out of control and go off like a hydrogen bomb? Nuclear scientists are entirely confident that the answer is no, an explosion could not occur. The total quantity of fuel in the reactor at any one time would be too small for an explosion to occur. If the temperature were to drop, the reaction would stop. In effect, the fusion would turn itself off. The situation is rather similar to that of a burning candle. If something goes wrong, the flame goes out — the candle does not explode.

Would there be a problem of environmental radioactivity? The answer here is yes, but the problem would be much less severe than that of wastes from fission reactors.

TVA's Browns Ferry Nuclear Power Plant is the world's largest nuclear power plant, with a total generating capacity of nearly 3½ million kilowatts. (Courtesy of U. S. Department of Energy)

9.9
THE FUTURE OF NUCLEAR POWER

In 1970, government officials in the United States estimated that 1200 large nuclear reactors would be built in this country before the year 2000. In 1978, officials estimated that fewer than 300 reactors would be built by the year 2000. What happened during those eight years? Has the promise of nuclear abundance been an illusion? We only need to review some of the problems of the nuclear issue to understand why power plant production has slowed down.

(1) Nuclear fuels are expected to become more expensive in the near future.

(2) The capital costs of nuclear power plant construction are much higher than the costs of fossil fuel plants.

(3) Many citizen groups oppose the construction of nuclear power plants in their areas. These objections have led to lengthy court battles and other delays.

(4) The problem of nuclear waste disposal has yet to be solved to everyone's satisfaction.

9.10
CASE HISTORY: THREE MILE ISLAND AND THE CHINA SYNDROME

The Three Mile Island nuclear plant in Middletown, Pennsylvania, is a subsidiary of the General Utilities Corporation. Starting at 4 A.M. on March 28, 1979, a series of mishaps occurred within the plant. The sequence of events is enormously complicated, with some 40 different events having been identified during subsequent investigations. We will therefore try to highlight the key factors involved in the accident and address two major questions:

(1) Was the accident caused by human error or did it come about because the plant was incorrectly designed?

(2) Was there ever a serious risk that the accident might have gone out of control, causing a meltdown ("China Syndrome") that would endanger people living over a wide area (the State of Pennsylvania)?

The Three Mile Island nuclear facility. (Photo courtesy of *Philadelphia Inquirer*/Chuck Isaacs)

The operation that seems to have started the trouble was a routine one—the changing of a batch of water purifier in a piping system. For a nuclear plant, this job is considered to be as routine as, say, changing the oil filter in an automobile. But a problem developed: Some air got into the pipe, causing an interruption in the flow of water. Of course, the back-up systems in a nuclear plant are designed to respond automatically to such an event, but instead several other things went wrong:

• Two spare feedwater pumps were supposed to be ready to operate at all times. But the valves that control the water from these pumps were out of service for routine maintenance, so the spare pumps could not deliver water. The controls for these valves were provided with tags to indicate they were being repaired. The tags hung down over red indicator lights that go on when the spare pumps are not feeding water. Since the lights were obscured by the tags, the operators did not see that

they were lit and did not realize that no water was flowing.

• As a result, pressure built up in the reactor core. Relief valves in the primary coolant loop then opened automatically (as they should have) to let out superheated steam. *But then the valves failed to close*, causing a dangerous drop in pressure. This malfunction is considered to be the crucial failure in the entire sequence.

• When the emergency core cooling system came on automatically, the pressure gauges in the control room gave a false reading, leading operators to think the water level was still above the fuel rods. It wasn't. Instead, bubbles of gas from below were pushing the water up, leaving part of the core exposed.

• The primary and the emergency cooling pumps, which should have been left on, were turned off twice by operators misled by the faulty pressure gauges.

The net result of this confusing series of mishaps or errors was that the nuclear core overheated. The temperatures inside the reactor vessel climbed off the recorder charts. For 13½ hours the situation was very unclear. It seemed that the reactor core was partially exposed above the cooling water and that there were voids or perhaps bubbles in the system. Decisions at this juncture were crucial. In the words of Nuclear Regulatory Commission (NRC) Chairman Joseph M. Hendrie,

We are operating almost totally in the blind; [my] information . . . is nonexistent and—I don't know—it's like a couple of blind men staggering around making decisions.

In fact, subsequent interpretations indicate that the entire core *was* exposed for some time. This means that only steam, not liquid water, was circulating through the core to remove excess heat.

The "void" or "bubble" that caused the problem was something entirely unexpected—it was a 1000 cubic foot (28,000 liter) volume of hydrogen gas. The plant management as well as the federal regulators was utterly unprepared for this possibility. Where did the hydrogen come from? There are two possibilities, both of which probably played a part. One is **radiolysis**, which is a chemical change produced by radiation. In other words, the radioactivity in the core chemically decomposed the water and produced hydrogen. The other possibility is that water can react chemically with certain metals, which also produces hydrogen. The metal tubes holding the uranium fuel are made of zirconium, which reacts with water when the temperature gets high enough. The net result, in the words of Roger J. Mattson, director of the division of system safety, was that "We saw failure modes the likes of which have never been analyzed."

The superheated steam released to the atmosphere early in the emergency was radioactive, and for this reason Pennsylvania's Governor Thornburgh ordered the evacuation of children and pregnant women from the area near the plant. Others left of their own accord. As the problem subsided, the evacuees returned, with misgivings one can only imagine.

Let us now return to our two questions:

(1) *Was it operator error or faulty design of the plant?* It is assumed that critical components (such as the relief valves that failed) are exhaustively pretested by the Nuclear Regulatory Commission. Operators certainly should not be blamed for problems related to faulty equipment or gauges. But what about the next mishap in the sequence, when the operators turned off the cooling pumps when the core was partially exposed by bubbles below? A scientist on the investigating team commented, "Well, if you were filling a bucket with water, and it started to overflow, wouldn't you turn the water off, even if there was a big bubble below that you couldn't see?" Of course, hindsight always shows how operators could have done better, but the scientific consensus seems to be that the Three Mile Island accident was mainly a failure of the design of the plant, not of the operators' performance.

(2) *Was there ever a serious risk of a complete meltdown?* During the event, some certainly thought so. At one point NRC Chairman Hendrie said in his excitement:

. . . When I say 6 to 12 hours once things begin to go and you figure it's going to go, you know, that there's nothing else you can press or pull in the way of switches and going to have to let it run its course and the best thing to do is to just get away, this could take several hours which is consistent with this, four hours, three to four hours, at least, to work its way through the vessel.

This confused language says that a meltdown might have happened, and, if it did, there would not have been enough time to get everyone out. But later analysis indicated that, when the feedwater flow stopped, the fission reaction was automatically shut down (by the control rods) within 12 seconds. Of course, the radioactivity continued— that cannot be shut down. Furthermore, as mentioned above, the core actually *was* completely exposed for some time. This evidence is both frightening and, to some with another point of view, comforting. It is frightening because all the reassurance of the nuclear industry that "it can't happen" came to naught.

Many people consider that their confidence and trust have been betrayed. On the other hand, the fact that an exposed core did not melt down may mean a nuclear lant is more resistant to catastrophe than its severest critics believe. No answer clear enough to convince everyone has emerged, and the nuclear debate continues.

CHAPTER SUMMARY

9.1 THE NUCLEAR CONTROVERSY

There is considerable debate about the desirability of nuclear power systems.

9.2 RADIOACTIVITY

When a nucleus splits apart spontaneously, energy is released. This process is called **radioactivity.** The time required for half of the atoms in a sample to decompose spontaneously is called the **half-life.** Atoms with a short half-life emit dangerously high levels of radiation.

9.3 BIOLOGICAL EFFECTS OF RADIATION

High doses of radiation can be lethal within a few days to a few weeks. Low level exposure causes cancer and birth defects.

9.4 NUCLEAR FISSION REACTORS

In a fission reactor, a neutron strikes the nucleus of a uranium-235 atom. The nucleus breaks apart, releasing energy, two or three neutrons, and fission products.

A reactor is designed so that some, but not all, of the released neutrons strike other uranium-235 atoms. The reaction then continues at an even, controlled rate.

9.5 SAFETY IN NUCLEAR FISSION REACTORS

A nuclear reactor cannot blow up like a bomb.

A **loss of cooling accident** would allow the reactor core to overheat, melt, and release its contents. There is wide disagreement on the probability of such an accident.

"Routine" radioactive emissions do occur to some extent. They are generally held down to levels comparable to background radiation.

9.6 DISPOSAL OF RADIOACTIVE WASTES

Fission products must ultimately be disposed of as radioactive wastes. The problem of ultimate storage has not yet been solved. Future large-scale use of nuclear reactors will not be possible unless some disposal site that everyone considers safe is found.

9.7 BREEDER REACTORS

Breeder reactors extend the life of our nuclear fuels, but they also increase the danger of nuclear power systems.

9.8 NUCLEAR FUSION

Energy can be released by the fusion of hydrogen isotopes. The lowest temperature envisioned for a practical reactor is about 40 million degrees Celsius. However, no solid material can withstand such heat. Therefore the reaction will probably be contained in a magnetic field.

A fusion reactor would not pose any danger of explosion because the total quantity of fuel present at any time would be very small. There would be some hazardous radioactive wastes, but these problems would be very much less than those of nuclear fission reactors.

9.9 THE FUTURE OF NUCLEAR POWER

Nuclear power plant construction has lagged in recent years because fuels and construction costs are high, citizen groups oppose the "nukes," and the waste disposal problem has not yet been solved to everyone's satisfaction.

KEY WORDS

Atom
Nucleus
Proton
Neutron
Isotope
Radioactivity
Half-life

Mutation
Nuclear fission
Chain reaction
Nuclear reactor
Loss of cooling accident
Breeder
Nuclear fusion

1. **Branching chain reaction.** You are not going to attempt to make a homemade atomic bomb, but this experiment will illustrate the principle of one. You will need about three dozen wooden matches, some glue, a little aluminum foil, and a pencil. Lay the matches down on the foil in the pattern shown in the sketch. Run a layer of glue over each row of matches, and set a pencil in the top layer of glue. Do not let any glue touch the match heads. Let the assembly dry overnight, then pick it up carefully and remove any foil that might be sticking to it. Now, using the pencil as a handle and holding it at the far end, suspend the entire assembly over a sink or wash basin partly filled with water, and light the bottom end of the match first. Hold the assembly vertically, so that it burns uniformly upward. Describe the results. What would happen if every neutron released 'from fission of uranium-235 were to strike another uranium-235 nucleus? (See Fig. 9–3 on page 181.) Discuss.

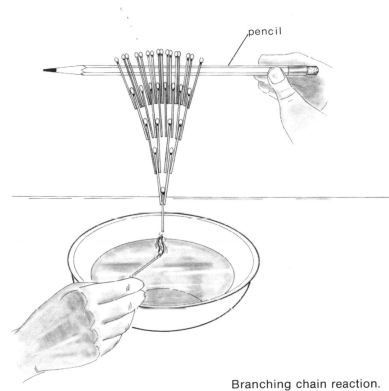

pencil

Branching chain reaction.

PROBLEMS

1. **Half-life.** Define half-life. Iodine-137 has a half-life of 24 seconds. If you had one gram of iodine-137, how much would remain after 24 seconds? 48 seconds? 96 seconds?

2. **Half-life.** Explain why atoms with short half-lives emit more radiation per second than atoms with long half-lives.

3. **Radiation.** A fission waste product from a nuclear reactor has a half-life of 108 days. Imagine that a sample of this material was stored in a temporary holding tank for 108 days. Would the sample have decomposed completely and be safe to handle? What about 108 weeks? 108 years? Explain.

4. **Health.** Outline the types of damage to the body that can result from exposure to high-energy radiation.

*5. **Health.** Give two reasons why a 70-year-old woman does not face so serious a problem regarding the health effects of radiation as a 17-year-old.

6. **Health.** Imagine that a person is exposed to an unknown quantity of radiation. If the person does not become ill in one month, would it be safe to say that he was not affected by the radiation? Discuss.

7. **Nuclear reactor.** Could a person operate a home furnace by throwing chunks of uranium ore into a large tank of water and then using the hot water in a home radiator? Discuss.

8. **Nuclear reactor.** Write the reaction that occurs when a neutron strikes a uranium-235 nucleus.

9. **Nuclear reactor.** Why is it impossible for a nuclear reactor to blow up like an atomic bomb?

10. **Loss of cooling accident.** What is a loss of cooling accident? How could it occur?

*Indicates more difficult problem.

11. **Reactor safety.** Is it possible to be certain what the risks of a nuclear power plant are? Discuss.

12. **Nuclear safety.** Describe any specific series of events that would cause all of the safety features of a nuclear plant to fail and a radioactive cloud to be released to the atmosphere.

13. **Routine radioactive emissions.** If there are no accidents, does some radiation leak out from nuclear power plants? If there are radiation leaks, are they harmful? Discuss.

14. **Radioactive wastes.** Explain why the waste products from a nuclear reactor are dangerous to humans. Why is it difficult to dispose of them? Discuss.

*15. **Radioactive wastes.** Sand-like radioactive leftovers from uranium ore processing mills, called "mill tailings," have been used to make cement for the construction of houses in Colorado, Arizona, New Mexico, Utah, Wyoming, Texas, South Dakota, and Washington. These tailings contain

---

*Indicates more difficult problem.

radium (half-life 1622 years). When radium decomposes, one of the fission products is radon. Radon is a radioactive gas with a half-life of 3.8 days. Are the following statements true or false? Defend your answers.

(a) Since radon has such a short half-life, the hazard will disappear quickly. Old tailings, therefore, do not pose any health problems.
(b) Continuous ventilation that would blow the radon gas outdoors would decrease the health hazard inside such a house.

16. **Breeder.** What are the liabilities and benefits of breeder reactors? Discuss.

17. **Fusion.** What is a nuclear fusion reaction? How does fusion differ from fission?

18. **Fusion.** Suppose that someone claims to have found a material that can serve as a rigid container for a fusion reactor. Would such a claim merit examination, or should it be ignored as a "crackpot" idea not worth the time to investigate? Defend your answer.

19. **Fusion.** Compare and contrast some potential environmental problems of a fusion power system with those of a fission power system.

QUESTIONS FOR CLASS DISCUSSION

1. Prepare a class debate. Have one side argue in favor of nuclear power systems and the other side argue against them.

2. Let us suppose that the establishment of a nuclear reactor system in the United States were to result in the deaths of a certain number of people each year from cancers that they would not have contracted if the radiation levels were lower. What effects of nuclear power plants can you imagine that might *increase* the general level of health?

3. At some time in their lives, most people have experienced a set of weird, totally unpredictable accidents. Review some strange series of events that have happened to members of your class. What do these experiences tell you about safety in nuclear power plants? Discuss.

4. Suppose that you were forced to decide to: (a) live near a nuclear power plant, or (b) reduce your personal energy consumption by 20 percent. Which alternative would you pick? Discuss with your classmates.

BIBLIOGRAPHY

There are many books on atomic energy and nuclear engineering; many of them assume previous training in physics and chemistry. The following two texts, however, present somewhat more elementary introductions.
Alvin Glassner: *Introduction to Nuclear Science.* New York, Litton Educational Publisher, Van Nostrand-Reinhold Books, 1961.

Samuel Glasstone: *Sourcebook on Atomic Energy.* New York, Litton Educational Publisher, Van Nostrand-Reinhold Books, 1958.
(Glassner's book is based on a short course given at Argonne National Laboratory since 1957 and presupposes only one year of college physics. Glasstone is more comprehensive and offers more introductory matter.)

The pro-nuclear viewpoint is very clearly presented in:

Bernard L. Cohen: *Nuclear Science and Society*. Garden City, N.Y., Anchor Press/Doubleday, 1974. 268 pp. Paperback.

Anti-nuclear viewpoints are given in:

John J. Berger: *Nuclear Power — The Unviable Option*. Palo Alto, Calif., Ramparts Press, 1976, 384 pp. Paperback.

Union of Concerned Scientists: *The Nuclear Fuel Cycle*. Revised Ed. Cambridge, Mass., Massachusetts Institute of Technology, 1974, 291 pp. Paperback.

For more advanced books, refer to the following:

George I. Bell: *Nuclear Reactor Theory*. New York, Van Nostrand-Reinhold Books, 1970, 619 pp.

Peter John Grant: *Elementary Reactor Physics*. New York, Pergamon Press, 1966. 190 pp.

M. M. El-Wakil: *Nuclear Energy Conversion*. New York, Intext Educational Publishers, 1971. 666 pp.

Two excellent books that integrate social and technical aspects of the problems of nuclear energy are:

David Rittenhouse Inglis: *Nuclear Energy: Its Physics and Its Social Challenge*. Reading, Mass., Addison-Wesley Publishing Co., 1973, 395 pp.

Henry Foreman, ed.: *Nuclear Power and the Public*. Minneapolis: Univ. of Minnesota Press, 1970. 272 pp.

Specific discussion of nuclear hazards may be found in:

Geoffrey G. Eichholz: *Environmental Aspects of Nuclear Power*. Ann Arbor, Mich., Ann Arbor Science Publishers, 1976. 681 pp.

Problems of sabotage and terrorism are considered in:

Mason Willrich and Theodore B. Taylor: *Nuclear theft: Risks and Safeguards*. Cambridge, Mass., Ballinger, 1974, 252 pp.

The Rasmussen Study is published under the following title:

Reactor Safety Study. U.S. Nuclear Regulatory Commission, October 1975. 198 pp. Copies can be purchased from the National Technical Information Service. Springfield, Virginia. A separate *Executive Summary* (12 pp.) is also available.

Finally, the excellent periodical *Bulletin of the Atomic Scientists* carries many articles on the problem.

Newborn lamb, a few minutes old, on the Très Piedres Ranch in Arboles, Colorado.

10

AGRICULTURAL SYSTEMS

10.1
INTRODUCTION

Primitive people collected food by hunting and gathering. Hunters dragged meat back to their dens. Wandering nomads competed with other animals for plant food. Life was very hard during drought or flood. When agriculture first started to become effective some 6000 years ago, many people were freed from the tasks of searching and hunting. Artists, scientists, engineers, and builders could work their trades and then buy their food.

The history of agriculture includes both successes and failures. Today, many millions of people live in cities and purchase food raised in the country. In many areas there is unprecedented ease and freedom from want. The total harvest of food has increased steadily for thousands of years. But at

Primitive agriculture. (Courtesy of American Museum of Natural History)

Following is a story about a buffalo hunt in Montana around 1890. "His heart's what I'm aiming for, but bein' weak and trembly from hunger I notice the sight wavin' when I pull the trigger . . . When I look up there's a red blotch on his side but it's too high and far back . . . He comes for me snortin' and gruntin' . . . and the next thing I know I'm amongst his horns. Lucky for me I get between 'em, an grabbin' a horn in each hand I'm hanging on for all there's in me, while the bull's doing his best to bear break my hold . . . I hear the bark of Bad Meat's gun. The bull goes over and the fight's mine.*"

―――――――

*Charles M. Russell, *Trails Plowed Under*, Doubleday and Co., New York.

Plants compete for space, light, water, and nutrients in a natural grassland.

Sun
1000 cal

10 cal

1 cal

FIGURE 10–1. Comparison between meat production on the range and in a feedlot. The feedlot system produces more meat for humans but consumes fossil fuel energy.

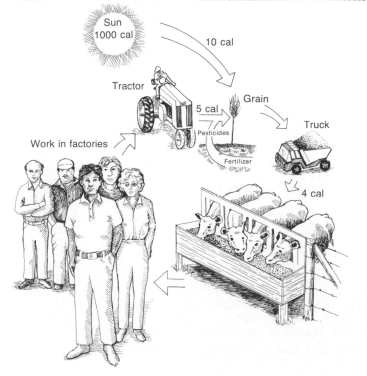

Sun
1000 cal

10 cal

Tractor

Work in factories

5 cal

Pesticides

Grain

Fertilizer

Truck

4 cal

198

Cattle in a feedlot.

the same time, more than half of humanity is hungry and malnourished.

10.2
THE FLOW OF ENERGY IN INDUSTRIAL AGRICULTURE

A wild oat plant growing in an unfarmed prairie must compete successfully with its neighbors for sunlight and moisture. The plant will survive if it is tall, sprouts early, or grows deep roots. The energy a wild plant needs to grow a tall stalk or a deep root comes only from the Sun. On the other hand, an oat plant growing on a farm is carefully tended. The farmer fertilizes and waters it when necessary, removes weeds, and loosens the soil around its roots. The farmer then expects to benefit from his efforts. Domestic crops generally produce more food than wild plants do. For instance, a variety of Oriental rice developed in the late 1960's produces more grain than wild Oriental rice. But the new breed is also less resistant to disease and more dependent on fertilizer and irrigation than older strains of rice. The new seeds require care, and care requires energy. Energy is needed to plant, to weed, to control pests and disease, and to irrigate.

In modern farming, even beef production is dependent on high energy input. Remember that in a natural system an animal must eat 10 Calories of plant matter to produce 1 Calorie of meat. Some of these "lost" 9 Calories are used by the animal to search for food. In a feedlot, a farmer grows hay or grain, harvests it, and brings the food to the cattle. Various chemicals are added to the feed to encourage rapid weight gain. The animals do not wander around in search of grass. They simply stand in front of a feed bin and eat. As a result, feedlot beef grow faster on less food than range cattle do. But the farmer must invest energy to encourage this rapid growth. Thus, a feedlot is more efficient in converting sunlight to meat than a natural system is, but the feedlot requires large quantities of fossil fuels.

Modern agriculture requires so much energy that the price and even the availability of food is closely linked to the price of gasoline. When the cost of petroleum rose abruptly in 1973, many farmers in the less developed nations could afford neither fertilizers nor

Cattle on the range, from the painting "Jerked Down" by Charles M. Russell. (Courtesy of the Thomas Gilcrease Institute of American History and Art, Tulsa, Oklahoma)

fuels. As a result, many people starved. In the industrialized nations, the fuel shortage led to higher food prices. Because increasingly severe fuel crises are projected for the future, it is important to understand the role of energy in modern agriculture. Modern farming depends on four fundamental techniques: (a) chemical fertilization, (b) mechanization, (c) irrigation, and (d) the chemical control of disease, insects, and weeds. All four techniques require large inputs of energy. The first three topics will be discussed in the following sections. Chemical pest control is the subject of Chapter 11.

10.3
CHEMICAL FERTILIZATION

Farmers fertilized their crops with manure, straw, or dead fish long before they understood the chemistry of fertilization. Today, mined and manufactured fertilizers are used extensively. The total energy requirement of the fertilizer industry is a major portion of agriculture's large demand for fossil

fuel. Recall from Chapter 2 that *nitrogen* is needed by all plants. The atmosphere is nearly 80 per cent nitrogen. But atmospheric nitrogen is not usable by most plants. Rotted manure or other types of animal or plant wastes contain nitrogen in a form that can be used by growing crops. Alternatively, bacteria on the roots of certain plants called legumes (such as peas, beans, and alfalfa) are known to "fix" atmospheric nitrogen. A third way to add nitrogen to the soil is by using manufactured fertilizers. Atmospheric nitrogen can be converted to fertilizer according to the following equation:

Atmospheric nitrogen + hydrogen + energy → ammonia (fertilizer)*

At the present time, great quantities of ammonia are produced in this manner. Nitrogen fertilizers will be plentiful as long as energy is available. If energy sources become limited, the future pro-

---

*The energy input needed to make ammonia fertilizers refers to the total energy required for the process, and not to the thermodynamics of the reaction.

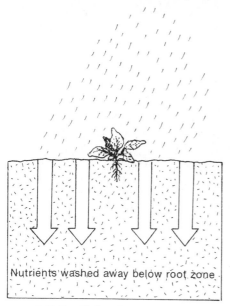

A Soil with low humus content

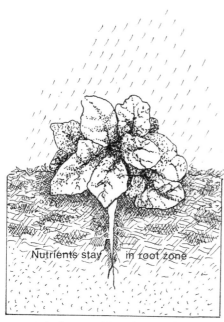

B Soil with high humus content

FIGURE 10–2. Humus helps retain nutrients in the soil.

ductivity of our present agricultural systems will be seriously threatened.

Farmers also need to add *minerals* to the soil. If the minerals cannot be recycled by the reuse of dead plant or animal matter, they must be dug out of a mine. Thus phosphorus, potassium, calcium, magnesium, and sulfur are all found in the Earth. Of these five minerals, all but phosphorus are plentiful in the Earth's crust. Known geologic deposits of phosphate ores are limited. Without an adequate supply of inexpensive phosphorus, highly productive agriculture would be impossible.

What Is Fertile Soil? If a person were to spread fertilizer on a sandy beach and then plant vegetables, the garden would not be expected to grow well. Much of the fertilizer would dissolve in rainwater and be carried away from the plant's roots. The plants would not receive enough nutrients to survive.

Fertile soil is a complex mixture of many materials. Small pieces of ground rocks and minerals are mixed with many types of plant and animal matter. Leaves, stems, and other dead plant tissue lie partially or almost completely rotted in the earth. The organic components of healthy soil are called **humus**. When rain falls on fertile soil, much of the moisture is held in place by humus. Nutrients are retained. Thus fertilizers added to rich soil will not be washed away.

Humus can be maintained if manure, straw, or other plant wastes are regularly added to the soil. If a field is fertilized only with chemicals, the humus will gradually decompose. Then water will run off more quickly. The flowing water will carry valuable fertilizers away. This loss of nutrients represents a waste of energy and ore. Soil fertility will decrease.

Organic Farming. Manufactured fertilizers have raised agricultural yields across the world. But in spite of this fact, many agricultural scientists question the use of inorganic fertilizers and advocate "organic" gardening and farming. What is meant by the term "organic farming"?

Originally, "organic" meant derived from a living organism, and "inorganic" meant not derived from a living organism. In the early 1800's, scientists discovered that certain organic compounds could be synthesized from nonliving materials. Then the word "organic" came to mean chemical compounds with a certain characteristic chemical bonding.

Mechanization in agriculature. *Left*, Combine. *Right*, tomato picker. (Photos courtesy of Rorer-Amchem, Inc., Ambler, Pennsylvania)

Recently, the word "organic" has recaptured some of its old meaning and added some new ones. Organic, as in the terms *organic gardening* or *organic foods*, means materials that are raised naturally and are not highly processed or manufactured. Thus manure contains valuable nitrogen fertilizers. These are called organic fertilizers. Nitrogen fertilizers that are manufactured in a factory are called inorganic. Organic farmers use only natural fertilizers and pesticides and do not rely on manufactured or highly processed chemicals.

In the United States and Canada billions of tons of manure, slaughterhouse wastes, and other organic "garbage" are produced each year. Most of this material is bulldozed into dumps. Only a small portion is now used as fertilizer. If more of this material were used as fertilizer, soils could be enriched considerably.

Before the fuel crisis of 1973, many farmers found that it was cheaper to buy chemical fertilizers than to handle bulky, heavy organic fertilizers. But when the price of fuel rose, the price of fertilizers rose. Now, organic farming is economically attractive. The change in position teaches us an important lesson. Economics often reflects the social attitudes of the times. If we calculate the true price of wasteful practices, we find that in most instances it pays to preserve the environment.

Recycling not only aids the farmer, but benefits the community as well. If manure and sewage are discarded, some of the wastes wash into streams and groundwater and pollute them. If they are used as fertilizer, however, valuable fuel and mineral reserves are

Harvesting wheat by hand in northern India.

conserved, and the soil is enriched. Although the cost of economic externalities is not reflected in the farmer's account book, it is paid for by our society.

10.4
MECHANIZATION IN AGRICULTURE

High farm productivity is impossible without good seeds, fertilizers, and pest control programs. On the other hand, tractors do not improve the yields. For example, most Japanese farms vary in size from about 0.4 to 2 hectares (about 1 to 5 acres). Many farmers own small, two-wheeled rotary tillers, but very few own four-wheeled tractors. Yet the yields per hectare in Japanese agriculture are the highest in the world. However, these yields require great quantities of fertilizers and much labor. Therefore, the total auxiliary energy demand of Japanese farming is high, even though very little heavy equipment is used.

A person tending a small vegetable or grain patch can work very precisely. Rows can be planted close together, and every corner of the field can be planted. The tractor is much clumsier. Rows must be spaced to leave room for wide tires, and the corners of fields are often left untouched. Automated harvesters are not nearly as accurate as humans are. As a result, large machinery actually leads to a *decrease* in yields. For example, semi-automated broccoli harvesters in California leave 10 to 20 percent of the food unpicked. Mechanical tomato pickers travel through a field once and pick all the tomatoes that are ripe at that time. Tons of vegetables are left to rot after the picker has passed.

What would happen if American farmers were to use more manual labor and fewer machines? In other words, suppose that the factory workers who built mechanized tomato pickers were to move to the country and pick tomatoes. This type of system would be similar to Japanese farming. It is unreasonable to believe that masses of urban workers would migrate to the country to pick tomatoes and broccoli. Such a movement would involve tremendous social and economic adjustments.

Navajo farmer George Blue-eyes works far harder than his neighbors because he tills his fields entirely by hand. But traditional farming methods limit moisture loss and erosion and bring him consistently higher yields per hectare. (Photo by Janet Bingham)

10.5
IRRIGATION

Irrigation is almost as old as agriculture itself. The ancient Egyptians, Babylonians, Chinese, and Incas all brought water from nearby rivers to increase the yields of their crops. Today a large portion of the world's crops of

A small boy clears an irrigation ditch in Peru.

Irrigation system in the United States.

vegetables and some grains depend on irrigation. With imported water, marginal farmland has become more productive, and even former deserts are being farmed. Despite these successes, irrigation leads to some environmental problems.

When rainwater falls on mountainsides, it collects in small streams above and below ground. As it flows downward it filters over, under, and through rock formations. The water dissolves various mineral salts present in the rock and soil. Therefore, river water is slightly salty. In most cases, you can't taste the salt, but it is there. If this water is used for irrigation, farmers are bringing slightly salty water to their fields. When water evaporates, the salt remains. Thus, over the years, the salt content of the soil increases slowly. Because most plants cannot grow in salty soil, the fertility of the land decreases. In Pakistan, an increase in salinity (saltiness) decreased soil fertility alarmingly after one hundred years of irrigation. In parts of what is now the Syrian desert, archeologists have uncovered ruins of rich farming cultures. However, the land lying near the ancient irrigation canals is now too salty to support plant growth. California farms now produce about 40 percent of the vegetables consumed in the United States. Here, too, salinity is threatening productivity.

10.6
AGRICULTURE AND RURAL LAND USE

People need food to survive, and therefore large expanses of rural land must be devoted to farming. But people also need fuel, transportation systems, places to live, and centers for manufacturing and distribution. Mines, roads, pipelines, factories, homes, stores, and warehouses all require space. What happens when a prime site for a new harbor is also the mouth of a river in which fish come to breed? Or suppose a coal mine lies under a wheat field? Perhaps a vegetable farm would be more profitable if the land were used as the site of a shopping center. When such conflicts arise, who decides how the land is to be used?

In most nations today, economic factors control these decisions. In the last example cited above, if a developer offers the vegetable farmer enough money, the farmer may sell. The developer will then bulldoze the topsoil away and cover the land with concrete and asphalt. Bit by bit, many prime agricultural areas have been converted to nonagricultural uses. What will happen in the future? Will continued conversion of agricultural land to other functions eventually lead to worldwide famine? If so, is there any way to reverse current trends and stabilize the system?

Legal and economic systems vary from nation to nation. In most regions of the United States, land is taxed according to its market value. If the land is valuable, the taxes are high. Imagine, then, what happens in some agricultural communities. One farmer sells fields to a manufacturing corporation, and a factory is built. The presence of the factory boosts land prices in the surrounding areas. Rising land prices lead to rising taxes. The high taxes then become a great burden to neighboring farmers. As profits from farming decline and the market value of farmland skyrockets, many more farmers sell their land. The spiral continues. Eventually an agricultural region is converted into an industrial one. If agricultural land were taxed differently from land used for other purposes, such a situation might be prevented. In fact, in some areas taxes on agricultural land are lower than taxes on commercial or residential land.

The South Platte River basin is potentially one of the richest agricultural regions in Colorado, yet much of the farmland has been preempted by commercial interests.

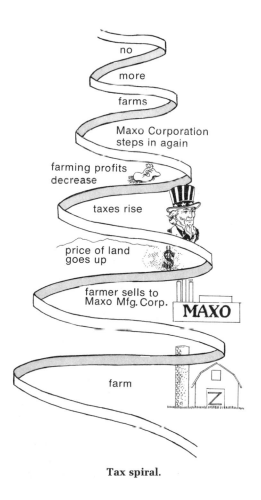

no more farms

Maxo Corporation steps in again

farming profits decrease

taxes rise

price of land goes up

farmer sells to Maxo Mfg. Corp.

MAXO

farm

Tax spiral.

In other situations (discussed in Chapter 7) valuable fuel or mineral deposits lie under fertile farmlands. If the ore is mined and the farmlands destroyed, the world community may be trading a short-term gain for a long-term loss. In the near future we must ask ourselves whether we are facing the issues satisfactorily. Perhaps significant changes are necessary.

10.7
WORLD FOOD SUPPLY

Worldwide food production has been in a nip-and-tuck race with population growth in recent years. There were many famines in the late 1950's and early 1960's. During the latter part of the 1960's, significant agricultural development led to increased crop yields and a reduction of famine (see Section

A hungry girl searches through a pile of garbage for something to eat in a slum in New Delhi.

10.8). Gains were slight, however. When a serious drought struck India, Pakistan, and North Africa between 1972 and 1974, millions again starved. In 1975 and 1976, when the rains returned to India and Africa, the situation improved. Bumper crops of grain in North America during the next two years further reduced the world food problem. Despite these gains, serious famine still threatens the world. A flood or drought in any major food-producing region could again cause widespread starvation.

How can one even think about, let alone describe, the human pain and misery that results from hunger and famine? So we ask ourselves why, in modern times, must people die for want of enough to eat? There are no simple answers to this question. Certainly, population growth is part of the problem. If the population is too high, there may not be enough to eat. But the world food problem is much more complicated than a simple balance between population size and food. Other factors are important as well, as described below:

(a) Political and Economic Barriers to Food Production. The United States is one of the richest nations of the world. It is one of the few countries that regularly export large quantities of food. Yet there are people in this country who go hungry. In the United States and in most other nations as well, many people go hungry because resources are not distributed evenly. There is certainly enough fuel, fertilizer, and land to feed everyone in the world today. Some people just don't get what they need. Some examples follow:

- In 1974 floods destroyed many rice fields in Bangladesh. Millions of poor farming families starved. Yet in other parts of the same country harvests were good. Approximately four million tons of rice were stockpiled in the central cities. The rice did not reach the hungry because they were "too poor to buy it."*

- In 1978 a large American food corporation owned 23,000 hectares (57,000 acres) of prime farmland in Guatemala. Of this land, only 3650 hectares (9000 acres) were planted. The rest was left sitting unused. Yet thousands of peasants went hungry because they could not grow the food that they needed. Even people who worked on the plantation barely had enough to eat. Pay was low and most of the food grown was exported.

- Great famines took a large toll of human lives in Africa in the early 1970's. Yet, throughout the famine years large quantities of cotton were grown in the drought area and exported to Europe.

- In 1973 and 1974 fuel shortages existed throughout the world. As a result, fertilizer shortages became acute. Most developed countries responded by restricting fertilizer exports. Wealthy people could buy fertilizers for their ornamental lawns and gardens, but many poor farmers could not purchase the fertilizer they needed to feed their families.

- In many places around the world starving people live near large plan-

---

*Steve Raymer. "The Nightmare of Famine," *National Geographic Magazine*, July, 1975.

tations of coffee, tea, and tobacco. In other situations food crops such as bananas are exported to foreign countries while once again local peasants have little to eat.

Given enough fertilizer, fuel, and machinery, the crop yields in the Ganges River delta in India could be increased sixfold. The impoverished Sudan in southeast Africa could be a bread and cereal basket for the world. The rice paddies of the Philippines could feed the hungry throughout the South Pacific. But great quantities of fuel and machinery are being used for warfare, and the remaining resources are not divided evenly. It is not easy to change present economic and political patterns. But such changes would reduce suffering considerably.

(b) Overconsumption. Part of the problem of uneven distribution of resources arises from overconsumption. The average person in North America uses 900 kilograms of grain per year. Only 90 kilograms are consumed directly as rice, bread, or cereals. The other 810 kilograms are used as animal feed to produce meat, milk, and eggs. In India, by contrast, the average per-

According to the United Nations Conference on Disarmament, the nations of the world spent one billion dollars *per day* on arms and warfare in 1977. If this money had been spent on fertilizers and irrigation systems, many people could have lived happier and healthier lives.

son uses only 180 kilograms of grain per year. Many people in wealthy societies are overweight and would be healthier if they ate less. Even more food could be conserved if people whose diet consisted largely of meat would eat more grain instead. Most North Americans could cut their meat consumption in half and not suffer any loss of health.

(c) Climate. Farmers have always been at the mercy of the weather. If the temperature and rainfall are favorable, food will be plentiful. But frost, flood, or drought often leads to famine. In recent years there has been grave concern that world climates are changing. In many regions, cool temperatures, monsoon failures, and frequent drought conditions have been occurring.

If world climate continues to under-

The effects of drought in the Sahel region of north Africa. (Courtesy of Alain Nogues-Sygma)

207

go change, the human race may indeed be in a grave situation. Global climate was unusually warm from 1900 to 1950. During that time the population rose considerably. Dryland farming has been extended right to the edge of most of the world's deserts. Mountainside pastures creep up the slopes of the major mountain ranges. Grains are grown in northern Canada and Russia nearly into the Arctic. Farmers are already pushing into the great jungle basins. There is only limited room for further expansion. The problem that causes great concern is this: Food production and population have been pushed to the limit during an unusual period of particularly mild climate. If the climate should become less favorable, farms in marginal areas will fail. Many more people will starve. No one is willing to predict that unfavorable climate change *will* occur, but most scientists say that such climate change is likely.

10.8
INCREASES IN CROP YIELDS AND THE GREEN REVOLUTION

Next time you are walking in the country, look closely at the wild grasses growing in meadows and fields. Notice that each plant has a slender stalk, a few leaves, and a small cluster of seeds at the top. These seeds are rich in starch, protein, and vita-

The wheat stalk on the right supports much larger quantities of grain than the native grass shown on the left.

Workers planting rice at the International Rice Research Institute, Los Baños, Laguna, Philippines.

mins. But they are so small that they are not harvested for human consumption.

Modern varieties of grain produce much more food for humans, but the farmers must pay for this increase in production. Fields must be fertilized and watered, and pests must be kept away. In recent years, scientists have developed seeds that produce spectacularly high yields. These new "wonder seeds" can produce so much food that people have called their introduction the "**Green Revolution**." For example, in Mexico in the 1940's wheat fields averaged 750 kg/hectare. Farmers planting the new strains of seeds in the 1970's averaged 3200 kg/hectare. In India and Pakistan, mass shipments of Green Revolution wheat in the late 1960's raised the wheat harvest between 50 and 60 percent during a period of two growing seasons.

But these seeds require even more care and energy than grain varieties

that had been planted twenty years ago. A "wonder seed" by itself does not produce large quantities of food. The new grain varieties planted in unfertilized, unirrigated soil produce less food than traditional grains. In addition, the new grain varieties are less resistant to insect pests and various diseases than traditional plants. As a result, farmers who invest in seed must also invest in fertilizers, irrigation systems, and pesticides.

We see that if a farmer has enough money to buy the new seeds and care for them properly, high yields can be realized. But if the fuel or money is not available for the needed fertilizer, pesticide, and water, the Green Revolution may provide no help. In fact, in some cases the poorest farmers may actually be harmed by the Green Revolution. If more grain is grown by wealthy farmers the price of food will drop. The poor who cannot afford to plant the new seeds find that they grow the same

This gladioli plantation is being grown on land reclaimed from the Negev desert in Israel. (Courtesy of Israel Government Tourist Office)

amount of food as they did in past years, but they receive less money for their harvest when it is time to sell.

The consequences of the Green Revolution have been mixed. Food production in the developing nations has increased. Starvation has decreased, and millions are living with a new-found freedom from hunger and want. On the other hand, many rural people have sunk into deeper poverty. Homeless and untrained rural poor are migrating in large numbers to the city slums.

Shrimp fishing near Galveston, Texas. (Courtesy of National Oceanic and Atmospheric Administration)

10.9
OTHER APPROACHES TO INCREASING FOOD PRODUCTION

Increases in Crop Acreage

Most of the prime fertile land in the world is currently being farmed. Therefore, future expansion of crop acreage will have to be in areas such as dry prairies, deserts, swamps, and jungles. The farming of these areas will be possible only if massive quantities of energy and money become available. Thus, the southern Sahara could be farmed if irrigation projects were built. Huge regions of southeast Africa could be productive if their swamps were drained. Agriculture in Bangladesh would be improved if rivers were dammed to increase irrigation during droughts and reduce flooding during periods of heavy rainfall. Unfortunately, the nations that need food most desperately can least afford such technical improvements.

Food from the Sea

People have often regarded the sea as a boundless source of food, especially protein. In truth, most of the central oceans are biological deserts. Many of the more productive coastal areas have been so heavily fished that fish populations have declined. As a result, world fish production has been dropping steadily in recent years. If fishermen were careful, productivity could be increased. But most probably the sea can provide no more than 15 percent of the world's needed protein.

Pollution and urbanization are also threatening the world's fisheries. As increasing quantities of sewage, oil, pesticides, and other chemicals enter the ocean, plankton and fish are being poisoned. In particular, disasters such as the Nantucket oil spill in December, 1976, are having a devastating effect on world fish populations. (This subject will be discussed further in Chapter 13.)

In many nations, the biological productivity of coastal bays, marshes, and river mouths has been destroyed by industrialization. The destruction of an estuary is ecologically disastrous. As was mentioned in Chapter 2, estuaries are nursery grounds for many fish that spend their adult lives in deeper waters.

10.10
THE FOOD WE EAT — CHEMICAL ADDITIVES

Almost every food available in modern supermarkets contains small amounts of chemicals that are not natural to the food itself. Fruits and vegetables contain pesticide residues. (Pesticide pollution will be discussed in the next chapter.) Cattle are fed artificial growth compounds before they are slaughtered. Some of these chemicals remain in the meat when it is sold. Almost all pre-prepared "convenience foods" contain a variety of additives.

There are several different catagories of food additives.

(a) Preservatives are used to prevent spoilage.

(b) Vitamins and minerals enhance the nutritional value of foods.

(c) Artificial colors, flavors, and sweeteners alter the taste or appearance of various products.

(d) Other additives, such as emulsifiers, are used to blend foods. For example, the oil in natural peanut butter separates and floats to the top of the jar. Emulsifiers are added to peanut butter so that the oil will not separate out.

(e) Many chemicals are added to animal feed to encourage the production of meat, milk, or eggs.

As we see, there are a wide variety of food additives. All of them have arisen in response to modern ways of life. Homemakers no longer bake bread every morning; instead many shop once a week. Since bread without preservatives will go moldy in a week, chemical preservatives are added to prevent spoilage. Similarly, many people do not have time to cook an elaborate breakfast before work. So they pour some cereal out of a box, add milk, and eat an instant meal. The vitamins and minerals in the packaged foods add to these people's dietary intake. Hormones added to animal feed speed growth and reduce the cost of many foods. Other additives do not improve the quality of the food we eat. For example, natural peach ice cream is a dull pink. However, with the

Most packaged foods contain a variety of food additives.

simple addition of some coloring, a bright, almost "day-glo" color can be achieved. The color does not improve the taste, flavor, or nutritional value of the ice cream. But advertising companies have conditioned us to expect highly colored foods and to buy their clients' products.

The benefits of food additives are balanced by certain risks. Many of these materials are synthetic chemicals. Recall that a synthetic chemical is a chemical that is manufactured by industrial processes and not produced by any organisms. Of course, once a molecule is produced, it doesn't matter where it came from. For example, the chemical vanillin, which imparts a vanilla flavor, comes from the vanilla bean but can also be made by a chemical factory. It is the same substance from either source. But many synthetic chemicals *are not* in this category. They are made *only* by industrial processes and are therefore foreign to the bodies of living organisms, including people. Here lies a danger. Since such chemicals are strange to the body, it is reasonable to suspect that the body cannot deal with them satisfactorily. They can cause trouble. One possible effect is that they may cause cancer. The actual outbreak of cancer does not occur at once, but may be delayed for years. A delay of 15 or 20 years is by no means unusual.

Here, then, is a real problem. How can we judge whether a particular synthetic chemical is a carcinogen? It isn't easy. We do not conduct experiments on human beings for two reasons. First, it is immoral to do so. Second, even if we would consider such a wrongful act, it would not be very helpful to have to wait 15 or 20 years for results. We could experiment by feeding concentrated food additives to humans, but this is something we should not even think about.

So the next step is to experiment on animals. The usual choices are small animals (mice, rats, guinea pigs, cats, dogs, rats, and small monkeys). Of course, a rat is not a person. However, a rat will dine quite happily on the food that people eat and can be poisoned by chemicals that are harmful to people. (People shouldn't eat rat poison either.) We can't wait 20 years to see whether a rat that eats a synthetic chemical will get cancer. For one, the rat doesn't live that long. So we do to the rat what we would never do to a human being — we feed the rat a *concentrated* diet of the suspected chemical. If the rat gets cancer, the chemical is a carcinogen. Can we then set up some sort of an equation, such as, "If the rat gets cancer from the concentrated chemical in one month, then a person would get cancer from a dilute mixture of the chemical in 20 years?" No, we can do no such thing. All we can say is that if the rat gets cancer, then the chemical is a carcinogen for the rat. Since rats and people are both mammals, we can also say that the chemical is *probably* a carcinogen for people. Period. That is about as far as we can go. We can't say much about the chances that one person will get cancer, nor how long it will take.

Let us say that experiments with animals show that a certain synthetic chemical that is used as a food additive is a carcinogen in its concentrated form. Should the chemical be banned?

Even here there are many people who would hesitate to do so because of their concern for animal rights. Nevertheless, most people assume that we, as humans, take priority over animals in such matters. As a result, much experimentation uses animals as subjects.

Now the argument starts. If the food additive is a synthetic color, you might take the position that its use is not very important. So what if your soft drink isn't a bright red? It will quench your thirst anyway. Ban the additive, you might say. What about a synthetic flavor that is cheaper than the natural one? An example would be a synthetically flavored pancake syrup instead of the real maple syrup that comes from maple trees. Real maple syrup is quite expensive. Many people cannot afford it and hate to eat their pancakes dry. Well, now the question becomes a little more difficult. Should people be allowed to take a risk of cancer in later years in exchange for syrupy pancakes now? Or should the synthetic flavor be banned? If the argument involves a synthetic sweetener such as saccharine or cyclamate, one might argue that banning them turns more people to sugar. Then they become overweight and suffer from diseases common to overweight people. Synthetic preservatives are even more difficult to judge. If you can do your own organic farming and freeze food for the winter, you are fortunate. But if you live in a city, it is considerably more expensive to buy foods that contain no preservatives than foods that contain chemical additives.

The problem of what to do about such questions is a matter of considerable controversy. If we knew how synthetic chemicals affected humans, it would be easy to establish a policy of use of these materials. In the absence of definite knowledge, it is difficult to establish laws regulating the use of these chemicals.

10.11
FORESTS AND OPEN SPACES

A noted environmentalist, Garrett Hardin, once asked the question, How much is a redwood tree worth? Suppose a person bought a seedling for $1 and planted it in her yard. It would take approximately 2000 years for the tree to grow to maturity. If the tree was then cut and sold for lumber, the investor would realize a profit of about ½ percent interest per year. That isn't very much, considering that most savings banks offer at least 5 percent interest per year. Should we use this calculation to argue that redwood trees have no value? Of course not! Since the beginning of time people have recognized that places or things can have a value that cannot be expressed monetarily. People would lose a great part of their heritage if the beautiful places on this Earth were destroyed for money. Most governments in the world have recognized this human need and have set aside national parks and forests as recreational areas. As cities and suburbs become increasingly crowded, more and more people visit these parks every year. But at the same time, many commercial loggers and miners view forest and desert areas for their wealth of timber or ore. Many questions arise. For example, there are at least two ways to cut the timber from a forest. A natural forest will usually contain some tiny seedlings, small and medium sized saplings, and large, mature trees. If loggers selectively cut only the large, fully grown timber, they can harvest much valuable wood without upsetting the forest ecosystem. It is expensive to search through a forest and pick out only the best trees. Loggers find it much more profitable to cut all the trees in a region. This process is called **clearcutting**. Small logs are used to manufacture paper, and large ones are sawn into boards. The forest is reduced to rubble. When all the trees are removed and the brush and grasses are dug up by the heavy logging machinery, the forest soils are exposed to erosion. Rains fall on open hillsides and wash the topsoil away. A new forest will eventually grow back, but the forest soils become partially destroyed. Soil scientists believe that in some regions the soils will become impoverished after two or three generations of tree growth and clearcutting. New forests may not grow back for years to come.

Surely the clearcutting issue is a classic confrontation between immediate profit and long term environmental preservation. It is very expensive to build a home today. Part of this expense is the rising cost of lumber. Lumber costs can be reduced if clearcutting is allowed on government lands. But recreational areas and the future productivity of the soils will be destroyed.

Agriculture — A Short Overview

It is tragic to realize that millions are starving around the world, that some

Area in the foreground is a 15-year-old Douglas-fir plantation that is well established. In the background is a recent clearcut, with mature timber on both sides. (Courtesy of U.S.D.A. and the Forest Service)

healthy foods are needlessly altered by additives, and that many forests and farms are being destroyed by carelessness and by industrialization. But let us remember that people have the knowledge and resources to improve the human condition greatly. If our technology is used wisely and justly, there could be enough to eat for everyone.

10.12
COMPARATIVE CASE HISTORIES: FAMILY FARMING IN INDIA, JAPAN, CHINA, PERU, AND THE UNITED STATES

India

A woman follows a cow down the path. She collects the manure, dries it,

Broad view of a clearcut logging operation in Willamette National Forest. (Courtesy of U.S.D.A. and the Forest Service)

213

Harvesting wheat in India. (Courtesy of the Agency for International Development)

and uses it as fuel for cooking. She has no alternative sources for heat because there are virtually no trees in the area and coal or oil is too expensive. Nor can her family afford to purchase chemical fertilizers. Therefore, when manure is used as a fuel for cooking, there is less fertilizer for next year's harvest. A loan to purchase fuel or fertilizer would help, but even that is unavailable. There appears to be no immediate relief. A continuous downward spiral develops, and her family sinks into deeper poverty. With adequate money and technology, one family's fields could be made to yield food for five families. But in her case there is barely enough for anyone.

Japan

A poor farming family in Japan lives on 0.8 hectare (2 acres) of hilly land. Twenty percent of the land (0.16 hectare, or 0.4 acre) is devoted to growing the paddy rice needed to feed the family. Another small plot of 0.16 hectare is used for growing upland rice, which is made into New Year's cakes and sold at a profit. The remainder of the land is given over to mulberry leaves. These are used to feed thousands of silkworms. The family owns a single-room mud hut, which they share with the silkworms. They also have a two-wheeled power tiller, which they use to prepare the soil, and a small motor scooter. The equipment is mortgaged. The cash from the silk and rice cake production is used to buy fish meal and soybeans and to repay loans on the mortgage. In their spare time the adults weave baskets of bamboo strips to supplement their income. Though they are poor, they maintain high crop yields by intense fertilization, so there is always enough to eat.

China*

A mainland Chinese family has two children, whose grandparents remember periodic famine and mass starvation. The family lives in association with a government controlled agricul-

Farming in Japan. (Courtesy of Japan National Tourist Organization)

* Since this picture of China is taken from reports of observers who visited the country on a government-sponsored tour, and since free travel through mainland China is not allowed, we are not sure whether this picture reflects the situation throughout most of the country.

214

tural community called a **commune**. The commune administration works with the national government to assure that adequate supplies of fertilizer and fuel are available every year. Excess grain is sold through the commune administration. New strains of short-stalk rice are planted, fertilized, and irrigated. When possible, two or three crops are grown successively in a single field every year. There is great effort toward efficient farming. In many places maximum yields are possible only if planting and harvesting are done by hand. The people work hard, and there are few frills in life, but there is enough to eat.

Peru

I was hiking along a mountain trail in the Andes in Peru. As I entered the small town of Ancash, I was greated by a happy, drunken, noisy group of people. It was harvest time and the day of the Indian New Year. Above us, fields of barley and wheat yellowed on the steep mountain pastures, and skinny cows and sheep grazed on the rocky slopes that were too steep to plow.

"Come for dinner, *amigo*, it is a lucky time — the feast of the year."

Chickens and rabbits scurried from under foot as we entered the adobe courtyard where the feast was to be held. Wooden plows and hand sickles hung along the walls. No one in the town owned a tractor. There weren't enough dishes for everyone to eat at the same time, but as a guest I was served on the first round. The feast consisted mainly of thin, salty, barley soup with a small piece of pig intestine. In addition, a small plate of boiled corn was passed around and everyone took a handful. The farmers could not afford to eat the meat they raised, but sold it instead in town. Fruits and vegetables were scarce, and only corn was served this evening. We returned to the house as darkness fell. There was a wooden table in the center of an otherwise bare, windowless room. A large doll, surrounded by memorial candles, lay on the table. The doll symbolized the death of the old year. We sat on the dirt floor and the farmers sang chants ushering in the start of the new.

Farming a steep mountainside in the Andes in Peru.

United States

I am resting near my home in the Maine countryside. It is a sunny day in August, but dark clouds in the western sky warn of rain. By United States' standards, central Maine is a marginal agricultural area. Individual farms are small. An average farm is 100 to 200 hectares. Farm income is low.

A new four-wheel-drive pickup pulls into the driveway and a woman in her early sixties steps out.

"Hi, neighbor. do you think you could help us get the hay in before the storm? If it rains on the cut hay, it'll rot. We can't pay much — two dollars an hour — but we need the help."

"Well, I won't work for two dollars an hour, but if you want to lend me a tractor to pull stumps after hay season's over, I'll work."

The deal is made. When we arrive at the farm, two generations of family are already on the job. All pitch in and work hard. The hay trailer was homemade in 1932, and one wheel threatens to collapse, so it is driven slowly. The tractors, too, are old and perform erratically. I ask why they don't invest in some automated hayloading equipment.

"Not enough money. Last year we borrowed heavily for new milking equipment and repairs on the barn (built in 1915). As long as the price of milk stays low, we can't afford any more machinery for a few years at least."

Baling hay in the United States.

As the storm approaches, we all work faster. Each time the wagon is loaded, it is driven slowly to the barn, unloaded, and returned to the fields. The conveyor belt that carries the bales to the interior of the barn jams repeatedly. The women are constantly running back and forth with pitchforks to release hay caught in the belt, while the men load and stack the hay. After four loads, the job is done, minutes before the rain falls.

I am invited to stay for dinner in their new prefabricated mobile home. We all watch color TV and then sit down in the modern all-electric kitchen to a plentiful supply of steak, instant "heat'n eat" french fried potatoes, and a large salad from the garden.

CHAPTER SUMMARY

10.1 INTRODUCTION

Agriculture began about 6000 years ago. There have been many successes and many failures. Today many people are well-fed but many others are hungry.

10.2 THE FLOW OF ENERGY IN INDUSTRIAL AGRICULTURE

Modern agriculture requires large quantities of energy to produce high yields.

10.3 CHEMICAL FERTILIZATION

Nitrogen fertilizers can be manufactured from the nitrogen in air if energy is available. Minerals such as phosphorus and potassium are mined from the earth.

Organic matter in the soil decomposes to form **humus**. Humus holds moisture and nutrients and aids the growth of plants.

10.4 MECHANIZATION IN AGRICULTURE

The use of large machines does not improve the yield per hectare, but it greatly increases the amount of food that can be raised by a single person. Gasoline for tractors accounts for increased energy consumption.

10.5 IRRIGATION

With imported water, marginal farmland has become more productive and even former deserts are being farmed. Overly extensive irrigation can threaten agricultural productivity by increasing the salinity of the soil.

10.6 AGRICULTURE AND RURAL LAND USE

People need food to survive, but they also need fuel, transportation systems, places to live, and manufacturing and distribution centers. Often conflicts of land use arise. According to current social, legal, and economic structures it is often more profitable to convert farmland into industrial or residential uses.

10.7 WORLD FOOD SUPPLY

Even today, many people are starving. If everyone is to have enough to eat we must solve problems related to: (a) overpopulation, (b) political and economic barriers to food and fertilizer distribution, (c) over-consumption. In addition, (d) climate affects food production, and many trends indicate that global climates are deteriorating.

10.8   INCREASES IN CROP YIELD AND THE GREEN REVOLUTION

In the mid 1960's new varieties of wheat and rice were developed. The potential yields of these new varieties are spectacular. A wonder seed does not produce large quantities of food by itself. If this production is to be realized, the plant must be fertilized heavily, watered, and protected from disease and insects. Green Revolution seeds help only those farmers with sufficient money and knowledge to practice modern farming techniques.

10.9   OTHER APPROACHES TO INCREASING FOOD PRODUCTION

Massive quantities of energy and capital are needed to expand crop acreages in semi-hospitable regions, but often farmers cannot afford such expenditures.

Food from the sea is limited, and many fish supplies may be depleted by over-exploitation.

10.10   THE FOOD WE EAT — CHEMICAL ADDITIVES

Chemical additives are used to: (a) prevent spoilage, (b) increase the vitamin or mineral content of food, (c) add color or flavor, (d) emulsify foods, (e) increase the production of meat, milk, or eggs. No one knows for certain whether food additives cause cancer or not.

10.11   FORESTS AND OPEN SPACES

Forests and open spaces provide recreation for many people. They are also valuable sources of timber and sometimes of ores. With proper management, it is possible to enjoy and preserve wilderness and still harvest its riches. But it is more profitable, on a short term, to reap the harvests without consideration for the ecology of the region.

KEY WORDS

Humus
Green Revolution

Clearcutting

TAKE-HOME EXPERIMENTS

1. **Sand and soil.** For this experiment you will need a small shovelful of dry sand or sandy gravel. Mix one portion of the sand with an approximately equal weight of dry rotten leaves, coffee grounds, grass clippings, or other organic matter. Place about 100 g of the unmixed portion of sand in one small dish and 100 g of the sand-organic mixture in the other. Add about 20 ml of water to each, and then weigh both dishes accurately. Place them side by side on a table and weigh them every half-hour for several hours. Does the weight change? Why? Which sample loses water faster? Why? Discuss the implications of this experiment for agricultural practice. (Note: 1 fluid ounce of water = about 28 g.)

2. **Sand and soil.** Fold two pieces of filter paper as shown in the accompanying sketch, and place one in each of two funnels. Fill one funnel with dry sand and another with dry rich topsoil, but be sure that the level of the dirt is below the level of the filter paper. Carefully weigh two small glasses or beakers and place one under each funnel. Now, using another container, dissolve 3 g of table salt in 100 ml of water. Pour 50 ml of this solution

into each funnel and collect the water that drips out in the two preweighed beakers. Carefully heat each container until all the water has evaporated and a residue of salt remains. Weigh the beakers. Which beaker contains more salt? Explain.

3. **Erosion.** Find a place near your school where you can observe the effects

1. Fold filter paper into quarters, as in A, B, and C.

2. Open filter paper to form cone, with one quarter segment separated from the others, as in D. Insert cone in to glass funnel (E).

How to Make a Filter

of erosion. Describe what is happening, and, if possible, photograph the site. What agent (wind, ice, water, or other) is causing the breakdown and removal of materi-

al? Collect samples of the rock or soil as it appears before being disturbed, and then collect samples of the sediment that is being carried away. Compare the textures of the two materials. In your example is the erosion directly harmful to humans and their activities? Discuss.

PROBLEMS

1. **Industrial agriculture.** Explain why a natural prairie containing wild oats will produce less food for humans than a farmer's field will.

2. **Energy.** List the energy inputs that are added to a wheat field in industrial agriculture.

3. **Industrial agriculture.** Why is it possible to raise more beef in a feedlot than in an open range? Discuss.

*4. **Natural vs. agricultural ecosystems.** Briefly discuss the relative importance of each of the following characteristics to a plant species existing (a) in a natural prairie; (b) in a primitive agricultural system; (c) in an industrial agricultural system: (i) resistance to insects, (ii) a tall stalk, (iii) frost resistance, (iv) winged seeds, (v) thorns, (vi) biological clocks to regulate seed sprouting, (vii) succulent flowers, (viii) large, heavy clusters of fruit at maturity, (ix) ability to withstand droughts.

5. **Industrial agriculture.** Discuss some problems in food production that would occur if there were serious oil shortages.

6. **Industrial agriculture.** What are the major techniques of industrial agriculture? Under what circumstances of use or misuse may each of these methods lead to a loss of fertility of the land?

*7. **Fertilizers.** Imagine that you saw the following sign at a farm supply center, "SALE — Oxygen and carbon fertilizers. This fertilizer contains all the oxygen and carbon your plants will need! Guaranteed! Only $5.00 for a 50 kilogram bag." Would you consider the offer to be a good buy? Why or why not?

8. **Fertilizers.** Do you expect that there may be shortages of nitrogen needed to manufacture nitrogen fertilizers? Will nitrogen fertilizers be scarce in the future? Discuss.

9. **Humus.** Briefly discuss some of the properties of humus.

*10. **Fertilizers.** Assess the advantages and disadvantages of using manure as a fertilizer. Do you feel manure should be used to fertilize crops? (Remember that a field fertilized with manure may stink, but such odors are not harmful to health.)

*11. **Mechanization.** One hundred years ago most farmers used horses or oxen to plow fields and harvest crops. Today tractors are commonly used. Imagine that the fossil fuel crisis became so severe that farmers were forced to rely on animal power once again. Would the total food production increase, decrease, or remain constant? Defend your answer.

12. **Mechanization.** A large combine cuts and threshes wheat efficiently. Does it improve the yield of grain produced per acre? Why or why not?

13. **Irrigation.** Discuss some advantages and disadvantages of irrigation systems.

14. **Rural land use.** In general, how are decisions about rural land use made today? Discuss how these decision making processes may affect world food supply.

*15. **Rural land use.** Outline some approaches to legislation that might prevent agricultural land from being converted to other uses.

16. **World food supply.** Explain why people are starving today even though there are ample resources to feed everyone.

17. **Climate.** Explain why a small shift in climate could seriously affect world food supplies.

18. **Green Revolution.** Discuss the advantages and disadvantages of Green Revolution grains.

19. **Green Revolution.** Explain why the Green Revolution has not helped the very poorest farmers.

---

*Indicates more difficult problem.

20. **Agriculture in marginal lands.** Discuss some problems inherent in farming deserts; jungles; temperate forest hillsides; the Arctic.

21. **Food from the sea.** Jacques Cousteau has said, "The shores of the rivers are the roots of the oceans." What do you think he meant?

22. **Food additives.** Discuss some advantages and disadvantages of using additives in food.

23. **Food additives.** Explain why it is impossible to determine the exact risk of a given food additive.

24. **Clearcutting.** What is clearcutting? How is a forest affected by clearcutting?

25. **Clearcutting.** It is cheaper to clearcut a forest than to cut selectively. What factors are omitted from such an economic evaluation?

QUESTIONS FOR CLASS DISCUSSION

1. Many authors have pointed out that if all the fertilizer used in the United States for lawns and ornamental gardens were shipped to farmers in the developing nations, many millions of people could grow more food to feed their starving families. Imagine that you would normally fertilize your lawn, but this year you chose not to so that there would be more fertilizer for the poor. Would the fertilizer that you chose not to buy actually get shipped to people in the developing nations? Would your action do any good? What policy, private or public, would you recommend? Defend your position.

2. Water from the Colorado River in the southwestern part of the United States is diverted both for irrigation and for domestic use. Some is allocated locally and some is piped to California. In addition, the coal and oil shale companies in the Southwest need the water for industrial expansion. However, environmentalists feel that the level of the river should remain high to preserve natural ecosystems.

A serious snow drought during the winter of 1976–1977 greatly lowered the water level in the Colorado River. Discuss the impact of the drought on the various groups that want to use the water. If water were to be rationed, which group should receive the greatest supply? Defend your selection.

3. Interview a farmer in your region. Ask the following questions. Is farming profitable? Why does he or she choose to be a farmer? Would it be more profitable to sell the land and retire? What can be expected to happen to the land in a generation or two? Report on your results.

4. List the advantages and disadvantages of each of the five categories of food additives: (a) preservatives, (b) vitamins and minerals, (c) artificial colors and flavors, (d) emulsifiers, (e) animal growth additives. Would you recommend that any of these additives be banned? Discuss with your classmates.

BIBLIOGRAPHY

Two sets of articles on food and agriculture are:
Scientific American, the entire issue of September, 1976.
Philip H. Abelson (ed.): Food: Politics, Economics, Nutrition, and Research. (A compendium of articles from Science.) Washington, D.C., American Association for the Advancement of Science, 1975, 202 pp.
Two excellent books that discuss the power base of industrial agriculture are:
William J. Jewell: Energy, Agriculture and Waste Management. Ann Arbor, Mich., Ann Arbor Science Publications, 1975, 540 pp.
Howard T. Odum: Environment, Power, and Society. New York, Wiley-Interscience, 1971, 331 pp.

A controversial book dealing with the politics of food distribution is:
Francis M. Lappe and Joseph Collins: Food First. Boston, Houghton Mifflin Co., 1977, 466 pp.

The reader who wishes to investigate the Green Revolution and food in the future should refer to:
Lester R. Brown: By Bread Alone. New York, Praeger Publishers, 1974. 272 pp.
Lester R. Brown: Seeds of Change: The Green Revolution and Development in the 1970's. New York, Encyclopaedia Britannica, Praeger Publishers, 1970. 205 pp.
Francine R. Frankel: India's Green Revolution. Princeton, N J., Princeton University Press, 1971. 232 pp.

Two excellent books on agriculture and food resources are:
John Harte and Robert H. Socolow: The Patient Earth. New York, Holt, Rinehart and Winston, 1971. 364 pp.
Kusum Nair: The Lonely Furrow: Farming in the United States, Japan, and India. Ann Arbor, University of Michigan Press, 1970. 336 pp.

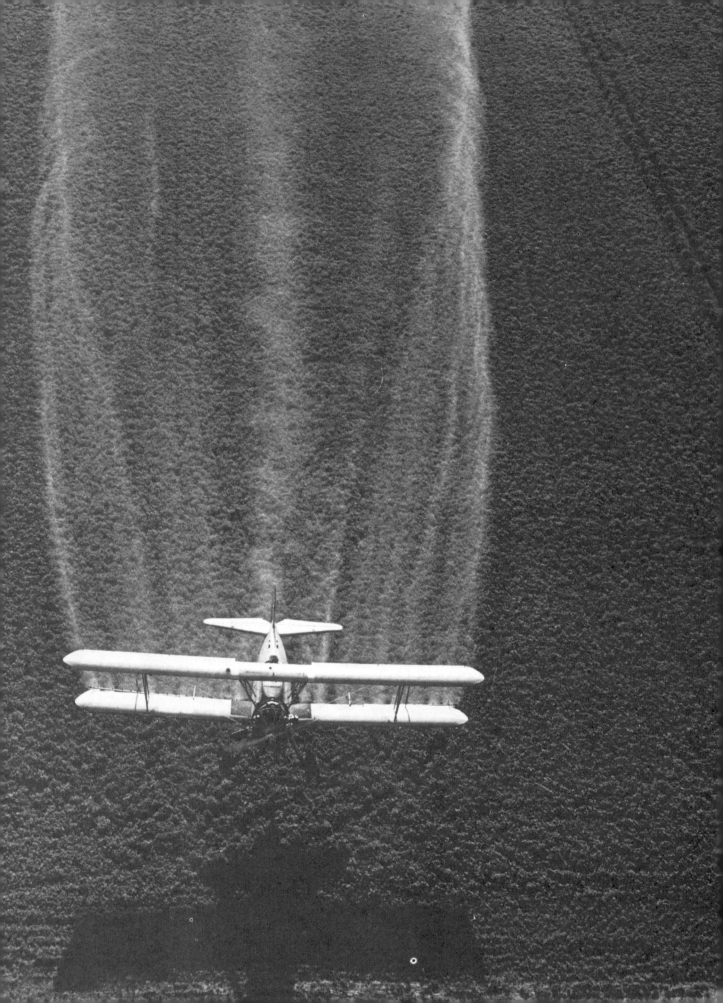

11

CONTROL OF PESTS AND WEEDS

11.1
COMPETITION FOR HUMAN FOOD

Throughout all ages, farmers have had difficulty protecting their crops from small animals and disease organisms. If a cow wanders into a field to eat the corn, the farmer can chase her away. An insect, a field mouse, the spore of a fungus, or a tiny root-eating worm is more difficult to deal with. Since these small organisms reproduce rapidly, their total eating capacity is very great. Small pests may also be carriers of disease. Malaria and yellow fever, spread by mosquitos, have killed more people than have all wars.

Not all insects, rodents, fungi, and soil microorganisms are pests. Most do not interfere with people, and many are directly helpful. Millions of small animals live within a single cubic meter of healthy soil. Most are necessary to the process of decay and hence to the recycling of nutrients. Fungi, too, are essential to the process of decay in all the world's ecosystems. Rodents are part of natural ecosystems. Thus squirrels help to spread pine seeds, and lemmings provide a stable food for almost all carnivores in parts of the Arctic. Many carnivorous insects eat various pest species and therefore are directly helpful to humans. In addition, bees are essential to the life cycle of most flowering plants. In their search for food, bees transfer pollen from flower to flower and thus fertilize the plants.

Pests have lived side by side with people for millions of years. At times pest species have bloomed and brought disease and famine. But most of the time, natural balance has been maintained, and humans have lived together with insects in reasonable harmony.

The periodic invasion of some African villages by driver ants is a fascinating illustration of insect ecology. Many disease-carrying rodents and insects live in the village houses and pose a constant threat to the human population. Occasionally, millions of large driver ants invade the villages. They chase away the inhabitants, and eat everything that remains. When the people return, they find that their stored food supply is gone. But so are all the cockroaches, rats, and other pests — everything has been eaten.

In modern times, people are no longer willing to accept these natural cycles. Perhaps even more critical is

Recall that a carnivore is an animal that eats other animals. There are many carnivorous insects that eat other insects.

Rat flea

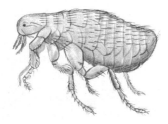

221

Body louse

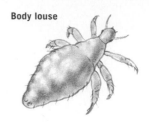

the fact that the human population is now so large that tremendous quantities of food are needed. One way to increase crop yields is to reduce competition from insects.

11.2
INSECTICIDES — A GREAT DEBATE

A farm is a carefully controlled environment. Fields are plowed, planted, and harvested. Usually only one species of plants is cultivated in a large field. Yet a great number of different species of animals exist within these artificial systems. Scientists studying a cabbage field in upstate New York found 177 different species of insects. Of these only 5 species were significant pests. Many were predators that ate the pest species. Some ate other types of food besides cabbages.

Agricultural systems are subject to the normal checks and balances of a natural ecosystem. If left alone, pest species will usually be kept under control by their enemies. According to one reliable estimate, insects ate 10 percent of the food crops in the United States in 1891.* At that time very few pesticides were being used. The pest populations were controlled by insect predators, parasites, and disease. In 1970, after years of intensive use of pesticides, crop losses rose to 13 percent. Does this mean that chemical sprays have increased the pest problem? Would farmers be better off if no pesticides were used at all?

No one knows the answer to this question. Times and agricultural practices have changed. As we will see later in this chapter, some scientists feel that chemical pesticides are harmful to the environment and increase pest problems. Others disagree strongly and feel that without pesticides humans would face mass starvation. Let us look at the problem in more detail.

The modern pesticide era started in about 1940 when a chemical called **DDT** was discovered to be a potent insecticide. DDT was far cheaper and more effective against almost all insects than the previously known chemicals. The use of DDT led to dramatic early successes. It squelched a threatened typhus epidemic among the Allied army in Italy. Anti-mosquito programs saved millions from death from malaria and yellow fever. Pest control, leading to increased crop yields all over the world, saved millions more from death by starvation.

Enthusiastic supporters of DDT predicted the complete destruction of all pest insects within the foreseeable future. Paul Müller, the chemist who first discovered its insecticidal properties, received a Nobel Prize. But within 30 years the promise of insect-free abundance had been broken. The "miracle" chemical that was to have achieved it had fallen from grace. On January 1, 1973, all interstate sale and transport of DDT in the United States was banned

FIGURE 11–1. Many insects are predators. Here a praying mantis is eating a Satyrid butterfly. (Courtesy of Emeritus Professor Alexander B. Klots, Biology Department, the City College of the City University of New York.)

*A.W.A. Brown: *Ecology of Pesticides*, p. 10 (see bibliography).

except for use in emergency situations.

A few decades ago, many people believed that chemical pesticides would liberate people from the whims of a changing environment. These chemicals were viewed as a road to plenty. But pest insects have not been removed from the face of the Earth. In fact they have survived and thrived despite an intense and continued program to annihilate them.

Many other pesticides besides DDT have been in use during the past few

TABLE 11–1. COMMON CHEMICAL INSECTICIDES*

Compound Class	Designation
Organochlorides (also called chlorinated hydrocarbons)	Aldrin
	Chlordane
	DDD
	DDT
	Dieldrin
	Endosulfan
	Endrin
	Heptachlor
	Lindane
	Toxaphene
Organophosphates	Diazinon
	Malathion
	Parathion
Carbamates	Sevin (carbaryl)

*Chemical formulas of various pesticides are shown in Appendix C.

years. Some of the most common of these are listed in Table 11–1. Note the three classes of compounds — **organochlorides, organophosphates, and carbamates**.

> The chemical formulas of some important pesticides are given in the appendix.

11.3
THE ACTION OF CHEMICAL PESTICIDES — BROAD–SPECTRUM POISONING

Most chemical pesticides are broad-spectrum poisons. That means that they kill all insects — pests and predators — in a spray area. Let us retrace what may happen after a spray is applied. In a typical example, 90 percent of all insects in the area are killed immediately. After a few days or weeks, insect populations start to grow back. Pest species then find an extremely hospitable environment. There are very few predators and lots of food. Therefore, the populations of these animals grow quickly. If populations of predatory insects do not grow as rapidly, the pest problem can grow even worse than it was originally. In some cases, the farmer would have suffered smaller losses by doing nothing at all.

Of course, pesticides often do more good than harm. There is obviously a balance here. At times, pesticides have destroyed pest populations and saved crops. At times, the same chemicals

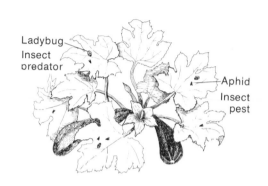

FIGURE 11–2. Effects of broad-spectrum pesticides. *A*, Before the spray application the ladybugs, a predatory insect, control the population of insect pests. *B*, Immediately after the spray application nearly all the insects are killed. *C*, If the predator population does not grow back quickly after the spray, the pest species may pose a more serious problem than they did originally.

Ladybug
Insect predator

Aphid
Insect pest

Trade winds .1-.3 ppb

Rain .1-.3 ppb

Fat of man
6-12 ppm

Rivers and lakes
.001-.2 ppb

Soil 2-10 ppm

Fat of cows .5 ppm

Ground water .001-.2 ppb

Ocean depths .001 ppb

FIGURE 11–3. Physical dispersal of insecticides with some average values
for DDT levels.

have destroyed predator populations and led to increased crop losses.

Another major problem with chemical insecticides is that many insects become resistant to the poisons. Plants and animals have adapted to a changing environment for millions of years. They have adapted to changing climates, food supplies, and predators. Recently, many strains of insects have evolved which are genetically resistant to a particular chemical. For example, certain species of bedbugs have become resistant to DDT. A colony of these animals were raised in the laboratory on cloth impregnated with DDT. They lived, grew, and mated, and the females laid eggs normally. The young, born on a coating of DDT, grew up and were healthy.

By 1945 at least a dozen species had developed some resistance to DDT. By 1976 the number had increased to over 200 species. About 35 of these resistant species carry disease and about 80 others are serious agricultural pests.

Many insects are also resistant to other chemical pesticides. Since resistant parents tend to pass this characteristic to succeeding generations, the old pesticides become ineffective. It is easy to see that if a pest species becomes resistant and the predators do not, extremely serious problems may arise.

There have been a great many cases throughout the world where pesticide use has increased pest damage. In the San Joaquin valley in California, chemical manufacturers sold the organophosphate Azodrin for use against the cotton bollworm, a pest of cotton plants. After several heavy spray dosages, it was discovered that the predator populations were killed more effectively than the pest species. Losses to the bollworm grew more severe. Sales representatives recommend that more pesticide should be used. (Of course this would lead to greater profits for the manufacturer.) An independent research team at the University of California determined that if no pesticides

"There I was, coming in low at a hundred and seventy-five m.p.h. with forty-two acres of broccoli to the left of me, eighteen acres of asparagus to the right of me, eighty acres of carrots straight ahead! Power lines all around! My target: seven acres of badly infested garlic smack in the center...." (Drawing by Dedini; © 1972. *The New Yorker Magazine, Inc.*)

were used *at all*, bollworms would consume approximately 5 percent of the crops. After three spray applications, so many predators had been killed that 20 percent of the crop was destroyed.

11.4
WHY WAS DDT BANNED?

All the insecticides listed in Table 11–1 are broad-spectrum poisons. They all kill nonpest species and have led to ecosystem disruption. Why, then, have DDT and other chlorinated pesticides been banned, while other types of pesticides are still used freely?

If an organophosphate is sprayed on a field, it will decompose rapidly in the environment. For example, when parathion, an organophosphate, is sprayed onto loamy soil, over 98 percent will decompose after four months. These materials are said to be **biodegradable**. The organochlorides are not biodegradable. They persist for long periods of time in the environment. For example, in some soils DDT or Aldrin has been detected fifteen years after a single spray application. Such persistence leads to a series of severe ecological problems:

(a) If a field is sprayed once, predators are subject to continuous exposure to poisons for many years. If a field is sprayed many times, the insecticide accumulates in the environment. Therefore, all the problems outlined in the previous section may persist for many years.

(b) Organochlorides are transported throughout the environment. Some contaminated soil particles are carried toward the oceans by creeks, streams, and rivers. Others fly high into the atmosphere and ride wind currents for thousands of kilometers. Eventually the dust is trapped by falling rain and brought back to earth. If a chemical decomposes rapidly in the environment, it will not have time to travel long distances. But organochlorides are long-lived and have time to be transported around the globe. Today, DDT can be

The abbreviation for parts per million is ppm. If a compound is present in a concentration of 1 ppm in water, that means that for every million grams of water, there is one gram of pesticide. Perhaps that doesn't sound like very much. It all depends on how we express the numbers. If there is 1 ppm of DDD in water, that means that there are 2×10^{15}, or 2,000,000,000,000,000, molecules of pesticide in one gram of water. DDD concentrations as low as 0.009 ppm can kill some species of trout.

DDT was banned in the United States in January, 1973. Yet production and use of DDT still continues in other parts of the world. In fact, more DDT is used today than before 1973. Therefore, global pollution problems still exist.

found in all major rivers of the world. There are traces in the ocean, in the fat of penguins in Antarctica, in people's bodies — just about everywhere. Pesticides are partially responsible for killing trout in the Great Lakes and in the Mississippi River system. They disrupt the growth of phytoplankton in the ocean. They kill microorganisms in the soil. They have been known to disrupt forest ecosystems. In short, if a persistent chemical poison is sprayed on one field it not only may affect the ecological balance of that field, but may also travel into other ecosystems. It may

continue to upset ecological balance in many regions for years to come.

(c) Organochloride chemicals are transported and concentrated in biological systems. Plants and animals are able to wash many poisons out of their systems. Most wastes are removed from the body in water solutions of urine, sweat, or pus. But organochlorides do not dissolve in water. Instead, they dissolve in fat. Therefore, they tend to remain in the body for very long periods of time. If an organochloride pesticide is sprayed onto a hay field and a cow eats the hay, the cow consumes small concentrations of poison every day. The pesticide then remains in the cow's body; each daily intake is stored. Therefore, the pesticide residues slowly accumulate in the animal. A person who then eats the cow's meat will consume the large concentrations of this accumulated chemical.

Clear Lake in north-central California is a popular recreational area. But the shallow, calm waters have been a breeding place for annoying colonies of mosquitos and gnats. In the 1950's government workers sprayed Clear Lake with DDD to control the insect pests. After the project was completed, the water contained 0.02 ppm of DDD.

Clear Lake, California.

Plankton that live in Clear Lake concentrated and stored some of the pesticide until there was 5 ppm in their tissues. Plant-eating fish ate the plankton and further concentrated the poison. As a result, their bodies contained 40 to 300 ppm of DDD. Some predatory fish and birds had as much as 2000 ppm of DDD in their tissues. Thus, the original concentration of the pesticide in the water seemed innocently low. But fish and birds were poisoned. The problem persisted for a long period of time. A year after the spray application, no pesticide could be detected in the water at all. Yet the poison still existed in the plants and animals.

The effects of high levels of insecticides on carnivorous fish, mammals, and birds have been particularly severe. In many cases the animals die soon after being poisoned. In other situations, small doses may not kill the animals directly, but they may upset the normal activities of the body and cause delayed death. In some cases, the adults may survive but the infants will die. For example, when mink were fed fish that were contaminated with DDT the adults remained healthy but 80 percent of the newborn infants died within a few days.

The evidence seems undeniable that DDT poisoning is responsible for the sharp decline in populations of many predatory birds. Birds that have been poisoned by DDT cannot use calcium properly. This has led to the production of thin-shelled eggs. Often these weakened eggs crack and break in the nest with resulting death of the unborn chicks. The populations of several species of birds — among them, the peregrine falcon, the pelican, and some eagles — are declining so rapidly that many conservationists fear that they will become extinct in the near future.

11.5
INSECTICIDES AND HUMAN HEALTH

Most chemical pesticides are poisonous to humans as well as to insects. The organophosphates, which have been used extensively in North America since 1973, are much more poisonous

FIGURE 11–4. Biological magnification of DDD (Clear Lake, California).

than the banned DDT that they replaced.

Since the mid-1940's, many thousands of people have been killed directly from severe pesticide poisoning every year. At present, more than half of these are children who are exposed to the toxic chemical through carelessness in packing or storage. Most of the others are workers who handle these materials in the factory or on farms.

When peregrine falcons are subjected to DDT poisoning, many eggs crack before hatching, and chicks often die prematurely. The two dead chicks and cracked egg in this nest all resulted from such poisoning. (Courtesy of the Peregrine Fund)

Migrant farm workers near Salinas, California. These people are not told of past spray applications and are sometimes exposed to high concentrations of pesticides.

The case of the workers in a kepone factory deserves mention. Kepone is an organochloride pesticide developed in the early 1950's. It was manufactured by a small corporation closely associated with Allied Chemical Company. Safety regulations at the factory were amazingly lax. Pesticide dust was scattered over floors, equipment, clothes, and even the employees' lunch area. In 1975, several workers began to complain of severe illness. Their muscles would tremble uncontrollably at times. People suffered blurred vision, loss of memory, and pains in their joints and chest. In all, 70 of the 150 employees were poisoned by the pesticide. In addition, some of the workers brought the pesticide dust home on their clothes and poisoned their families as well. In July of 1975, the plant was shut down and Allied Chemical paid millions in damage suits. But cash settlements do not make people healthy again.

It is relatively easy for most people to avoid exposure to large doses of insecticides. But it is impossible to avoid exposure to trace contaminants in food, in the air, and in drinking water. We all carry measurable quantities of insecticides in our bodies. What are the chronic (long-term) effects of these chemicals? Unfortunately, it is impossible to answer this question with certainty. Suppose there were two large groups of

people and one group ate contaminated food while the second group ate pesticide-free food. If we studied both of these groups for twenty or thirty years, then perhaps we could measure the long-term effects of pesticides. However, it would be morally unjustifiable to perform such an experiment. In addition, it would be impossible to control the diets of large groups of people over many years. Instead we must rely on studies of small groups of people and of laboratory animals. Such studies are always uncertain. In addition, most experimental subjects are examined for only a few years, although it is known that cancer may appear ten to twenty years after exposure to a foreign poison.

Scientists have fed various pesticides directly to people. They have studied men and women who work in pesticide factories or with these chemicals in the field. None of these studies has *proved* that pesticides cause cancer. Extensive experiments have been conducted with rats, mice, and other laboratory animals. Some of the results have shown that various pesticides induce cancer in certain species of animals. These experiments suggest that the same pesticides may induce cancer in humans. But other experts have questioned the validity of the experiments. In some instances extremely high doses of pesticide were used. In others, the strains of rats and mice that were studied showed an abnormally high susceptibility to cancer without exposure to pesticides. All in all, it can be safely said that after a decade and a half of study, no one is certain whether pesticides do or do not cause cancer in humans.

11.6
OTHER METHODS OF PEST CONTROL

Broad-spectrum pesticides have been effective in many situations. But they have serious flaws as well. They destroy populations of predators, upset ecosystems, and poison people and animals. In addition pesticides may possibly cause cancer. What are the alternatives?

A female screwworm fly—the one that lays the eggs and does all the damage. (Courtesy of U.S.D.A.)

Use of Natural Enemies

Pesticides have caused problems where they have poisoned natural predators. Why not use the opposite treatment. Instead of poisoning predators, why not import them? This type of approach has been quite successful in many cases.

The Japanese beetle was imported to North America from the Orient, and has become a serious agricultural pest. In Japan, the beetle population is controlled by natural enemies. These predators do not live in North America. One of the natural predators is a type of wasp. A female wasp paralyzes a Japanese beetle grub and attaches her eggs to it. When the young wasp hatches, it eats the grub as its first food. These wasps lay their eggs only on Japanese beetle grubs. They do not attack other species. In the United States, spray applications against the Japanese beetles have caused serious ecological disruptions. In recent years, people have realized higher levels of success simply by importing natural enemies.

Insects can also be controlled through the use of certain strains of bacteria and viruses. A virus effective against the cotton bollworm and corn earworm has been manufactured as a pesticide in the United States. It has been approved by federal regulatory agencies and is now ready for mass production. Viral strains that combat several other species of pests should be commercially available in the near future.

There are many advantages to importing enemies of pests. Because these agents are living organisms, they reproduce naturally, and one application can last for many years. Most insect parasites and disease organisms are very specific and do not interfere with the health of large animals. No harmful or questionable chemicals are introduced into the environment.

Sterilization Techniques

Pests can be controlled without killing them directly if the adults of one generation are sterilized. The sterilized adults cannot produce healthy babies, so the pest population will soon die off.

Screwworm fly

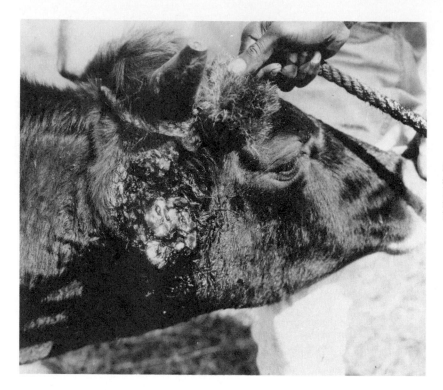

Screwworm larvae infestation in ear of a steer. An untreated grown animal may be killed in 10 days by thousands of maggots feeding in a single wound. (Courtesy of U.S.D.A.)

The screwworm, the parasitic larva of the screwworm fly, is a serious pest that has killed many cattle in the southeast United States. Some years ago, the United States Department of Agriculture initiated a program to raise male screwworm flies, sterilize them, and release them in their natural breeding grounds. If a sterile male mates with a normal female, she will lay eggs, but they will not hatch.

For several years, millions of sterilized males were released annually to mate with healthy females living in the area. Initially the program was a spectacular success. Screwworm infestations were completely eliminated from 1962 to 1971. But in 1972 problems

---

SCREWWORM WAR

It was billed as the world's largest biological warfare program, with daily bombings out of Mission, Tex., and use of an atomic weapon. The target was the screwworm fly.

In May 1977, pilots flying C-45's were dropping 400 million male flies a week in Mexico and infested parts of Texas. Each fly had been rendered sterile by radiation from a cesium 137 isotope. Sterile males breeding with fertile females would produce zero population, the war planners said.

The screwworm fly, which is metallic green, is a bit larger than a housefly. But unlike the housefly, its larvae feed on living tissue. They can kill a mature steer in 10 days. In 1976 they killed an invalid elderly woman in San Antonio who was unable to help herself.

Once before the Federal Department of Agriculture had been tricked into believing that the screwworm fly was all but wiped out. In 1970, Texas reported only 92 cases. In 1972, following a fly assault from Mexico, the total was 90,980.

So far this year, according to the Texas Animal Commission, the line is being held in Texas but the screwworm fly has opened new fronts in New Mexico, Arizona and California. The commission reports 57 cases of cattle infestation in Texas, 163 in New Mexico, 913 in Arizona and 26 in California. Air drops of sterile flies have been raised to 500 million a week and extended to New Mexico.

Dr. Carl R. Watson, a staff scientist of the commission, observes: "This is a war. You may lose a battle here and there, but we won't lose the war."

Dr. H. Q. Sibley, the commission's director, predicts: "I'd say that we'll eventually eradicate the screwworm fly in the continental U.S.A. and Mexico, and hope we can do it sometime in '81, '82 — along in there." From the *New York Times, June 18, 1978.*

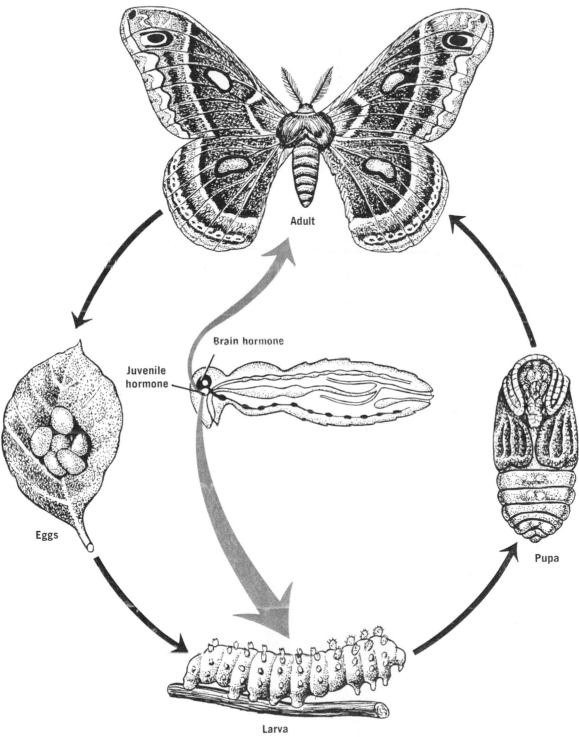

Adult

Brain hormone

Juvenile hormone

Eggs

Pupa

Larva
The role of hormones.

arose. Wild females no longer chose to mate frequently with sterilized males. The screwworm population rose once again. Further research is in progress. But this example reminds us that insect control is a complex and ever-changing problem.

Control by Hormones

In many cases, natural hormones are expensive to manufacture. Scientists have been using **hormone mimics**—compounds that are similar to the natural hormones, but not identical. These chemicals replace the natural compound and serve the same purpose.

Many insects begin their life in some larval stage and later change (metamorphose) into a mature adult. As an example, a caterpillar is a larva that later matures to become a moth or butterfly. The changing stages in an insect's life are controlled by chemicals called **hormones**. If insects are artificially sprayed with specific hormones, their life cycle can be upset. Eventually such spray applications can kill the animals. Therefore, hormones can be used as insecticides. It is likely that widespread use of these chemicals would produce minimal environmental insult. They are biodegradable and active only against specific insects. They are not poisonous to large animals. With the aid of careful timing, pests can be destroyed without killing their predators.

Sex Attractants

In many species of insects a virgin female signals her readiness to mate by emitting a small amount of chemical sex attractant. The males detect (smell) very minute quantities of these chemicals and follow the odor to its source. Attempts have been made to bait traps either with a chemical or, more simply, with live virgin females. In this program, only tiny quantities of natural chemicals are released into the air. Environmental disturbances are practically nonexistent. However, effective control is realized only if a great number of traps are used. Some successes have been registered by the confusion technique. In one application millions of cardboard squares were impregnated with sex attractants and dropped from an airplane. There were so many cardboard squares that the males became confused and could not find females.

Use of Resistant Strains of Crops

Some plants are naturally resistant to pests because they synthesize their own insecticides. For some time, plant breeders have been actively developing more resistant strains of various crops. They have succeeded in a number of instances. A variety of alfalfa that is resistant to the alfalfa weevil has been developed. Various strains of cereal crops resistant to rust infections are also available.

The successes achieved have been encouraging, but it should be remembered that the technical problems are difficult. The new plant variety must produce high yields as well as maintain the natural resistance to diseases. Throughout biological time, the genetic defense of one species traditionally has been met by genetic changes in the attacking organisms to neutralize the defense. Imagine that scientists develop a new variety of plants that are resistant to a species of insects. Soon the insects may evolve to overcome the plant's defenses. Then a new variety of plants must be developed. The seesaw battle may continue forever.

There are four groups of people involved in the pesticide controversy.
1. Farmers want to grow as much food as possible so as to earn a living.
Illustration continued on opposite page

Integrated Control

If there is any lesson to be learned from the DDT era it is that insects are not easy to control. They evolve resistance to chemicals, evade enemies, and continue to survive. Many scientists feel that we should never again rely on a single approach to insect control. Rather, we should use as many control measures as possible in an integrated and well-planned manner.

11.7
POLITICS AND ECONOMICS OF PESTICIDES

There are four groups of people involved in the pesticide controversy. (a) Farmers want to grow as much food as possible so as to earn a living. (b) Pesticide manufacturers want to sell pesticides and realize a profit. (c) Government agencies have the responsibility to regulate and control dangerous chemicals. (d) Consumers want a plentiful supply of food at reasonable prices, but they want to avoid environmental pollution that endangers their health. Any pesticide policy results from a balance of these four viewpoints.

Let us think for a minute of an "ideal" pesticide. Perhaps it would be a virus or bacterium that would attack a specific insect pest. These microorganisms would infect only one pest species. They would not attack nontarget species. No foreign chemicals would be introduced into the environment. Moreover, bacteria and viruses reproduce themselves. One spray application would protect a field for years to come. With these encouraging advantages, one may wonder why insect enemies are not used more frequently.

One problem is that this type of control measure acts too slowly. A chemical pesticide kills pests within a few hours. A disease organism requires weeks or months to act. Although effective pest control can eventually be realized, results cannot be expected immediately. Many farmers cannot afford large losses in a single year and are reluctant to trust biological control methods.

Another problem is that such an ideal pesticide program would not be likely to be highly profitable to the

2. Pesticide manufacturers want to sell pesticides and realize a profit.

3. Government agencies have the responsibility to regulate and control dangerous chemicals. (Courtesy of U.S.D.A. Soul Conservation Service)

4. Consumers want a plentiful supply of food at reasonable prices, but they want to avoid pollution that endangers their health.

manufacturer. Many millions of dollars are needed to develop, produce, test, and market a new product. If the product can be used against many pest species, the company can expect large sales. But if the insecticide is effective against only one species of pests, sales will be much less, and profits will be lower. In addition, recall that parasites and disease organisms reproduce naturally. Therefore, a farmer can buy one application of the pesticide and guarantee protection for many years. Obviously, sales of a truly effective pest program will be much lower than sales of chemicals that must be used every year. Therefore, it is not surprising that industry has traditionally been more interested in broad-spectrum poisons than in biological control. Perhaps government support of truly effective pesticide measures would be a valuable environmental goal.

In 1962, companies in the United States sold 70 million kilograms of DDT. In the same year, Rachael Carson published a very popular book called *Silent Spring*. She outlined some of the dangers of chlorinated pesticides and suggested that exposure to DDT might increase the occurrence of cancer in human beings. Environmental groups began to pressure the government to ban certain pesticides. After an extensive review of the situation, all sales of DDT in the United States were banned in 1973 except for emergency situations. During the next five years, several other pesticides and herbicides were banned.

The bans on various pesticides have been both praised as farsighted and damned as shortsighted. One critic of the bans has been Dr. Norman E. Borlaug, the winner of the 1970 Nobel Peace Prize for his work in developing high-yield wheat strains. He has emphatically denied that such chemical pesticides significantly contribute to the deterioration of the environment. Furthermore, he fears that other pesticides will soon be banned, leading to failures in agriculture and possible mass starvation.

On the other hand, many support the ban on DDT and other pesticides. They argue that any substances as biologically active, persistent, and foreign to natural food webs as the organochloride insecticides can be *presumed* to be harmful to human health. They point out that cigarette smoke and asbestos dust cause cancer several decades after exposure. Perhaps it would be slightly better to be cautious now than to risk a large scale epidemic of cancer in the next decade. Supporters of the ban also point out that so many insect species had developed genetic resistance to many chlorinated pesticides by the early 1970's that they were losing their usefulness anyway. They point out that government restrictions may serve as a drive to develop other, more environmentally sound methods of pest control.

11.8
HERBICIDES

Have you ever grown a garden? If you have, you know how much work is involved in hoeing and picking weeds. Yet if the weeds are not killed, they will choke out your flowers and vegetables and reduce yields. Farmers have battled weeds for as long as they have battled insect pests. People have known for centuries that certain chemicals may kill plants. The challenge in modern times is to develop chemicals that will kill weeds but not kill valuable crops. Such formulations are called **herbicides**.

Herbicide development and production has been quite successful in recent years. A hired laborer spends an average of 10 to 25 hours to hoe the weeds between the rows of one hectare of cotton. Today an airplane can spray hundreds of hectares in a few hours for $5 to $15 per hectare. Agriculture in North America and Europe depends on small inputs of labor and a heavy use of machinery and chemicals. Certainly herbicides are an important part of such a program. As a result, in 1978 approximately 400 million kilograms of herbicides were manufactured in the United States.

By now the student should be aware that anytime large quantities of synthetic chemicals are spread throughout the environment, serious ecological problems are likely to arise. Herbicides are not nearly as persistent in the soil as organochloride insecticides are. But

Many workers with hoes are needed to clear a field of weeds, whereas it is much cheaper to spread herbicides. But chemicals pollute the environment.

they pose difficulties nevertheless. Some herbicides kill earthworms and other organisms that contribute to a healthy soil. Others have been known to kill certain insect predators. Of course, if the predators are killed, pest populations will rise. Then, farmers are likely to use more pesticides along with the herbicides, resulting in a spiraling pollution problem.

The herbicide 2,4,5-T has been used in the United States by the Forest Service to kill brush that interferes with the growth of valuable timber. 2,4,5-T is poisonous to humans. But perhaps even more frightening is the fact the commercial preparations of this herbicide are contaminated with a chemical that causes birth defects. When the government sprayed forests in Minnesota, many farmers in the area suffered from headaches, nausea, dizziness, and diarrhea. In addition, an abnormally high number of women had miscarriages in the region within a few months after the spray application. In April, 1978, 2,4,5-T was placed on a list of dangerous chemicals.

Agricultural scientists are faced with a dilemma. If herbicides are banned directly, the price of food will rise sharply. Farmers might not find the laborers needed to work long hard hours in the fields picking weeds. Then food production would decrease. Therefore, the use of herbicides continues. However, chemical pollution of the environment

Early in 1979 the EPA used its emergency powers to temporarily ban all domestic uses of 2,4,5-T.

is also increasing, and no one knows how this pollution is affecting human bodies and world ecosystems.

11.9
CASE HISTORY: THE BOLL WEEVIL

Many of the most serious agricultural pests have been imported from foreign regions. When they enter a new environment, there are relatively few natural predators and population growth is rapid.

One time I seen a Boll Weevil, he was settin' on a square.
Next time I seen the Boll Weevil, he had his whole darn family there;
He was lookin' for a home
He was lookin' for a home.

The farmer take the Boll Weevil, put him on the ice;
Boll Weevil said to the farmer, You's treatin' me mighty nice;
And I'll have a home.
And I'll have a home.

Folk ballad

The boll weevil is a small insect — about 0.65 cm (¼ in) long—which first migrated to the United States from Mexico in about 1880. Weevils feed on the buds and bolls of young cotton plants, thereby destroying the valuable fiber. Since there are few natural enemies of the weevil in the United States, the population of these insects has grown rapidly at times. If left uncontrolled, a light infestation of 10 weevils per hectare can increase to 10,000 weevils per hectare within a single growing season. Such a large population of weevils can completely consume an entire crop. Weevils are now responsible for 200 to 300 million dollars worth of crop losses in the United States annually.

In the early 1900's, farmers used cultural controls to combat the weevils. Cotton was planted and harvested as early in the year as possible. Then the stalks and leaves of the plants were burned immediately after the harvest. The fires would destroy a large number of weevils that lived in the fields. If the farmer was lucky, an early crop could be grown and harvested the following year before the pests could repopulate.

This nip-and-tuck race with growing weevil populations was often unsuccessful, and many crops were completely destroyed. In the 1930's, a new chemical pesticide using arsenic was introduced. But it was so highly poisonous to nontarget species that it was largely replaced by DDT in the 1940's. By 1960, many DDT-resistant strains of boll weevil had evolved. Therefore, when DDT was banned in 1973, many farmers had already been using organophosphates for several years.

Broad-spectrum pesticides were effective in partially controlling the population of weevils. But many secondary problems were raised. Cotton is not the only crop grown in the South. Tobacco, corn, peanuts, soybeans, and vegetables are also commercially important. Boll weevils have few natural enemies in the United States. But the insect pests of other major crops have traditionally been controlled by natural enemies. When cotton fields were sprayed heavily with broad-spectrum chemicals, adjacent regions were contaminated also, and many natural predators were killed. As a result, epidemics of tobacco and corn pests became a problem for the first time.

Obviously, a new strategy was necessary. A Special Committee on Boll Weevil Eradication was formed with government support. A five-pronged attack on this insect pest was outlined. Farmers were asked to (a) spray lightly with chemical pesticides during the growing season, (b) spray again in the fall just before the weevils hibernate for the winter, (c) destroy all plants, leaves, and stalks just after harvest to kill any remaining insects, (d) set traps in the early spring baited with a sex attractant to catch and destroy females emerging from hibernation, and (e) release sterile male weevils to prevent fertile matings of any remaining females.

A pilot project was started in 1971 to see if such a program could wipe out the boll weevil forever. The results of the project were encouraging. Weevil infestations were generally reduced in the target area. But even with such a careful and complete program, total destruction was not realized. Research teams found that 95 percent control was feasible. But it is extremely difficult to destroy every weevil in a region. Despite an intense campaign, small pest populations survived in several safe havens. One farmer who was cheating on his income taxes did not tell authorities about a hidden field. Weevils survived there and later migrated outward. Another farmer claimed that pesticides were killing his chickens and also refused to cooperate. Many owners of roadside stands and restaurants kept small plots of cotton to attract Yankee tourists. For the most part these smalltime "growers" did not participate in the program, so more weevil breeding populations survived. Finally, wild cotton plants grow throughout the South, and many weevils breed in small cotton patches in the woods and swamps, far from agricultural areas.

The United States Department of Agriculture (USDA) seriously contemplated a billion-dollar program to annihilate the boll weevil once and for all throughout the United States. But in view of the problems outlined above, the program was abandoned, and the weevil war continues to this day.

Boll weevil on a cotton plant. (Courtesy of U.S.D.A.)

Cotton flower. (Courtesy of U.S.D.A.)

Open cotton boll. (Courtesy of U.S.D.A.)

CHAPTER SUMMARY

11.1 COMPETITION FOR HUMAN FOOD

Small herbivores have lived side by side with people since the human race has evolved. Some of these organisms are pests, some are directly helpful, and all are part of natural ecosystems.

11.2 INSECTICIDES — INTRODUCTION AND GENERAL SURVEY

There are three major classifications of chemical pesticides — organochlorides, organophosphates, and carbamates. The use of DDT, an organochloride, led to some dramatic successes, but now the chemical is banned in many areas.

11.3 THE ACTION OF CHEMICAL PESTICIDES — BROAD-SPECTRUM POISONING

Broad-spectrum pesticides kill predators as well as pest species and thus disturb natural controls. In some cases, pest species have evolved to be resistant to the poisons, and the populations of these animals have bloomed. The problem of resistance of insects to poisons can be compounded if the pest becomes resistant and various predators do not.

11.4 WHY WAS DDT BANNED?

DDT and other organochloride insecticides were banned because:
(a) They are not biodegradable and persist in the environment for long periods of time. (b) They are easily transported throughout the environment. (c) They are concentrated in biological systems and poison many species of large animals.

11.5 INSECTICIDES AND HUMAN HEALTH

Organophosphates are much more poisonous than organochlorides. After a decade and a half of study, proof for or against the carcinogenicity (cancer-producing quality) of pesticides in humans is still lacking, but the very fact that these materials are potent poisons makes caution advisable.

11.6 OTHER METHODS OF PEST CONTROL

(a) Natural enemies such as predators, parasites, and disease organisms are used. (b) Sterile animals (usually males) are released for nonreproductive mating. (c) Hormones can be used as sprays to interrupt the life cycles of insects, eventually causing death. (d) Sex attractants are used to bait insect traps. (e) Resistant strains of crops are used to replace more susceptible ones. (f) The most effective controls include an integrated approach using many different techniques.

11.7 POLITICS AND ECONOMICS OF PESTICIDES

Species-specific biological control methods are environmentally sound but are not highly profitable to the manufacturer. Farmers are wary of them because their action is slower than that of chemical pesticides. Therefore, they are not widely used. Pesticide bans have been both praised as farsighted and damned as nearsighted.

11.8 HERBICIDES

During the past ten years, the use of herbicides for chemical control of unwanted plants has been increasingly popular. Some commonly available herbicides are known to cause birth defects. Others are suspected carcinogens. Weed control is necessary in agriculture, but indiscriminate use of herbicides may lead to ecological imbalance.

KEY WORDS

DDT
Organochloride
Broad-spectrum poison
Organophosphate
Biodegradable
Carbamate

Hormone
Herbicide

Insect pests. Identify one species of insect pest that inhabits your area. Does this pest carry disease, cause economic loss, or is it merely an annoyance? Try to control the pest in a small, well-defined area without using chemical sprays. You can use baited traps (available at many hardware stores), screens to isolate the target area, fly swatters, imported predators, or any other nonspray techniques. Comment on the cost and effectiveness of your control program.

PROBLEMS

1. **Insects and people.** What are the harmful effects and the benefits that insects bring to people? How did people cope with insect problems before the introduction of modern insecticides?

2. **Broad-Spectrum poisons.** What is a broad-spectrum poison? Discuss some advantages and disadvantages of this type of control program.

3. **Broad-spectrum poisons.** Explain how use of a pesticide may actually increase the pest problem. Does pesticide use always cause an increase in the pest problem? Discuss.

4. **Genetic resistance.** What is genetic resistance? How does it lead to problems in pest control?

5. **DDT ban.** Why was DDT banned while organophosphates were not?

6. **Transport of pesticides.** Describe three ways in which DDT is transported through the environment by natural means. Why is an organophosphate less likely to be transported?

7. **Wrong targets.** Explain how soil run-off from agricultural systems can be deadly to fish in nearby streams.

8. **Pesticides in aquatic systems.** Explain why carnivorous fish are generally more susceptible to low levels of pesticides in the water than are herbivorous fish.

9. **Pesticides in aquatic systems.** If no measurable quantities of DDT were found in the water of a pond, would that necessarily mean that the aquatic ecosystem was unpolluted by DDT? Explain.

10. **Sublethal doses.** An amateur ecologist studying wildlife populations before and after a heavy spray application deter-

mined that since no animals were directly killed by the spray, no harm had resulted. Would you agree? Explain.

11. **Birds.** Do you think that some bird species might become resistant to DDT? Would you think that resistance might save the birds from extinction? Defend your answers.

12. **Health effects.** Discuss the short-term and possible long-term effects of pesticides on human health.

13. **DDT ban.** Outline the arguments for and against the DDT ban.

14. **Alternatives.** What methods of pest control are available as alternatives to the use of chemical sprays?

15. **Insect predators.** Discuss some advantages and disadvantages of the use of insect predators for pest control.

16. **Sterilization.** Explain how sterilization programs function. Why are they not always effective?

*17. **Control by hormones.** What is a hormone? Why can hormones be used as pesticide sprays? Why are they minimally disruptive to the environment?

*18. **Control by hormones.** Explain why it is important to spray with hormones at a precise time. If a spray application were successful on May 1 of one year, would it be safe to assume that farmers could spray successfully on that date every year?

19. **Integrated control.** Explain why one indiscriminate spraying with DDT could destroy the effectiveness of an integrated control program.

_____

*Indicates more difficult problem.

*20. **Integrated control.** Birds often become major pests in vineyards because they eat the grapes. (a) Which of the following control programs would you recommend for bird control: (i) spreading poison; (ii) broadcasting noise from a loudspeaker system to scare them; (iii) shooting; (iv) covering the vineyard with some fencing material? Discuss. (b) Do you think that it might be wise to initiate research directed toward: (i) developing a sterilization program against the birds, or (ii) developing new strains of grape which would be unpalatable or poisonous to birds? Discuss.

21. **Pest control and social attitudes.** Explain how some peoples' jobs (for example, of a crop duster pilot, a pesticide salesperson, a government inspector, a farmer, or

*Indicates more difficult problem.

a textbook writer) may affect their opinions on pesticide use.

22. **Economics of pesticides.** Discuss the statement: "Pest control techniques that were truly effective would be a disaster for the pesticide industry."

*23. **Herbicides.** Ragweed is considered to be a major plant pest because its pollen causes misery to hay-fever victims. In natural systems ragweed is characterized as an early successional plant. (An early successional plant is one that grows in areas that have been stripped of vegetation.) A few years ago, the State of New Jersey initiated a program to eliminate ragweed. Thousands of acres of roadways and old fields were sprayed with herbicides. Why do you think this program failed?

24. **Herbicides.** Outline briefly some of the parallel problems that exist between herbicide use and insecticide use.

QUESTIONS FOR CLASS DISCUSSION

1. Prepare a class debate. Have one side argue that a pesticide should not be banned until it is proved to be carcinogenic. The other side should argue that even a suspected carcinogen should be banned.

2. Sometimes an integrated control program may not become effective until the second season of its application. Moreover, crop losses during the first season may actually be greater than average. Do you feel that it would be good policy for

the government to subsidize such losses in an effort to improve environmental quality? Defend your answer.

3. A worm that thrives in the core of an apple does not usually eat a large proportion of the apple, but people do not like wormy apples anyway. Discuss the social and economic factors influencing the choices between wormy apples and apples that might contain pesticide residues.

BIBLIOGRAPHY

Several recent and authoritative books on pesticides are:

A. W. Brown: *Ecology of Pesticides.* New York, John Wiley & Sons, 1978. 523 pp.

W. W. Fletcher: *The Pest War.* New York, John Wiley & Sons, 1974. 218 pp.

Rizwanul Haque, and V. H. Freed (eds.): *Environmental Dynamics of Pesticides.* New York, Plenum Press, 1975. 387 pp.

C. B. Huffaker (ed.): *Biological Control.* New York, Plenum Press, 1971. 511 pp.

David Irvine, and Brian Knights (eds.): *Pollution and the Use of Chemicals in Agriculture.* Ann Arbor, Mich., Ann Arbor Science Publishers, 1974. 136 pp.

Fumio Matsumura: *Toxicology of Insecticides.* New York, Plenum, Press, 1975. 503 pp.

Robert L, Metcalf and William H. Luckman (eds.): *Introduction to Insect Pest Management.* New York, John Wiley & Sons, 1975. 587 pp.

A few older general references include:

W. W. Kilgore and R. L. Doutt (eds.): *Pest Control.* New York, Academic Press, 1967. 477 pp.

David Pimental: *Ecological Effects of Pesticides on Non-target Species.* Washington, D.C., U.S. Government Printing Office, 1971. 219 pp.

U. S. Department of Health, Education, and

Welfare: *Report of the Secretary's Commission on Pesticides and Their Relationship to Environmental Health*. Washington, D.C., 1969. 677 pp.

The book that started much of our current concern about pesticides, and a more recent sequel to it, are the following:

Rachel Carson: *Silent Spring*. Boston, Houghton Mifflin Co., 1962. 368 pp.

Frank Graham, Jr.: *Since Silent Spring*. Boston, Houghton Mifflin Co., 1970. 333 pp.

12

AIR POLLUTION

12.1
INTRODUCTION

Most of us have had direct and personal experiences with air pollution. Perhaps you were swimming in the ocean outside of New York City or Los Angeles and watched a red-yellow-brownish cloud collect over the city and drift slowly toward the beach. Maybe you were driving from the country into some city and suddenly felt a slight irritation in your eyes and throat as you approached the metropolis. Many people have at one time or another sat on a hilltop overlooking some commercial-industrial center such as Denver, San Francisco, Los Angeles, or Montreal and seen a dark haze hanging close to the ground.

Air pollution has grown side by side with industrial development. For years a picture of a factory with a tall chimney pouring smoke into the air stood as a symbol of progress and growth. But in recent times, people have grown increasingly aware of the harmful effects of pollution, and this former symbol of progress has now become a symbol of environmental insult.

12.2
THE ATMOSPHERE

The Earth is surrounded by a gaseous blanket that is called the **atmosphere**.

The atmosphere consists of a mixture of various gases and small particles collectively called **air**. Natural air differs from place to place around the globe. The air in a tropical rain forest is hot and steamy. People travel to the seaside to enjoy the "salt air." Visitors to the Smoky Mountains in Tennessee view the bluish hazy air. On a cold night in the Arctic the air feels particularly dry and "pure."

Actually it is hard to say what pure air really is. Air is not a single chemical compound such as water. Natural air is composed of a variety of gases mixed with water and solid particles of dust, pollen, sand, and other materials. Dry air is roughly 78 percent nitrogen, 21 percent oxygen, and 1 percent other gases. A more detailed breakdown is given in Table 12–1. Most samples of natural air contain some water vapor as well. In a hot, steamy, jungle, air may contain 5 percent water vapor, while in a dry desert or a cold polar region there may be almost none at all.

If you sit in a house on a sunny day you may see a sunbeam passing through the window. Look at right angles to the sunbeam and you will see tiny specks of dust suspended in the air. There are many sources of airborne dust. Sand, dirt, pollen, bits of cloth, hair, and skin are all suspended in air. There are living particles as well, such as bacteria and viruses. All these materials are natural components of air.

243

(Drawing from O'Brian; © 1972 *The New Yorker Magazine, Inc.*)

12.3
SOURCES OF AIR POLLUTION

Some harmful substances are introduced to the atmosphere by natural processes. Volcanoes erupt and discharge great masses of dust and sulfurous gases. Forest fires release huge quantities of gases, smoke, and soot. Therefore, there were pollutants on Earth even before modern technology. However, these have generally been isolated events.

Of course, massive pollution from industrial activities is a recent development. Furthermore, the resulting con-

TABLE 12-1. GASEOUS COMPOSITION OF NATURAL DRY AIR

	Gas	Concentration (By Volume) *percent*
"Pure" Air	Nitrogen, N_2	78.09
	Oxygen, O_2	20.94
	Inert gases, mostly argon, (9300 ppm) with much smaller concentrations of neon (18 ppm), helium (5 ppm), krypton and xenon (1 ppm each)	0.93
	Carbon dioxide, CO_2	0.03
	Methane, CH_4, a natural part of the carbon cycle of the biosphere; therefore, not a pollutant although sometimes confused with other hydrocarbons in estimating total pollution	0.0001
	Hydrogen H_2	0.00005
Natural Pollutants	Oxides of nitrogen, mostly N_2O (0.5 ppm) and NO_2 (0.02 ppm) both produced by solar radiation and by lightning	0.00005
	Carbon monoxide, CO, from oxidation of methane and other natural sources	0.00003
	Ozone, O_3, produced by solar radiation and by lightning	0.000002

taminants are likely to be emitted to the air in regions where many people live. Therefore, the effects, even if small on a global scale, may be locally very severe.

What are the human sources of air pollution? The following paragraphs will summarize the major categories.

Stationary Combustion Sources

Certainly the burning of fuel for heat is the oldest form of "civilized" air pollution. The blackened walls of caves in ancient dwellings tell us that even primitive peoples must have breathed polluted air in their homes. Since the Industrial Revolution, people have burned coal, natural gas, and petroleum as sources for heat and work.

Coal is largely carbon, which, when it burns completely, produces **carbon dioxide**. Petroleum consists mainly of compounds of hydrogen and carbon,

Forest fires are a source of natural air pollution. (Courtesy of U.S. Forest Service)

FIGURE 12-1. Approximate gaseous composition of natural dry air.

other gases
1%
oxygen
21%
nitrogen
78%

Electric power plants such as this one near Farmington, New Mexico, are often significant sources of air pollution.

which, when they burn completely, form **carbon dioxide** and **water**. Carbon dioxide and water do not pollute the air. Therefore, under "ideal" circumstances, the simple burning of fuels would not cause pollution. But the situation is never ideal. Fossil fuels contain other chemicals besides carbon and hydrogen, and the combustion is never complete. Coal is always mixed with small quantities of minerals from the Earth's crust. When the coal burns, some of these minerals fly out of the chimney. The soot thus produced is called **fly ash**.

Coal and oil are derived from living organisms that were trapped between layers of rock and sediment millions of years ago. Since sulfur is an essential element of life, all deposits of coal and oil contain some sulfur. When the fuel burns in air, the sulfur burns also. **Sulfur dioxide** and **sulfur trioxide** are produced. These chemicals are probably the most significant air pollutants because they are harmful to humans and are difficult to control. High sulfur dioxide concentrations have been associated with major air pollution disasters that have been responsible for numerous deaths. Sulfur trioxide reacts rapidly with atmospheric moisture to produce **sulfuric acid**.

sulfur trioxide + water → sulfuric acid

Sulfuric acid corrodes metals, destroys living tissue, and deteriorates buildings.

Nitrogen, like sulfur, is common to living tissue, and therefore is found in all fossil fuels. This nitrogen, together with a small amount of atmospheric nitrogen, combines with oxygen when coal or oil is burned. The products are mostly **nitrogen oxide** or **nitrogen dioxide**. Nitrogen dioxide is a reddish-brown gas with a strong odor. It therefore contributes to the "browning" and the smell of some polluted urban atmosphere.

We stated earlier that complete combustion of fuels produces carbon dioxide and water. Not all furnaces and engines are totally efficient, and often fuels are not burned completely. The incomplete combustion of carbon and gasoline produces a variety of pollutants. Coal generally yields considerable quantities of **carbon monoxide**, a gas that is colorless, odorless, and nonirritating, yet very poisonous. Accidental deaths occur, for example, from carbon monoxide escaping into a room from a faulty gas heater. Gasoline generates a wide variety of products of incomplete combustion. Some are gases, others are particles. Particles that consist mostly of carbon are called **soot**. Many of these substances are known carcinogens.

Transportation Systems

By the early 1900's, many industrial cities were heavily polluted. The major sources of pollution were no mystery. The burning of coal was number one. Other specific sources, such as a steel mill or a copper smelter, were readily identifiable. The major air pollutants were mixtures of soot and oxides of sulfur, together with various kinds of mineral matter that make up fly ash. When the pollution was heavy, the air was dark. Black dust collected on window sills and shirt collars, and new-fallen snow did not stay white very long.

Imagine now that your grandfather had decided to go into the exciting new business of making moving pictures. Old-time photographic film was "slow" and required lots of sunlight, so he would hardly have moved to Pittsburgh. Southern California, with its warm, sunny climate and little need for coal, was more like it. A region of Los

Angeles called Hollywood thus became the center of the movie industry. Population boomed, and after World War II, automobiles became almost as numerous as people. Then the quality of the atmosphere began to deteriorate in a strange way. Four classes of effects were noted: (a) A brownish haze settled over the city area; (b) people felt irritation in their eyes and throat; (c) various vegetable crops became damaged; and (d) the side walls of rubber tires began to crack.

The air pollution "experts" of those days were found, of course, in the Pittsburgh area. They were called in to diagnose the problem. They looked for the same sources they knew so well — especially sulfur, but the facts did not fit the theory. Then in the early 1950's, A. J. Haagen-Smit, a chemist from California, proved conclusively that the mysterious Los Angeles smog is produced when automobile exhaust is exposed to sunlight.

The general principle of Haagen-Smit's experiments is easy to understand. He piped automobile exhaust into a sealed room equipped with sun lamps. The room contained various plants and pieces of rubber. The room was also provided with little masklike

Automobile pollution along a freeway in Denver, Colorado. (Photo by Mel Schieltz; courtesy of *Rocky Mountain News*)

windows that permitted people to stick their faces in and smell the inside air. With the automobile exhaust in the room but the sun lamps off, the room smelled like automobile exhaust, and

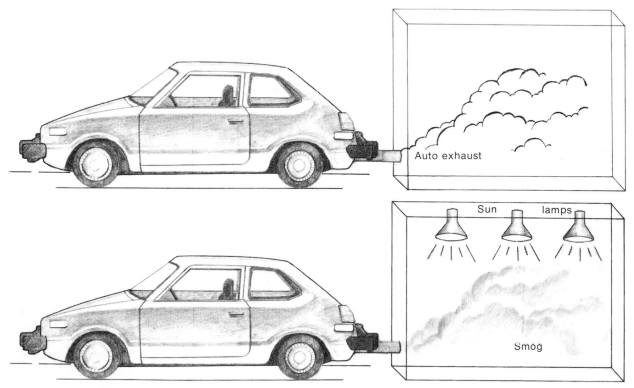

Smog is produced when automobile exhaust is exposed to sunlight.

The CF and I Steel mill in Pueblo, Colorado, contributes significantly to the air pollution of the region.

Hot air rises because it is less dense than cold air. Thus the heated air near a radiator moves toward the ceiling. Similarly, the air near the equator is heated by the sun, and it generally moves upward.

not like Los Angeles smog. But when the sun lamps were turned on, the symptoms developed. After a time, the air brought tears to the eyes, the plants showed the typical smog damage, and the pieces of rubber developed cracks. These were the crucial experiments that pointed the accusing finger at the automobile.

Chemical Manufacturing Industries

An automobile produces air pollution when it is driven. Many air pollutants are also released when cars (or any items) are manufactured. When metals are mined, the drilling, blasting, and digging operations spread dust into the air. More pollutants are released when ores are smelted to produce metal, and when metals are refined and molded into shapes. Air pollutants are released when paints, plastics, glass, and chemicals are manufactured. Oil refineries discharge unwanted smoke, soot, and gases into the air. There are many industrial sources of air pollution and all add to the problems to be discussed in the following sections.

12.4
METEOROLOGY OF AIR POLLUTION

We all know that air pollution exists. The questions that we must face are: (a) How does it affect the human con-

dition? (b) What can we do about it? Anyone who was lived in a polluted area knows that the pollution is worse on some days that it is on others. Even in Los Angeles or Tokyo, it is possible to wake up in the morning and breath relatively clean air. On other days, the air is so badly contaminated that it irritates people's eyes and causes headaches. No one can control the weather, but temperature and rainfall affect air pollution conditions considerably. Therefore, you must consider how pollutants are transported in the atmosphere, and how atmospheric conditions affect their concentrations. The science that deals with these and other atmospheric phenomena is **meteorology**.

It is generally colder on a mountain top than it is in a valley. Even near the equator, high peaks are covered with snow and ice, and mountain climbers must wear heavy clothing. If a person were to fly straight upward from sea level in a balloon under average meteorological conditions, he would find that the air temperature drops steadily with the altitude.

Suppose some foul gases rise out of a chimney into the atmosphere. Since the air from the chimney is generally

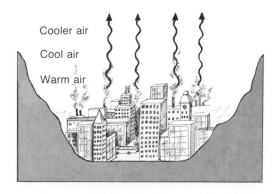

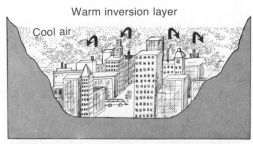

FIGURE 12–2. Atmospheric inversion. Inversions concentrate pollutants and cause the air to become unpleasant and unhealthy to breathe.

warmer than the air outside, it will rise rapidly at first. If the temperature of the upper air is much colder than the temperature of the lower air, the pollutants will continue to rise. In this way the polluted air is dispersed into the upper atmosphere and is diluted and carried away from the city.

Air pollutants do not always disperse so easily. Sometimes a layer of warm air accumulates just above a city (see Fig. 12–2). When this happens, polluted air from factories and automobiles cannot rise above the warm air layer into the upper atmosphere. Instead, the pollution remains trapped and rests low, near the ground. This condition is called **atmospheric inversion**. If an in-

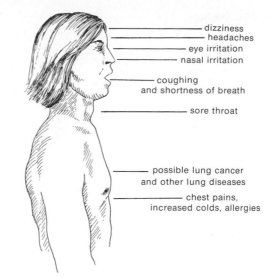

FIGURE 12–3. Effects of air pollution on the human body.

Three views of downtown Los Angeles. *Top*: a clear day. *Middle*: pollution trapped beneath an inversion layer at 75 meters. *Bottom*: pollution distribution under an inversion layer at 450 meters. (Photos from Los Angeles Air Pollution Control District)

version continues for several days, air pollutants may slowly accumulate. When this happens, pollution levels may increase to the point where the air becomes unhealthy to breathe.

12.5
EFFECTS OF AIR POLLUTION ON HUMAN HEALTH

Some environmental poisons can cause acute illness and even death. Others may be harmful, but the disease may take years or even decades to appear.

In a few situations, high air pollution levels have led to dramatic disasters. The case history at the end of this chapter describes the pollution in Donora, Pennsylvania, in 1948 that claimed 20 lives and caused over 6000 illnesses. In 1952, a series of inversions caused air pollutants to accumulate over London. Within a period of two to three weeks there were nearly 4000 more deaths in the area than would have normally been expected for that time of year. A much larger number of individuals became ill, chiefly with diseases of the heart and lungs. Such disasters are fortunately infrequent.

For many years, medical scientists have been studying the effects of long-term exposure to mildly polluted air. Very simply, they have been trying to find out whether people who live in polluted environments have a higher

"Thank you for not smoking."

Drawing by Booth; © 1977 *The New Yorker Magazine, Inc.*

rate of death and illness as compared with people who breathe clean air.

Various studies have shown that lung disease is up to four times as common in cities as in rural towns. But this fact alone does not *prove* that air pollution causes lung disease. There are a great many ways that urban and rural populations may differ. People in the country may smoke less, be in better physical condition, or live under less stress. Some of the better studies have tried to take these differences into account. The best evidence indicates that after considering all factors, people in the city have a higher probability of contracting lung disease than rural people do. It seems likely that air pollution is the major cause of this difference.

In addition, it is quite clear that even people who do not die of lung disease may suffer from discomfort or shortness of breath resulting from air pollution. Children living in highly polluted areas in the United States and Japan have been shown to have a higher rate of severe respiratory infections.

In another interesting study, scientists took several people who had a previous history of heart disease. They were taken for a ride along a freeway in Los Angeles during rush hour traffic. After a ninety minute drive the subjects were asked to perform various exercises. They all experienced pains in the chest and shortage of breath more quickly than they had before the drive. People who rode the freeways but breathed purified compressed air did much better on the exercise tests. These studies show conclusively that polluted urban air is directly harmful to the health of people with a previous history of heart disease. Most evidence also indicates that this air is harmful to healthy people as well.

12.6
EFFECTS OF AIR POLLUTION ON PLANTS, ANIMALS, AND MATERIALS

Air pollution has caused widespread damage to trees, fruits, vegetables, and ornamental flowers. In fact, the total annual cost of plant damage in the United States has been estimated at close to

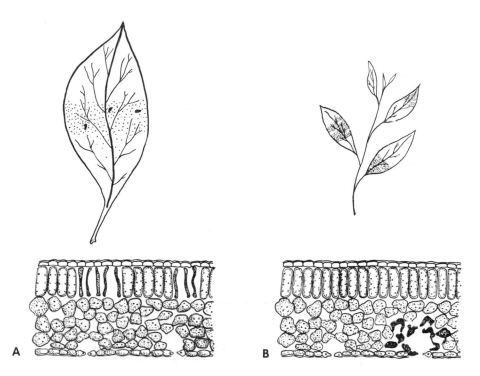

Effects of air pollution on plants. *A*, Ozone injury. Note the flecking or stippled effect on leaf. On sectioning, only the palisade layer of cells is affected. *B*, Smog-type injury. Note the change in position of the effect with the age of the leaf. On sectioning, initial collapse is in the region of a stoma. (From Stern: *Air Pollution.* New York: Academic Press, 1962)

A   B

one billion dollars. In several places in North America the pollution from large smelters has killed all brush and trees in nearby areas (Fig. 12–4).

We now know that there is a wide variety of plant damage caused by air pollutants. Photochemical smog bleaches and glazes spinach, lettuce, chard, alfalfa, tobacco, and other leafy plants. Ethylene, a compound that occurs in automobile and diesel exhaust, makes carnation petals curl inward and ruins orchids by drying and discoloring their sepals.

Countless numbers of North American livestock have also been poisoned by various air pollutants. When cattle eat compounds containing fluorine they develop abnormal growth of bones and teeth. They lose weight and often

A smelter is a place where metal ores are chemically processed to separate the metal from the rock.

Fluorine is released from a variety of industrial processes, notably many smelting operations.

FIGURE 12–4.   The Copper Basin at Copperhill, Tennessee. A luxuriant forest once covered this area until fumes from smelters killed all of the vegetation. (U.S. Forest Service photo. From Odum: *Fundamentals of Ecology*, 3rd ed. Philadelphia, W. B. Saunders Co., 1971)

FIGURE 12–5.   A cow afflicted with fluorosis.

become lame (Fig. 12–5). Arsenic poisoning, which is less common, has been transmitted by contaminated gases occurring near smelters.

Air pollutants also cause extensive damage to materials. Sulfides blacken paint so that people who live in polluted environments find that they must repaint their houses fairly frequently. Ozone, another air pollutant, causes natural rubber to crack. Recall that sulfur trioxide, an air pollutant, reacts with water and air to produce sulfuric acid. This acid corrodes metal and certain stones. In Athens, Greece, recent air pollution has caused widespread erosion of ancient carvings. Government officials have removed many stat-

Old post office building being cleaned in St. Louis, Missouri, 1963. (Photo by H. Neff Jenkins. From Stern: *Air Pollution*, 2nd ed. New York, Academic Press, 1968)

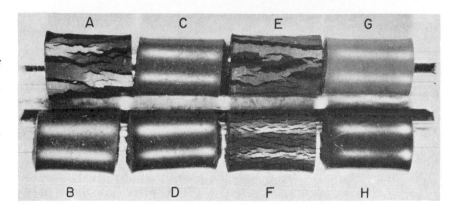

Effect of ozone exposure on samples of various kinds of rubber. *A*, GR-S; *B*, Butyl; *C* and *D*, Neoprene; *E*, "Buna-N"; *F*, Natural rubber; *G*, Silicone; *H*, "Hypalon." (Photo courtesy of F. H. Winslow, Bell Telephone Laboratories. From Stern: *Air Pollution*, 2nd ed. New York, Academic Press, 1968)

ues from their outdoor settings and sealed them in glass cases in museums. In the United States the cost of deterioration of buildings and materials from air pollution is estimated at several billion dollars per year.

12.7
AIR POLLUTION AND CLIMATE

There has been serious concern in recent years that air pollution may affect global climate. The climate at any place on Earth results from a balance of many factors. Only a small portion of the sunlight that reaches the upper atmosphere strikes the Earth directly. Most is absorbed or reflected by clouds, gases, and solid particles in the air. Obviously, if more sunlight is reflected, the Earth will cool. If more is absorbed, the Earth will grow warmer.

Dust in the upper atmosphere reflects light. Pollution makes dust. Therefore, one would expect that the effect of dust pollution is to cool the Earth. What evidence do we have to bear on the accuracy of this conclusion? So far, major volcanic eruptions far outstrip human activities in producing dust (but man is catching up). The most spectacular eruption in modern times was that of Krakatoa (near Java) in 1883. Its dust particles stayed in the atmosphere for five years. Summers seemed to be cooler in the Northern Hemisphere during this period, although the temperature differences were small.

What, then, of the dust injected into the atmosphere by people? As yet, com-

paratively little reaches the upper levels of the atmosphere, so there should be little global cooling. But dust particles can cause water vapor to collect into raindrops, so pollution may change rainfall patterns.

Certain molecules in the atmosphere absorb **infrared** radiation (heat). If more

Cerro Negro Volcano (Nicaragua) blanketed the countryside and the city of Leon (approximately 27 km away) with ash from October 23 to December 7, 1968. (Reprinted with permission of *Science & Public Affairs*, the Bulletin of the Atomic Scientists)

The atmosphere of the planet Venus contains large concentrations of carbon dioxide. This carbon dioxide absorbs so much heat that the average temperature of Venus is approximately 500°C.

sunlight is absorbed, the earth will grow warmer. Molecules of water, carbon dioxide, and ozone absorb infrared efficiently. When fossil fuels are burned in air, carbon dioxide is released. Our best estimate is that worldwide carbon dioxide concentrations have been increasing since the start of the Industrial Revolution. If this effect continues, it may someday cause a gradual increase in global temperature.

Dust cools the Earth. Carbon dioxide warms it. Is it possible, then, that the global temperature will remain the same? Our knowledge of climate is certainly too feeble to make any accurate predictions of this sort. Certainly, even a small change in global climate could affect the human condition significantly. If the Earth were to cool by even a few degrees, crop failures would become serious throughout the world. If a slight warming trend were to develop,

the glaciers and polar ice caps might melt and flood coastal cities.

Depletion of the Ozone Layer

The sun emits an entire spectrum of light, including visible, infrared, and ultraviolet. It is the ultraviolet (UV) radiation which, having high energy, tans our skin. Heavy doses of UV can cause burns and can increase the chances of skin cancer. Certain plants, such as tomatoes and peas, grow more slowly when they are exposed to high levels of UV light. If more of the UV light that reaches our upper atmosphere were to penetrate to the surface of the Earth, the risks of such damages would be increased.

Fortunately, most of the high-energy UV radiation is removed in the upper atmosphere by **ozone**. Therefore, the ozone layer serves as a protective blanket around the Earth. Various air pollutants can disperse into

Depletion of the ozone layer by chlorofluoromethanes.

the upper atmosphere and destroy the ozone. These pollutants include **Freon** compounds that are used as propellants in some aerosol spray cans, and nitrogen oxides produced by high flying jet aircraft. Some people claim that there is nothing to worry about. They say that the ozone layer will not be destroyed and it will continue to protect us. But others feel that pollution will destroy the ozone and upset world ecosystems. Whenever such uncertainties arise, it seems sensible to balance the potential gains from a certain activity against the risks. With' aerosol spray cans, the benefit of spray-on cosmetics or paints as compared with brush-on or roll-on products seems slight compared with the risk of global disruption. Responding to a variety of legal and social pressures, many manufacturers have recently been using other, less harmful propellants in spray cans.

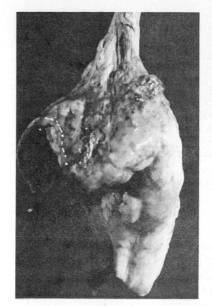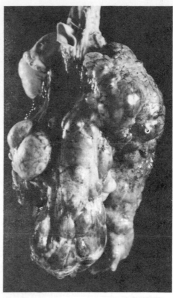

Pictured are actual human lungs. Normal lung (*left*) retains its size, shape, and elasticity. Lung with emphysema (*right*) is enlarged and misshapen. Cigarette smoking and air pollution contribute to lung diseases such as emphysema. (Photo by Glen Cuerden, Cuerden Advertising Design)

12.8
AIR POLLUTION CONTROL

Almost everyone complains at one time or another about air pollution. Obviously, no one feels that air pollution is a "good thing." You may ask yourself, "If people know that polluted air is harmful to humans and to ecosystems, why don't they force industries to clean their exhaust gases and drive automobiles that don't pollute the atmosphere? Then we could all breathe clean air."

Of course this question is extremely difficult to answer. There are many factors to consider. Some people would lose money if air pollution controls were enforced. Others would stand to gain. Many citizens are deeply concerned about the problem. Others don't care. A few years ago I met two young men who worked in an asbestos mine in Canada. They were carefree and relaxed and spent money freely. I told them that the asbestos fiber in the air near the mine would most probably lead to a fatal form of cancer within twenty years. They already know that they would probably die before they were forty-five or fifty. But their attitude was, "Where else can an unskilled person earn $15 per hour? We'd rather live it

up for twenty years and die early than live a mediocre life." Similarly, it is common knowledge that cigarettes are harmful to human health. Smokers are more likely to contract diseases of the heart and lungs than nonsmokers are. They suffer from shortages of breath and uncomfortable coughs. Yet millions of people smoke.

On the other hand, there has been slow progress toward pollution control. Thousands of scientists and engineers are working on pollution problems. In 1977, approximately $13 billion was spent in the United States on air quality control. In the following paragraphs we will recount two short stories. The first was written by my father, Amos Turk, who has been a chemist for over forty years, and has worked on air pollution control for the past thirty years. The second is an account of automobile legislation and exhaust control for the past twenty-five years.

My Father's Story

"I was a graduate student at Ohio State University during the years 1937 to 1940. I received a stipend of $50 per month (later raised to $55) from the American Petroleum Institute for work in synthesizing gallon samples of pure gasoline compounds. These samples were tested at the research laboratories of General Motors in Detroit, and now and then I would drive up to deliver the precious liquids. My professor was Albert Henne, who formerly worked with Thomas Midgley, the co-discoverer of tetraethyl lead, and those 'glorious' days were well remembered. Henne's cousin was vice-president of the Ethyl Gas Corporation and used to come around from time to time to hire chemists for his company. Henne had done considerable work on Freon compounds. One of the students in a nearby laboratory, Roy Plunkett, later went to work at the duPont laboratories, where he discovered Teflon. So here was a group of young chemists busily setting the stage for future environmental problems — smog from automobile exhausts; lead; Freon in the upper atmosphere; nonbiodegradable plastics— while outside the laboratory the economic depression was receding, and the storm of World War II was gathering. What did we think? We thought we were pretty good, or at least virtuous. What about health and safety? It was somewhat of a joke to us—we were *muy macho*. I remember one student who didn't want to work with a chemical that was known to cause blindness. He was kicked out of the lab, and that meant from the school, and I never saw him again. So, in large measure, the driving force for our work (and even for theoretical chemists) was to make *products* for people to use. This drive didn't make our chemistry incorrect, but it did influence the kind of chemistry we pursued.

"A whole generation later I went back to the General Motors Research Laboratory (now in new quarters), not to bring gasoline samples, but to lecture on diesel exhaust pollution. Hardly any of the old gang was left. I talked with young chemists about catalysts for cleaning up the pollution fragments of compounds like the ones I had made so many years ago. We discussed carbon monoxide, oxides of nitrogen, 'oxidants,' and sulfur compounds — their effects in the atmosphere and methods for controlling them.

"Of course, the General Motors Laboratory is not the national center for environmental studies, and their chemists are still very much interested in products. But no laboratory can now afford to ignore the environmental aspects of any of the products they make or modify. Does this influence the *kind* of chemistry that chemists do? Yes, it does. Ecosystems are sensitive, and environmental effects are subtle. These facts confront the chemist with problems involving trace concentrations, complex mixtures, delayed reactions, new pathways (often involving living organisms), and very large reacting systems, such as the Los Angeles Basin or the Ruhr Valley.

My professional colleagues now deal with problems such as analyzing and purifying air in a work space or in a community, or studying the complex chemical systems by which insects communicate with each other for sex attraction or for locating food sources. These chemists, too, feel they are doing the right thing, and, so far as I can tell, they are."

Automobile Pollution Control

Throughout the 1950's, automobile traffic and air pollution both increased steadily. By the early 1960's, it had become generally recognized that the automobile was a major source of air pollution. In 1963, the federal government passed the **Clean Air Act**. This legislation was aimed at reducing pollution both from automobiles and stationary sources such as power plants and chemical manufacturers. But the law was not particularly strict. Although some pollution control devices were added to cars, even those that passed the Clean Air Act emitted relatively large quantities of pollution. Seven years later, in 1970, an amendment to the Clean Air Act required that there be a 90 percent reduction of automotive pollution by 1975. To meet these strict standards, manufacturers would have to redesign certain engine components. Also, catalytic mufflers would be used. The automobile industry claimed that it was technically impossible to meet these demands in the time period allowed. Since the law was beyond the realm

---

Tetraethyl lead is an organic lead compound that is added to gasoline to improve engine performance. However, the lead is released into the atmosphere and causes pollution problems. Therefore, all automobiles manufactured after 1974 are required to operate on unleaded gasoline. Large quantities of leaded gasoline are still used for older cars.

A catalyst is a compound that speeds up a reaction, such as the oxidation of a chemical pollutant. A catalytic muffler destroys pollutants in the exhaust system of a car.

of reason, they argued that it could not be enforced. The president of General Motors said, "They can close the plants, they can put somebody in jail — maybe me. But we can't meet the control standards by 1975.*"

Congress agreed to allow the industries more time for development. Several rounds of arguments and delays occurred, until the deadline was moved to the mid-1980's.

Would it have been technically possible to develop and sell a clean air car by 1975? There are wide differences of opinion. As stated above, automobile executives said no, it would be absolutely impossible. Many scientists disagreed. They argued that engineers for automobile racing teams can make tremendous improvements in engine design within a few years. Drastic design changes cost money, but they can be achieved.

However, the major decisions for or against the provisions of the Clean Air Act were not made by scientists and engineers. The laws were debated by lobbyists, business people, and legislators. The automobile manufacturers

---

*New York Times Magazine, November 21, 1976, page 126.

Drawing by Sidney Harris.

The United States Senate. It is often difficult for legislators to evaluate opposing technical opinions. (Photo courtesy of the U.S. Senate Historical Society)

argued that: (a) it would be so expensive to build clean air cars by 1975 that car prices would increase significantly. Then people would buy fewer cars and automobile workers would lose their jobs. A national recession would result. (b) Air pollution control devices reduce gas mileage. Therefore, the energy crisis would worsen.

Both these arguments are subject to significant debate. For example, automobile industry spokesmen stated that it would cost approximately $1000 over the life of a car to reduce pollutants to the 90 percent level. Independent scientists concluded that the cost would be between $200 and $500. Furthermore, some engineers also claimed that truly efficient pollution control would require total new engine designs. These new engines would get better gas mileage and reduce energy consumption.

Congressmen aren't automotive engineers. It is extremely difficult for a nonscientist to read opposing technical opinions and decide which are correct. So the emotional statement "Clean air will cost the public JOBS and ENERGY" swayed the legislators.

In the meantime, significant gains toward clean air are being made. In 1960, an average new car produced 130 grams of carbon monoxide per mile. In 1978, the quantity was reduced to 15 grams. Oxides of nitrogen have also been reduced, but not quite so drastically. Development toward clean air is slower than some had hoped it would be, but progress is being made.

12.9
CASE HISTORY: THE DONORA SMOG EPISODE*

Donora is an industrial town of about 14,000 people, located about 48

---

*This section describes the situation in 1948, not at present.

Region of Donora, Pennsylvania.

Donora, as the fog thickens.

km (30 miles) south of Pittsburgh, Pennsylvania. In 1948, there was a large steel-and-wire plant and a zinc-and-sulfuric acid plant located in the town. Taken together, these two plants extended for some 5 km (3 miles) along the Monongahela River that runs through Donora. Across the river a series of hills rise sharply above the city. These hills tower far above the factory chimneys. As a result, any atmospheric inversion threatens to convert Donora into a basin that collects pollutants from the stacks along the river bank. Anyone familiar with the hill country of Pennsylvania and West Virginia knows that foggy days are common there, especially in the fall. When pollutants are trapped in the valleys by fogs and inversions, they blacken houses and kill trees.

One such polluted fog settled over the area during the last week of October, 1948. On October 26, the air seemed somewhat heavy and motionless, but the townspeople had seen many such fogs before. By the next day, however, there was such a dead calm that it was considered unusual, even for Donora. Streams of sooty gas from locomotives did not even rise, but hung motionless in the air. The visibility was so poor that even natives of the area became lost. Such dense fog is much worse than mere

darkness, which can be pierced by a flashlight beam. During such episodes, a driver cannot see the side of the road, nor even the white line that marks the center. If he leaves his car to explore the immediate area he may become disoriented and lost. Drivers were unable to find their way back to their cars even if they were only a few meters away.

The smog continued through Thursday, when it seemed to become thicker. People could smell the pollution, but the chemical "sensations" went well beyond the sense of smell. There was a distinct tickling, scratching, irritation, and even taste in the nose and throat. Sulfur dioxide, particularly, is a gas that gives a bitter-sweet taste to the back of the tongue. The fog had piled up in height as well, and the mills, except for the tops of their smokestacks, had become invisible.

Some illnesses had begun, rather gradually, on Tuesday. Their number had increased on Wednesday and Thursday, but they were yet too scattered to alarm the community. On Friday, however, the rate of illness soared. Altogether nearly half of the population of Donora developed some symptoms of illness before the episode was over. The symptoms included smarting of the eyes, nasal discharge, constriction in the throat, nausea, vomiting, tightening of the chest, cough, and shortness of breath.

Donora Calendar, 1948

SUN	MON	TUES	WED	THURS	FRI	SAT
Oct. 24	Oct. 25	Oct. 26 Fog settles in	Oct. 27 Fog thickens 810 illnesses	Oct. 28 Strong chemical irritations in fog	Oct. 29 Fog very thick 4700 illnesses	Oct. 30 Fog very thick 400 illnesses 17 deaths
Oct. 31 Rain starts in the afternoon 2 deaths	Nov. 1 Rains all day. Smog clears	Nov. 2	Nov. 3	Nov. 4	Nov. 5	Nov. 6
Nov. 7	Nov. 8 1 more death					

The first death came to a retired steelworker, Ivan Ceh, at 1:30 Saturday morning. The second occurred an hour later. By 10:00 a.m. eleven bodies lay in the undertaker's. Knowledge of the extent of the disaster gradually spread through the town. It was a long time before the citizens of Donora, accustomed as they were to smog, began to realize that a real tragedy was at hand. Help then began to come in from neighboring towns and hospitals (there was none in Donora) and an emergency-aid station was in operation by Saturday night. The death toll that day reached 17. Two more people died on Sunday, and at six o'clock on Sunday morning the factories finally began to shut down. By afternoon it started to drizzle. It really rained Sunday night and Monday, and then it was all over, except for the lingering illnesses and one more death, which occurred a week later.

There are two sequels to the Donora episode. One involved Donora itself and its people. The other involves all of us, everywhere.

After the killing smog lifted, teams of investigators from the U. S. Public Health Service and the Pennsylvania Department of Health came to Donora. They studied the medical records, the sources of the pollutants, and the local weather patterns. They found that of the nearly 6000 people who became ill during the air pollution episode, most did so because of severe irritation of the throat and lungs. Those who became sickest tended to be older than sixty and had pre-existing diseases of the heart and/or lungs. The investigators determined that the pollutants were emitted from the various manufacturing operations in the town. The meteorological study confirmed what everyone already knew, namely that "a definite relationship . . . existed between the concentration of contaminants and the atmospheric inversion."

Another important conclusion resulted from the Donora study. It appeared that pollutants from the factories reacted together in the atmosphere to produce poisons that were more deadly than the compounds originally released from the smokestacks. These "secondary effects" are often difficult to predict.

There have been deaths caused by air pollution before and since. But in the United States, Donora was a landmark with respect to public attitudes toward air pollution. The incident shocked people into the realization that: (a) air pollution can be deadly; (b) chemicals may react in the atmosphere to produce new compounds that are more poisonous than the original pollutant; (c) factories should be shut down during air pollution episodes before deaths occur.

CHAPTER SUMMARY

12.2 THE ATMOSPHERE

The gaseous blanket that surrounds the Earth is called the **atmosphere**. It is made up of particles and gases. Dry air contains roughly 78 percent nitrogen, 21 percent oxygen, and 1 percent other gases. Natural air also contains water and various gaseous and solid air pollutants.

12.3 SOURCES OF AIR POLLUTANTS

Major source categories are: (a) stationary combustion sources (such as power plants), which emit fly ash, sulfur dioxide, sulfur trioxide, nitrogen oxide, nitrogen dioxide, carbon monoxide, and carbon dioxide. (b) Mobile combustion sources (mainly automobiles), which emit pollutants that undergo chemical reactions in sunlight to produce **photochemical smog**; and (c) manufacturing industries, which generate a great variety of air pollutants.

12.4 METEOROLOGY OF AIR POLLUTION

An **atmospheric inversion** occurs when a layer of warm air accumulates above a layer of colder air. Pollutants cannot rise into the upper warm air layer, and so they become concentrated near the ground level.

12.5 EFFECTS OF AIR POLLUTION ON HUMAN HEALTH

High air pollution levels can cause acute illness or death within a few days. Lower concentrations increase the frequency of respiratory diseases. They also probably contribute to heart disease and cancer.

12.6 EFFECTS OF AIR POLLUTION ON PLANTS, ANIMALS, AND MATERIALS

Air pollution may kill or deform plants, injure animals, or erode buildings and other materials.

12.7 AIR POLLUTION AND CLIMATE

The possible climatic effects of air pollution are: (a) radiation may be reflected by airborne dust, thereby cooling the atmosphere; (b) radiation may be absorbed by moisture, or carbon dioxide, thereby warming the atmosphere; (c) airborne particles may change patterns of precipitation; (d) the ozone layer may be depleted by Freon and similar compounds and by jet exhaust.

12.8 AIR POLLUTION CONTROL

The Clean Air Act of 1963 set regulations for control of air pollution. Later, even stricter controls were set. It has been necessary to postpone some regulations to give industry more time to be able to conform to them. In recent years air pollution emissions have been reduced, but the process is slow.

KEY WORDS

Atmosphere
Air
Fly ash
Sulfur dioxide
Sulfur trioxide
Nitrogen oxide
Nitrogen dioxide

Carbon monoxide
Soot
Photochemical smog
Meteorology
Atmospheric inversion
Ozone
Freon
Clean Air Act

TAKE-HOME EXPERIMENTS

1. **Dustfall measurement of particulate air pollution**. Get about three half-gallon wide-mouthed jars, such as restaurant-size mayonnaise jars, and wash them out. If you live in a hot, dry climate, fill the jars about one-fourth full of distilled water. In winter or rainy season, fill them only to a height of about ½ to 1 inch. If you expect freezing weather, use a 50–50 mixture of water and rubbing alcohol. (Distilled water may be available from your school, automobile service station, or drugstore. "Deionized" water will also do. You may use rain water if you filter it first through a coffee filter or a paper towel.) Set the jars in *open* areas where you wish to measure dustfall, not under a tree or any part of a building. The jars should be elevated, preferably at least six feet, to avoid contamination from coarse windblown material, such as soil, that does not reflect general air pollution levels.

Leave the jars exposed for 30 days. Visit them at least once a week to replace any evaporated water by refilling the jars to the ¼-level mark. (If the water dries out completely, the test is invalid, because the wind may blow fine dusts out of the jar.) In rainy weather, check that the jar does not overflow. If the level approaches the top, stop the test.

Your next job is to evaporate all the water and collect and weigh the residue. There are several possible procedures. One is to heat the jar directly on a hot plate or electric stove, turned on to the *lowest* setting, and using a wire gauze or other convenient spacer between the jar and the hot surface. Even so, the jar may break. A better procedure is to pour the liquid, in stages, into a smaller open dish for evaporation. In the laboratory, you would use a previously weighed evaporating dish. In the kitchen, a clean frying pan can be used. Again, evaporate *slowly*. Make sure to remove *all* the solid matter from the jar, using a small rubber kitchen scraper. Do not overheat the solid residue at the end. Finally, weigh the residue, expressing the result in milligrams (mg).

2. **Observation of smoke shade.** When fuel burns inefficiently, some of the unburned carbon particles are visible as a black or gray smoke. The density of such smoke coming from the stack of a factory or power plant can be estimated visually by comparing it with a series of printed grids. These grids, first suggested by Maximilian Ringelmann in 1898, are formed from squares of black lines on a white background, as follows:

Ringelmann No.	Black (%)	White (%)
1	20	80
2	40	60
3	60	40
4	80	20

Select a suitable location for observing a smoke plume from a stack. Hold the chart in front of you and view the smoke while comparing it with the chart. The light shining on the chart should be the same as that shining on the smoke. For best results, the sun should be behind you.

Match the smoke with the corresponding Ringelmann smoke grid (1, 2, 3 or 4). Record your results and the time of your observation.

Calculate "observed smoke density" as follows:

Observed smoke density (average percent) for a number of observations
$$= \text{Ringelmann number} \times 20$$

Observed smoke density (average per cent) for a number of observations

$$= \frac{\text{sum of all Ringelmann numbers} \times 20}{\text{number of observations}}$$

Check the air pollution regulations in your community and compare them with your findings.

3. **Dispersal of gases in air.** Ask someone to pour a little fragrant liquid, such as cologne water or shaving lotion, into a saucer or other shallow open dish. The dish should be covered with plastic or aluminum foil and set in one corner of a quiet room. Now sit down in some other part of the room where you can keep busy with a quiet activity, such as reading a book. Ask your friend to remove the foil from the dish, while you note the time. Now read your book until you become aware of the smell, and note the time again. Repeat the experiment under various conditions, in different locations, perhaps with a fan blowing, and with different sources of odor. Note your findings with regard to the various factors that influence the dispersal of gases in air.

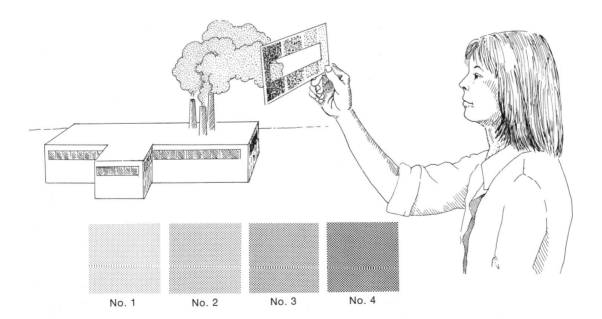

No. 1   No. 2   No. 3   No. 4

PROBLEMS

1. **The atmosphere**. List four gases and four types of solid particles that exist in natural air. Which of these can be found throughout the atmosphere and which are prevalent in some places but rare in others?

2. **Air pollutants**. List five gaseous and three particulate air pollutants, and identify their possible sources.

3. **Sources of air pollution**. Discuss the three major sources of air pollution. What types of pollutants are emitted from each source?

4. **Sources of air pollution**. Both coal and petroleum emit various pollutants when they are burned. List five pollutants that are emitted by both coal and petroleum. List one that is only released from the burning of coal.

*5. **Photochemical smog**. There are chemical pollutants in the air in most major cities that were not emitted by automobiles, electric generators, or industry. Trace the origin of some of these pollutants.

6. **Atmospheric inversion**. What is an atmospheric inversion? How can inversions affect the air we breathe?

7. **Meteorology of air pollution**. Explain why a city that is located in a valley is more likely to have serious air pollution problems than a city in an open prairie.

---

*Indicates more difficult problem.

8. **Air pollution and human health**. Discuss the available evidence linking air pollution with chronic (long-term) and acute (short-term) diseases in humans.

9. **Air pollution and human health**. Does the polluted air in most major cities generally make people feel acutely sick? Does it affect their health? Discuss.

10. **Sulfur dioxide and sulfur trioxide**. List some of the sources of sulfur dioxide and sulfur trioxide pollution. Discuss some of the harmful effects of these pollutants in the atmosphere.

11. **Climate**. How does manmade dust in the atmosphere affect climate?

*12. **Carbon dioxide and climate**. The worldwide carbon dioxide concentration has been increasing steadily since 1870. Carbon dioxide absorbs infrared (heat rays). Yet global temperatures have not increased over the past one hundred years. Discuss the apparent paradox.

13. **Ozone**. Explain the relationship between ozone in the upper atmosphere and ultraviolet light from the sun. How does ozone affect life on Earth? Discuss the relationship between Freon compounds and ozone in the upper atmosphere.

14. **Clean Air Act**. What is the Clean Air Act? How has it affected atmospheric air quality? Discuss.

QUESTIONS FOR CLASS DISCUSSION

1. We learned in Chapter 8 that solar collectors conserve our fossil fuel reserves. Solar collectors also emit no air pollutants. If people who use conventional home furnaces were charged for the economic externalities of air pollution emitted, solar collectors might become more desirable. Do you think that it would be fair to charge homeowners for the price of the air pollution that they produce? Discuss with your classmates.

2. Electric cars can be built to operate using a series of batteries. The cars emit no pollution when they are driven. However, the batteries must be recharged periodically, and the energy is derived from electric power stations. In general, electric cars are smaller, lighter, and accelerate more slowly compared with conventional automobiles. Outline some advantages and disadvantages of using electric cars.

3. The United States government has banned the use of DDT. Cigarettes have been proven to cause several serious diseases in humans. Do you think that cigarettes should be banned as well? Discuss.

BIBLIOGRAPHY

Three good texts on air pollution are:

Arthur C. Stern, Henry C. Wohlers, Richard W. Boubel, and William P. Lowry: *Fundamentals of Air Pollution*. New York, Academic Press, 1973. 492 pp.

Samuel J. Williamson: *Fundamentals of Air Pollution*. Reading, Mass., Addison-Wesley Publishing Co., 1973. 473 pp.

Samuel S. Butcher and Robert J. Charlson: *An Introduction to Air Chemistry*. New York, Academic Press, 1972. 241 pp.

Among the air pollution technical publications of the U. S. Environmental Protection Agency are two important series of documents on specific air pollutants. One series deals with "air quality criteria," which are established from our knowledge of the effects of air pollutants. The other series outlines "control techniques." The titles of the first series are:

Air Quality Criteria for Particulate Matter

Air Quality Criteria for Sulfur Oxides

Air Quality Criteria for Carbon Monoxide

Air Quality Criteria for Photochemical Oxidants

Air Quality Criteria for Hydrocarbons

Air Quality Criteria for Nitrogen Oxides

Two of the control documents are:

Control Techniques for Particulate Air Pollutants

Control Techniques for Sulfur Oxide Air Pollutants

For a very brief introductory text, refer to:

National Tuberculosis and Respiratory Disease Association: *Air Pollution Primer*. New York, 1969, 104 pp.

Various popular books that warn of the dangers of air pollution appeared in the 1960's. Some representative ones are:

L. J. Battan: *The Unclean Sky*. New York, Doubleday and Co., 1966. 141 pp.

D. E. Carr: *The Breath of Life*. New York, W. W. Norton & Co., 1965. 175 pp.

Howard R. Lewis: *With Every Breath You Take*. New York, Crown Publishers, 1965. 322 pp.

A moving human account of the Donora smog episode is given by:

Berton Roueché: *Eleven Blue Men*. Boston, Little, Brown, and Co., 1963; chapter entitled "The Fog."

Water from a "pure" mountain stream is good to drink even though it contains dissolved gases, minerals, and organic matter. Some of these added impurities are necessary additions to a person's diet.

13

WATER POLLUTION

13.1
INTRODUCTION

Travelers from North America are advised not to drink the water in the less developed nations. In many regions of the world, water purification systems are either inadequate or nonexistent. Millions of infants die at an early age from drinking polluted water, and even adults suffer from stomach illness. North Americans have long prided themselves on the fine quality of their drinking water. But in recent years, many people have questioned whether the drinking water is really so excellent. The tap water in several American cities has a strong smell and a bad taste. Local residents have turned to large scale consumption of bottled spring water. Chemists analyzing water supplies have raised disturbing questions. In 1977 and 1978, several articles published both in technical journals and local newspapers have raised fears about water supplies. For example, a headline in the *New York Times* read, "Safe U.S. Drinking Water Is No Longer A Certainty."* This chapter will review some of the problems of water pollution and purification.

13.2
THE NATURAL WATER CYCLE

Natural forces are responsible for most of the movement of water on

New York Times, April 9, 1978.

Earth. Water evaporates from the oceans or inland regions, travels with moving air currents, falls as rain, and flows back to the sea in streams and rivers. Some water seeps into the Earth and travels slowly through underground reservoirs. Large quantities are locked in glaciers and ice caps that inch slowly toward the sea. Plants absorb water and release most of it through leaf surfaces.

Now look at Figure 13–1. The total amount of water on Earth is about 1.35 billion cubic kilometers. Only about 1 percent of this vast amount, however, is in the form of inland waters such as lakes, streams, and underground storages. Vast quantities of water pass through the atmosphere, but there is only a small amount held there at any one time. Thus, water enters the atmosphere (through evaporation) and leaves it (falling as snow and rain) at approximately equal rates. This fact implies that if human activities upset the bal-

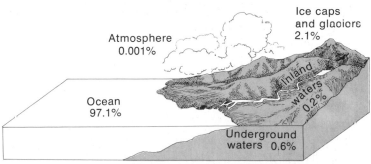

FIGURE 13–1.   Sources of water on the Earth.

ance of the small amount of water in the atmosphere, it may be possible to disrupt the entire water cycle.

As rain or snow falls to the Earth, various pollutants in the air are intercepted. By the time the water reaches the ground, therefore, it contains various impurities such as dust, carbon dioxide, microorganisms, and pollen. It may also contain traces of industrial pollutants such as sulfur dioxide, sulfuric acid, or pesticides. As the water runs over or through the soil, its load of impurities continuously increases.

Oil is one type of chemical waste, but it poses such specific problems that a separate section will be devoted to it.

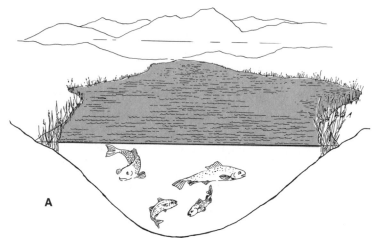

A healthy lake is one with a plentiful supply of oxygen, forming a balanced ecological system.

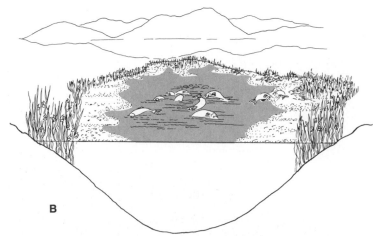

When nutrients are added, plants and microorganisms grow rapidly. They consume the available oxygen and the fish die.

FIGURE 13–2.   Pollution by nutrients.

These substances are fed into the ocean, where they accumulate. The ocean is thus the Earth's ultimate sink for water-borne impurities. The concentration of dissolved matter in the ocean rises very slowly, however, because its volume is so large.

13.3
WATER POLLUTION

I was once entertaining some guests in a small cabin in a backwoods portion in Maine. It was evening and growing dark as I went to the spring with a bucket to fill it with drinking water for the evening meal. When I returned to the lighted cabin, it was discovered that there was a small frog swimming about in the bucket. One of the guests was appalled and would not drink the water.

When is water pure and when is it not? From a chemist's point of view, pure water is a collection of water molecules, with nothing else in it. Using this definition, water that gurgles down a high mountain stream or bubbles up from a woodland spring is not pure. It contains carbon dioxide, some chemical salts that have been dissolved from rocks and minerals, microorganisms, and traces of organic matter. It may also contain small amounts of industrial chemicals from falling rain. But the water from most mountain streams and springs is nevertheless good to drink. Many of the mineral salts are necessary ingredients of human diets, and others are harmless. Our bodies are not poisoned by small quantities of organic matter. The frog in the bucket did not pollute the water appreciably. Small amounts of chemical pollutants seem to be unavoidable in our present society.

How, then, do we define water pollution? Water is polluted when it is unfit for its intended use. Drinking water is considered to be polluted when it is unfit to drink. River water is polluted when fish cannot live there. The water in a swimming pool is polluted when it is unsafe to swim in it.

Common types of water pollutants include aquatic nutrients, disease-carrying organisms, industrial wastes, and oil. Each of these will be considered in turn.

THE POLLUTION OF INLAND WATERS BY NUTRIENTS

All animals and most plants, even those that live under water, require oxygen to survive. On the surface of the Earth, oxygen is readily available. Oxygen is also dissolved in most bodies of water. But in general the supply under water is less plentiful than it is on land.

In a natural system, the growth of all organisms is controlled by the quantity of the nutrients available. The entire system is delicately balanced. If the supply of any essential ingredient is *increased* or *decreased*, the entire system may be upset. Imagine that there is a clean, cool, flowing river that supports a healthy population of trout and salmon. The fish share the stream ecosystem with populations of plankton, larger plants, microorganisms, small aquatic worms, and many other types of organisms. Now suppose that someone dumps some sewage into the stream. The sewage provides nutrients for plants and animals. Plankton and algae will grow faster. The fish that eat the plankton will be nourished as well. Yet the introduction of sewage may lead to severe ecological disruptions. In general, populations of smaller animals grow faster than populations of large animals because small animals reproduce more quickly. Thus the population of microscopic organisms, small worms, and larger "sludge" worms will expand. Algae will grow rapidly. All these organisms consume large quantities of oxygen. If the growth is rapid enough, most of the oxygen will be used up and fish may eventually suffocate and die.

It is important to understand that the sewage, by itself, does not kill fish. In fact, it nourishes them. But the sewage supports other forms of life that consume the oxygen. It is the lack of oxygen that kills fish. You don't have to read a book to be convinced that clear running water with trout and salmon is better for people than cloudy water that is slimy with plant growth and the home of sludge worms. When so many nutrients have been added to a body of water that fish die, the resulting condition is called **eutrophication.** Many lakes and rivers throughout the world are polluted in this manner.

Algae and Detergents — and How to Wash Your Clothes

Sewage is not the only material that can fertilize lakes and rivers and lead to eutrophication. Sometimes fertilizers that have been spread on farmer's fields wash into waterways. Fertilizers promote plant growth on land and in the

Persistence of soap suds in a wastewater treatment plant.

A

B

FIGURE 13–3.   The choking of waters by weeds. *A*, The dam on the White Nile at Jebel Aulia near Khartoum, Sudan. The area was clean when photographed in October 1958. *B*, The same area in October 1965, showing the accumulation of water hyacinth above the dam. (From Holm: Aquatic Weeds. *Science 166*:699–709, Nov. 7, 1969. Copyright 1969 by the American Association for the Advancement of Science)

water. If they are applied to a cabbage field, the cabbage grows well. If they are spilled into aquatic systems, algae and microorganisms grow well, leading to increased growth and depletion of oxygen. Then, as mentioned above, fish may die.

In recent years there has been considerable public discussion of laundry detergents and their role in water pollution. Many modern detergents contain phosphates. As we learned in Chapter 10, phosphates are an essential component of agricultural fertilizers. More

important, phosphates are frequently in short supply in natural waters. Without this nutrient, algae cannot grow in abundance. When phosphate detergents are discharged into waterways, they supply a needed nutrient and promote rapid growth of algae. Sometimes the results far exceed unsophisticated expectations. In many areas of the world, aquatic weeds have multiplied explosively (Fig. 13–3). They have interfered with fishing, navigation, irrigation, and the production of hydroelectric power. They have brought disease and starvation to communities that depended on these bodies of water. Water hyacinth in the Congo, Nile, and Mississippi rivers, the water fern in southern Africa, and water lettuce in Ghana are a few examples of such catastrophic infestations. People have always loved the water's edge. To destroy the quality of these limited areas of the Earth is to detract from our humanity as well as from the resources that sustain us.

How, then, should you wash your clothes? That depends on how you balance your concern for the environment with your requirements for effective cleaning. Environmental impact would be avoided if you used no soap at all, and simply washed in pure water. But your clothes wouldn't get clean. However, plain soap does not contain phosphate. A combination of soap and washing soda (also called sal soda) will clean clothes without disrupting the environment. Low phosphate or nonphosphate detergents are also acceptable alternatives. Another very effective action is simply to use less detergent. Manufacturers often recommend quantities appropriate (or even excessive) for extreme conditions. Frequently ½ of the recommended amount of detergent gives very good cleaning, and as little as ⅛ gives fairly good cleaning. As far as personal health is concerned, soap sanitizes clothes very effectively. Therefore, the use of bleaches or other antiseptic agents makes clothes cleaner, but not more sanitary. A final point is your personal standard of "effective" cleaning. The

various additives in modern detergents do contribute whitening and brightening effects, and you must therefore decide how important is is to have "dazzling" underwear.

13.5
WATER POLLUTION AND INFECTIOUS DISEASES

Sewage in an inland waterway may lead to depletion of oxygen and death of fish. But from a human standpoint, the pollution of waterways has a far more damaging effect. Sewage carries disease organisms. On a worldwide scale the pollution of water supplies is probably responsible for more human illness than any other type of environmental disruption. The diseases transmitted are chiefly due to microorganisms and parasites. Two examples will illustrate the dimensions of the problem.

Cholera is an illness carried by a bacterium. Victims of this disease contract intense diarrhea which results rapidly in massive fluid depletion. Most untreated patients die. In the past, the distribution of this disease was virtually worldwide. During the twentieth century, water purification systems in most developed nations have satisfactorily destroyed most disease bacteria, and epidemics of cholera or other similar diseases are rare. But cholera is still

"Notice how bright and white Brand X gets your clothing because of the harmful chemicals and enzymes it contains. Pure-O, on the other hand, containing no harmful ingredients, leaves your clothes lacklustre gray but protects your environment." (Drawing by Dana Fradon: © 1971 *The New Yorker Magazine, Inc.*)

In Srinagar, India, people live on houseboats along the lakes and rivers. Wastes are dumped directly into the muddy waters, and then people use the same water for washing and cooking. As a result, waterborne diseases are common.

common in Asia, and particularly in the area of the Ganges River in India. During the nine years from 1898 to 1907, about 370,000 people died from cholera. Thousands of Indians continued to die each year even up to the present. In 1947, a severe epidemic occurred in Egypt with about 21,000 cases, half of whom died. Other bacterial illnesses such as typhoid, as well as viral infections such as hepatitis, are also carried by polluted water.

Most Americans have never heard of schistosomiasis. This is actually a group of diseases caused by infection with one of three related types of worms (which worm you get depends on where in the world you live). The worm lives for part of its life in the bodies of certain snails. Then it leaves the snail and floats freely in polluted water where it infects people and other animals. Current estimates are that over 100 million people are infected with schistosomiasis. These cases are dis-

tributed throughout the African continent, parts of Asia, and in areas of Latin America. It is much more difficult to estimate the amount of suffering caused by this disease than for a disease like cholera. Cholera victims become intensely ill very quickly and generally die or recover soon theraften. Many victims of parasite infections are weakened for long periods of time but are not killed outright.

13.6
INDUSTRIAL WASTES IN WATER

Several years ago I toured a factory that recovered felt from old fur scraps. The company purchased smelly scraps of animal hides and then boiled them down in a huge vat of dilute sulfuric acid. The acid separated the valuable fur fibers from the bits of skin and flesh. The fur was then collected and used to manufacture felt products such as hats or the insoles of winter boots. The hot sulfuric acid solution of decomposed skin and animal fat was then dumped directly into the river. This small company was forced out of business about fifteen years ago by pollution control laws and a changing economy. But the pollution of waterways continues to this day. Large quantities of water pollutants are currently released into streams and rivers from paper production, food processing, chemical manufacturing, steel production, and petroleum refining

Some of these wastes are known to be poisonous. The effects of others are unknown. Some have existed in drinking water since ancient times. Many are quite recent, and new types of wastes continue to appear as new technology develops.

Many large American cities obtain their drinking water from nearby rivers. Consider the city of New Orleans. New Orleans is situated along the banks of the Mississippi River, a few kilometers from its mouth. Many thousands of factories lie along the river north of the city. Tons of foul wastes are discharged daily into the waters of the Mississippi. Some of this water is then withdrawn, purified, and used for drinking in New Orleans. If filthy, polluted water is

A fish kill caused by water pollution.

Industrial wastes in water.

completely purified, it is perfectly good to drink. The problem is that most purification processes aren't complete. To be sure, the disease-carrying bacteria are killed, and some chemical pollutants are removed — but not all of them. Heavy metals, pesticide residues, and small concentrations of hundreds or thousands of industrial chemicals are found in the drinking water of New Orleans. Many of these chemical pollutants are suspected carcinogens. For example, benzene is a common organic chemical used in a great many industrial processes. Some of this material is spilled into nearly all large rivers in North America. Benzene is known to induce cancer in mice and rats and is suspected to be carcinogenic in humans. It exists in the drinking water of New Orleans and many other American cities.

Many heavy metals are also found in drinking water. Two of the oldest water pollutants are lead and arsenic. Both of these metals are cumulative poisons. That means that they aren't washed out of your system. If a person ingests a little each day, the concentrations in the body slowly accumulate. If lead or arsenic is continuously present in drinking water, it may eventually lead to illness or even death.

Big waves are rolling over the wrecked tanker Amoco Cadiz while large amounts of oil are still pouring out. (Wide World Photos)

In January, 1978, the United States Environmental Protection Agency ordered that all major municipalities must start construction of more sophisticated water purification systems to remove chemical pollutants. Critics of the program argue that no one has really proved that the chemical pollutants in water poison people directly or cause cancer in later years. Therefore, it is said to be unreasonable to build these expensive treatment plants. But supporters of the program argue that the expense really isn't really that great. The new systems will cost approximately $.50 to $.75 per family per month. If the pollutants really don't cause cancer, people gain by drinking

When the Argo Merchant broke apart and sank off the coast of Nantucket in December, 1976, a detailed study revealed that 98 percent of the pollock and 64 percent of the cod eggs were destroyed. (Pollock and cod are types of fish.) Half of the zooplankton were killed in the spill area, and the total ecological costs were difficult to estimate.

better-tasting and better-smelling water. But if our present drinking water is harmful, the more complete purification will save millions of lives.

13.7
POLLUTION OF THE OCEANS

Oil

On March 16, 1978, the oil tanker Amoco Cadiz lost all her steering off the coast of Brittany in France. High winds blew the ship aground and within the next few days she gradually broke apart and spilled approximately 1.6 million barrels (220 thousand metric tons) of crude oil onto the beaches and fisheries of northern France. It is virtually impossible to estimate the damage caused by the spill. Brittany has long been a favorite vacation area. But until the oil is cleaned up, tourists will undoubtedly seek cleaner beaches. The ecological damages are even more severe. Over a million sea birds were killed within a few days of the disaster. The oil fouled their feathers and clogged their pores until the animals died. Nearly the entire oyster population was destroyed. Plankton, fish, and other sea animals were killed. Estuary systems were clogged. Moreover, oil residues are expected to remain in some tidal marshes for a decade or more.

The wreck of the Amoco Cadiz was one of the worst tanker spills in history. But it was certainly not an isolated

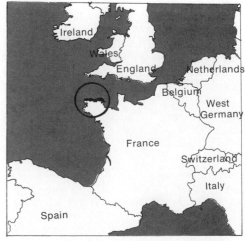

Location of the wreck of the Amoco Cadiz.

Oil-soaked gannet (a gull-like sea bird), Jones Beach, Long Island. (© Komorowski, from National Audubon Society)

event. Between 1969 and 1974, there were 500 tanker accidents that involved oil spills. More than one million metric tons of oil escaped altogether. In 1976, five major tanker accidents off the coast of the United States combined to spill another 35,000 metric tons of oil. Tanker wrecks are certainly dramatic examples of environmental pollution. But they are not the only sources of oil spills. Tanker captains often clean the holds of their ships with sea water. They then dump the oil wastes into the ocean, usually when the dark of the night matches the color of the oil. It has been estimated that 80 to 90 percent of the oil spilled in the sea comes from these small discharges, and only 10 to 20 percent from large spills.

Offshore drilling operations are also subject to accidents that result in the direct release of oil into the sea. Notable incidents of this type occurred off the coast of Santa Barbara, California, in 1969 and in the North Sea in 1977.

The total quantity of oil that finds its way into the sea each year is very large. It has been estimated that about one million metric tons of oil are spilled into the ocean each year from ships and oil drilling operations alone. But there are also many "mini-spills." Some examples are sludges from automobile crankcases that are dumped into sewers, routine oil-handling losses at seaports, leaks from pipes, and the like. Some oil aerosols also settle into the sea from the atmosphere. The grand total from all these sources is difficult to estimate, but it could well reach 10 million metric tons per year or more.

Shipwrecks have occurred ever since people first went to sea in ships. In ancient times these accidents were disastrous for the sailors, their families, and the ship owners. But today tanker wrecks and discharges are threatening the very life of the sea. Can anything be done? The sea has always been an international domain. No nation owns it, no nation can impose laws concerning it. Therefore, individual shipping com-

Cleaning oil soaked beaches in Northern France after the wreck of the Amoco Cadiz. (Wide World Photos)

panies can make decisions concerning the seaworthiness of their vessels. Often these decisions are based more on profit than on safety. For example, it is possible to build a ship with two rudders, two propellers, and two independent steering mechanisms. Then if a steering gear breaks, the ship will not flounder helplessly and be smashed to bits on the shore. But such careful design is expensive. A ship owner can realize higher profits if ships are built cheaply without such safety devices.

Governments cannot regulate foreign ships when they are far out to sea. But they can impose laws on vessels that sail into port. After the disastrous winter of 1976 when five tankers sank off the coast of the United States, many lawmakers suggested strict controls for all tankers landing in the country. But strict laws have not been passed. Meanwhile, oil continues to accumulate in the oceans. Ship captains report sighting large oil slicks daily in the North Atlantic Ocean.

Other Chemical Wastes

How does one dispose of highly poisonous chemical wastes, such as by-products from chemical manufactur-ing, chemical warfare agents, and pesticide residues? There is no easy answer. It is cheap and therefore tempting to seal such material in a drum and dump it in the sea. But drums rust, and outbound freighters do not always wait to unload until they reach the waters above the sea's depths. As a result, many such drums are found in the fisheries on continental shelves or are even washed ashore. It is estimated that tens of thousands of such drums have been dropped into the sea.

Of course, all the river pollutants enter the same sink: the world ocean. The organic nutrients are recycled in the aqueous food web. But the chemical wastes from factories and the seepages from mines are all carried by the streams and rivers of the world into the sea.

And where do the air pollutants go? Airborne lead and other metals from automobile exhaust, mercury vapor, and the fine particles of agricultural sprays ride the winds and fall into the ocean.

Is There an Overall Threat to Life in the Sea?

In ocean regions near large cities such as New York, pollution has killed most marine life in wide areas. These places have come to be known as "dead seas." Is it possible that the entire ocean may die?

It is, in general, very difficult to predict how complex ecosystems react to environmental pollution. Sometimes natural systems are amazingly stable and seem to be barely affected by pollution. Thousands of tons of crude oil have been dumped in mid-ocean and in a few months the residues seem to disappear quietly. In other situations ecosystems seem to be extremely fragile. Scientists are not sure what goes out of adjustment. But sometimes small concentrations of pollutants disturb entire ecosystems.

The poisons that accumulate in the oceans have no other place to go. The Earth's ocean is their ultimate resting place. Complex biological systems are usually able to adapt to changing conditions. But no one can guarantee that life in the sea will continue to survive the present contamination from chemical wastes.

Pollution may destroy fisheries and a valuable food supply for humans. Even greater disasters can be imagined. The Earth's oxygen is continuously replenished by photosynthesis. A large portion of that activity is carried out by the vegetation in the oceans. Some investigators have cautioned that the destruction of the phytoplankton might seriously reduce the oxygen content of the atmosphere.

13.8
ECONOMICS, SOCIAL CHOICES, AND POLITICS OF WATER POLLUTION CONTROL

Waste water systems in North America seem to revolve around the concept of "pass the pollution on downstream." Domestic wastes from toilets, bathtubs, and sinks are generally piped to a sewage treatment center. Here the dirty water is partially purified and then generally discharged into some convenient river. Certainly, the water flowing out of sewage treatment plants is not fit to drink. Thus, the river becomes polluted. Additional pollution originates from factories that dump wastes directly into streams and rivers. In many cases, the river is then used as a source of drinking water for people living downstream. Therefore, these communities must spend money to purify the water polluted by the people living upriver. On a national scale, we are discharging our wastes into the public waters and then spending billions of dollars to restore the water to a quality that is fit for drinking. In 1972, the Federal Water Pollution Control Act was passed in the United States. This law stated that:

(a) Whenever possible, water quality should be clean enough for swimming, recreation, and healthy growth of fish and wildlife by July, 1983.

(b) By 1983 there must be zero discharge of industrial and domestic wastes into US waters.

Industry responded strongly to the 1972 law. An early response was, "I flatly predict that zero discharge can't be done, won't be done, and that the people of American won't want it to be

Waste treatment plant, Charleston, West Virginia. Grit basin: GB, Primary clarifier: PC, Sludge pumps: SP, Sludge thickener: ST, Vacuum filters: VF, Chlorinators: C, Aero accelerator: AA, Aeration basin: AB, Secondary clarifier: SC, Pump station: PS. (Photo courtesy Union Carbide)

done because they won't want to pay the costs."*

Congress later passed a law postponing the 1983 deadline for one year. It now seems unlikely that the goal of zero discharge will be realized by 1985. One of the problems is that the quantity of water is so great. In the United States, cities, towns, agriculture, and industry use about 1½ trillion liters of water per day. If all this liquid were to be purified to the highest standards for drinking, the costs would, indeed, be tremendous. But there are many ways to use less water. For example, think of water consumption in the home. Considerable quantities could be saved simply by wasting less. Or, water could be reused. Under present prac-

*John T. Connor, Chairman of Allied Chemical Company, in an address to the Synthetic Organic Chemical Manufacturers Association in New York City.

A

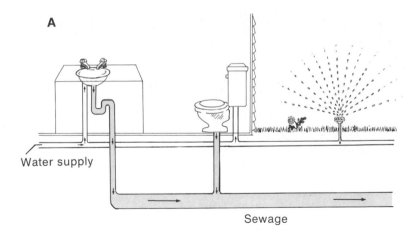

Water supply

Sewage

FIGURE 13–4.   Water use. *A*, Present, wasteful system — pure water is used for sinks, toilets, and lawns. *B*, More conservative system — wastewater from sinks is stored and used for toilets and lawn irrigation.

B

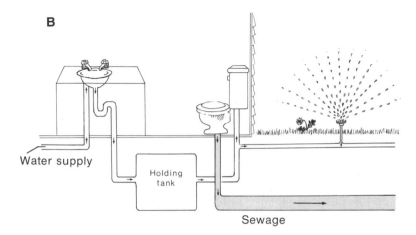

Water supply

Holding
tank

Sewage

The Metropolitan Sanitary District of Chicago transports its sewage and sludge by barge and piepeline (*left*) to a 15,000 acre (6000 hectare) site in Fulton County, Illinois, where it is spread on agricultural areas that had been left in poor condition by strip-mining (*right*).

Lewis Carroll's Mad Hatter, probably inspired by the syndrome of mercury poisoning in the British hatting industry.

tices, pure, clean, drinking water is used to flush toilets and irrigate lawns. If waste water from bathtubs and sinks were recycled through toilets and outdoor sprinklers, there would be considerably less demand on the purification systems. As another possibility, waterless toilets have been developed and could easily be used. Some of these burn wastes, others compost them for use as fertilizer.

It is also possible to treat the sewage and pipe it to farms and woodlands where it may be used as fertilizer. Large-scale use of sewage as fertilizer is practiced in locations in the Southwest and Midwest United States.

13.9
CASE HISTORY: MERCURY

Mercury has always been regarded with fascination and alarm. It is the only metal that is liquid at ordinary temperatures, and it is fun to play with. (But don't do it. Its vapor is poisonous, and at high temperatures it can vaporize rapidly enough to be deadly.) Some of its compounds have been used as agents of murder and suicide.

The mining of mercury has long been known to be hazardous to miners. Until the early part of the present century,

mercury was used in the manufacture of felt hats. The workers in hat factories suffered from tremors (the "hatter's shakes") and loss of hair and teeth. Many eventually died from the effects of the poison. Lewis Carroll's Mad Hatter was probably inspired by this industrial disease.

Until very recently, however, mercury was not considered a dangerous water pollutant. Although mercury is widely distributed on Earth, it generally occurs in only very small concentrations. Natural waters typically contain only a few parts per billion of mercury. Metallic mercury itself is poisonous in gaseous (vapor) form. But it is not particularly poisonous when taken by mouth as a liquid. Mercury has been used as a component of dental fillings without harmful side effects.

In the early 1950's, a plastics company operated near the shores of Minamata Bay in Japan. Large quantities of waste mercury compounds were dumped into the bay. At first no one was concerned, because, as mentioned above, mercury is not generally considered to be a water pollutant. But the

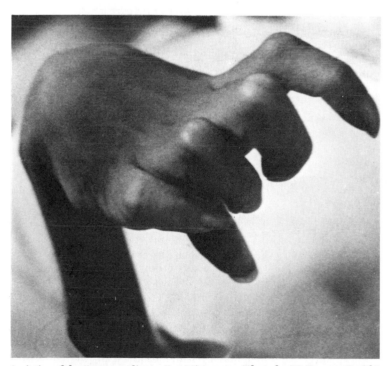

A victim of the "strange disease" at Minamata. (Photo by W. Eugene Smith)

Minamata: The factory, the dump-way, the bay, and on to the sea. (Photo by
W. Eugene Smith)

fishermen, their families, and their household cats that lived near Minamata Bay all became stricken with a mysterious disease. This disease weakened their muscles, impaired their vision, lead to mental retardation, and sometimes resulted in paralysis and death. After a thorough study, scientists unraveled the history of this disease. The mercury from the plastics factory settled into the mud on the bottom of the bay. Here certain species of bacteria ingested the mercury and converted it to organic mercury compounds. The organic mercury compounds are highly poisonous. Bacteria were eaten by larger microorganisms. These in turn were consumed by small animals and then by larger fish. During the process the mercury became concentrated in the tissues of the animals. In the next step the fish were eaten by humans and the humans were poisoned by the organic mercury compounds.

This case history reaffirms an important lesson. Often the environment has an amazing ability to cleanse and purify itself. Many poisons are discharged into the air, soil and water, and most slowly decompose and seem to disappear. But sometimes environmental poisons become concentrated and cause much more harm than would have been imagined.

CHAPTER SUMMARY

13.1 INTRODUCTION

There have recently been doubts about the quality of the drinking water in North America.

13.2 THE NATURAL WATER CYCLE

Water leaves the atmosphere as rain or snow, moves on the Earth in the form of ice or liquid water, and returns to the atmosphere by evaporation. At any one time, only a small fraction of the Earth's water is in the atmosphere; most of it is in the oceans, which serve as the ultimate sink for waterborne impurities.

13.3 WATER POLLUTION

The water in all natural lakes, streams, and springs contains many impurities such as aquatic organisms and their wastes, pollen, dust, dissolved salts, and gases. Water is considered to be polluted when it is unfit for its intended use.

13.4 THE POLLUTION OF INLAND WATERS BY NUTRIENTS

The important difference between life in water and on land is that oxygen is much less plentiful in water. When nutrients are added to water, small organisms reproduce very quickly and consume most of the oxygen. The fish suffocate and die. When so many nutrients have been added to a body of water that the fish die, the resulting condition is called **eutrophication.** Detergents, sewage, and agricultural fertilizers pollute waterways in this manner.

13.5 WATER POLLUTION AND INFECTIOUS DISEASE

Water pollution is responsible for more human illness than any other environmental influence, particularly because water carries microorganisms and parasites. Examples of illnesses from these sources are cholera and schistosomiasis.

13.6 INDUSTRIAL WASTES IN WATER

Many thousands of industrial chemicals exist in the drinking water of many large North American cities. Some of these chemicals are suspected carcinogens; others are cumulative poisons.

13.7 POLLUTION OF THE OCEANS

The most significant ocean pollutant is oil from tanker wrecks, discharges of oily ballast, leaks from offshore drilling, and runoffs from land. Oil may ruin shorelines, destroy marine life on the continental shelves, drift out to the open sea where it may be toxic to phytoplankton, or settle to the bottom. Economic factors tend to make tanker owners negligent in enforcement of safe practices. Other chemical wastes also contribute to ocean pollution. The total effect represents a threat to life in the ocean, and hence to life on Earth.

13.8 ECONOMICS, SOCIAL CHOICES, AND POLITICS OF WATER POLLUTION CONTROL

The Federal Water Pollution Control Act of 1972 proclaims national goals for clean water for the 1980's. The projected costs of meeting these goals are considered by some to be excessive. One possible remedy is to reduce the quantity of water that we use. Great savings could be realized by switching from the flush toilet to other forms that use less or no water, or by use of liquid sewage as fertilizer, or by other strategies.

KEY WORD

Eutrophication

TAKE-HOME EXPERIMENTS

1. **Dissolved solids in water.** Use a clean jar to get a sample of water you wish to test. Filter the water, using a coffee filter or paper towel. Now evaporate all the water and weigh the residue by means of the procedure described on page 262. Test the following types of water samples: rain water or distilled water; tap water, water from a natural source, such as a river, stream, or pond; sea water, if available. Add to this list any other sample you consider to be of interest.

2. **Water purification with activated carbon.** Food colors are usually available in small containers from which they can be dispensed in drops. Using such colors, prepare a set of lightly tinted solutions in ordinary drinking glasses, filled about ¾ full. Now stir a little activated carbon powder, of the type used for aquariums, into each glass. The carbon may be purchased from a pet shop, drugstore, or hobby shop. Place a saucer over each glass and allow the carbon to settle overnight. Note the effectiveness with which the colored impurities are removed.

Can you design a series of experiments to determine how much carbon is needed to remove a given amount of dye? Or to determine which dyes are easier or harder to remove'

3. **Phosphates.** Collect four liters of pond water along with a small sample of mud that lies along the bottom. (Pond water contains many small aquatic plants. If there are no convenient ponds in your area, purchase some algae or other water plants from a local fish supply store and add to conventional tap water.) Separate your sample into four equal portions and place each in a glass jar or aquarium. Leave one portion untouched, then add respectively ½ g, 1 g, and 2 g of high phosphate detergent to each of the other three. Report on the growth of the algae and other aquatic plants in each sample.

PROBLEMS

1. **Water cycle.** What forces control the movement of water through the biosphere?

2. **Water cycle.** Why is it conceivable that despite the great mass of water on Earth, efforts to control climate by such means as cloud seeding might affect the hydrological cycle?

3. **Water pollution.** List five types of impurities that may be found in "pure" natural spring water.

4. **Water pollution.** Explain how a river can be considered unpolluted even though the water may not be good to drink.

5. **Water pollution.** Explain how a nontoxic organic substance, such as chicken soup, can be a water pollutant.

6. **Water pollution.** Explain how an aquatic ecosystem can be disrupted by the addition of nutrients.

7. **Eutrophication.** What is eutrophication? Explain how it occurs and why it is hastened by the addition of inorganic matter such as phosphates.

*8. **Home laundry.** Write up a set of specific laundry instructions that outlines your personal decisions about pollution, hygiene, and washing effectiveness.

*9. **Water pollution.** A healthy person lives in harmony with bacteria in his digestive system. Why, then, should water that contains digestive bacteria be considered to be polluted?

10. **Impurities in water.** Imagine that a certain sample of water is absolutely free of all parasites and disease bacteria. Does this mean that the water is necessarily fit to drink? Discuss.

_____

*Indicates more difficult problem.

11. **Oil spills.** How does spilled oil affect ocean ecology? What types of organisms are most seriously affected?

12. **Oil spills.** Discuss some of the ecological consequences of the death of large numbers of sea birds.

13. **Oil spills.** Imagine that a large shipping company decided to build oil tankers that were carefully designed for very safe operation. List the advantages and disadvantages for the shipping company of such safe design. List the advantages and disadvantages for ocean ecosystems of such safe design.

14. **Ocean pollution.** What are the various categories of ocean pollutants? Where do they come from? Where do they go to after they reach the sea?

15. **Ocean pollution.** In its article on "Sewerage," the Eleventh Edition of the Encyclopaedia Brittanica, published in 1910, states, "Nearly every town upon the coast turns its sewage into the sea. That the sea has a purifying effect is obvious. . . . It has been urged by competent authorities that this system is not wasteful, since the organic matter forms the food of lower organisms, which in turn are devoured by fish. Thus the sea is richer, if the land is the poorer, by the adoption of this cleanly method of disposal." Was this statement wrong when it was made? Defend your answer. Comment on its appropriateness today.

*16. **Water purification.** All the water that enters a sewer must be purified before it is discharged. List five ways to reduce the amount of water entering a sewer from your home. Which of these proposals could be initiated immediately and which would involve expensive home improvements?

QUESTIONS FOR CLASS DISCUSSION

1. According to the President's Council on Environmental Quality, 40 billion dollars was spent on all types of pollution controls in 1977. This amounted to $187 per year for every person in the country. Do you feel that this amount is excessive? Too little? Discuss with your classmates.

2. Many of the Earth's ecosystems seem to be quite stable. They are not easily disrupted or, if they are, they recover readily. The world's oceans may be considered to be the most massive ecosystem of the Earth. It has an immense capacity for diluting and absorbing toxic substances. Discuss the ques-

tion of whether our concern about a threat to the oceans by human activities is exaggerated. Is international cooperation urgent? Is action needed at once, or within your lifetime, or "eventually," or not at all? Who should decide such matters?

3. Some people have suggested that North Americans place an unrealistic premium on cleanliness. If people took showers two or three times a week rather than once a day, and if clothes were laundered less often and with less detergent, then many pollution problems would be diminished. Discuss the issues of cleanliness and pollution with your classmates.

BIBLIOGRAPHY

The following texts are good sources of information on water pollution and its control:

Charles E. Warren: *Biology and Water Pollution Control.* Philadelphia, W. B. Saunders Co., 1971. 434 pp.

Metcalf & Eddy, Inc.: *Wastewater Engineering.* New York, McGraw-Hill, 1972, 782 pp.

John W. Clark, Warren Viessman, Jr., and Mark J. Hammer: *Water Supply and Pollution Control.* Scranton, Pa., International Textbook Co., 1971. 661 pp.

T. R. Camp: *Water and its Impurities.* New York, Litton Educational Publisher, Van Nostrand-Reinhold Books, 1963.

G. V. James and F. T. K. Pentelow: *Water Treatment,* 3rd Ed. London, Technical Press, 1963.

For a detailed study of eutrophication, refer to:

National Academy of Sciences: *Eutrophication: Causes, Consequences, Correctives.* Washington, D.C., National Academy of Sciences Press, 1969. 661 pp.

The following four books deal with the marine environment, although in different ways. The first is a general text. The next two analyze specific ship disasters. The last is a fascinating, well-written account of supertankers, written by an author who took a voyage on one.

National Academy of Sciences: *Beneficial Modifications of the Marine Environment.* Washington, D.C. 1972. 166 pp.

J. E. Smith (ed.): *"Torrey Canyon" Pollution and Marine Life.* London, Cambridge University Press, 1968.

Peter L. Grose and James S. Mattson (eds.): *The Argo Merchant Oil Spill.* National Oceanic and Atmospheric Administration, U.S. Dept of Commerce, 1977, 133 pp.

Noël Mostert: *Supership.* New York, Alfred A. Knopf, 1974. 332 pp.

The Minamata poisoning is graphically described in a photo-essay book by a husband-and-wife team who were very much apart of the action:

W. Eugene Smith and Aileen M. Smith: *Minamata.* New York, Holt, Rinehart and Winston, 1975. 192 pp.

A technical book on mercury contamination is:

Rolf Hartung and Bertram D. Dinman (eds.): *Environmental Mercury Contamination.* Ann Arbor, Mich., Ann Arbor Science Publishers, 1972, 349 pp.

A number of recent chemistry texts emphasize various environmental topics, including water pollution. One such book is:

John W. Moore and Elizabeth A. Moore: *Environmental Chemistry.* New York, Academic Press, 1976, 500 pp.

The following are cited as representative of popular books that deal with the crisis of water pollution:

D. E. Carr: *Death of the Sweet Waters.* New York, W. W. Norton & Co., 1966. 257 pp.

F. E. Moss: *The Water Crisis:* New York, Encyclopaedia Britannica, Praeger Publisher, 1967. 305 pp.

G. A. Nikolaieff (ed.): *The Water Crisis.* New York, H. W. Wilson Co., 1969. 192 pp.

14

SOLID WASTES

14.1
SOLID WASTE CYCLES

Today's landscape is not littered with huge mounds of dinosaur bones or ancient ferns. Waste products from living things have traditionally been recycled, and the chemicals of one organism's refuse have been reused by others. Occasionally, chemical elements are locked for long periods of time inside glaciers or geological deposits such as coal, but these deposits are eventually returned into the world's ecosystems.

Recycling is quite inefficient in modern industrial societies. An apple grown in an orchard in the State of Washington may be shipped to a city on the Atlantic seaboard. After someone eats it, the core is not left out to be consumed by decay organisms and returned to the soil. Rather, it is stored in a trash pail, picked up by a truck, and transported to a large garbage dump. Similarly, the feces of a person who has eaten the apple may ultimately be washed into a sewer, and the farmer in Washington must purchase manufactured fertilizers to grow more food.

This chapter will discuss the sources of solid waste, the extent to which they are recycled, and the problems involved in their disposal. Radioactive wastes were discussed in Chapter 9 and so will not be incuded here.

14.2
SOURCES AND QUANTITIES

Perhaps the most noteworthy characteristic of solid wastes is their variety. In our household garbage piles there are food scraps, old newspapers, discarded paper, wood, lawn trimmings, glass, cans, furnace ashes, old appliances, tires, worn-out furniture, broken

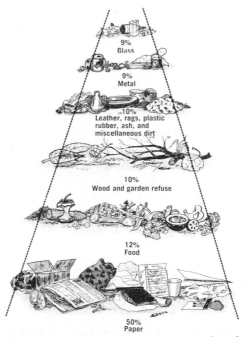

9%
Glass

9%
Metal

10%
Leather, rags, plastic rubber, ash, and miscellaneous dirt

10%
Wood and garden refuse

12%
Food

50%
Paper

FIGURE 14–1.   Composition of municipal trash in the United States.

toys, and a host of other items too numerous to mention. The total quantity of solid waste is large and increasing. In the United States municipal solid wastes averaged 1.2 kilograms per person per day in 1920. The quantity rose to 2 kilograms in 1965, and about 3 kilograms in 1975. In other words, the waste disposal system of an average city must accommodate about 85 kilograms of refuse per week for every family of four, and that is a lot of trash. In the year 1975 alone, about 70 million metric tons of municipal refuse accumulated in the United States. Residents of the United States generate more solid waste than any other people. In contrast to the 3 kilograms per person per day in the United States, residents of Australia produce 0.8 kilogram per person per day. The average resident of India produces only about 0.2 kilogram per day.

Wealth is responsible for most of this trash. Packaging materials amount to about one fifth of the municipal refuse. Over 50 percent of the cost of soft drinks is for the bottle, and almost half of the cost of many other items lies in their packages. Yet Americans have the money to buy these products. But even without wasteful packaging, Americans discard over twice as much as Australians. For example, more food is discarded, cars are junked sooner, clothes are patched less often, and tires are recapped less frequently in North America than in Oceania.

Municipal sources contribute only a fraction of the types and amounts of solid wastes discarded in the United States. Agricultural activities, for example, produce over 1.8 billion metric tons of wastes each year. About three quarters of this is manure. The balance includes forest slash from logging

Compacted means to be packed close together. In most garbage trucks large hydraulic presses pack the trash so more can be hauled into the truck. Small trash compactors for use in the home are also available.

operations, discarded fruits, slaughterhouse wastes, pesticide residues, and plant parts such as corn cobs, leaves, and stems. Mining operations produce about 1.35 billion metric tons per year. Most of this material is rock, dirt, sand, and slag that remain behind when metals are extracted from the Earth. Excess rock and dirt differ from any of the other wastes mentioned in this chapter. The problems associated with strip mining, acid mine drainage, and mine reclamation were discussed in Chapter 7.

Look about the room you are in and note the number of manufactured objects. Each different kind of object was produced by a series of industrial operations, and some solid wastes were generated at each stage of production. Whenever raw materials such as metals, fossil fuels, or agricultural products are used to manufacture airplanes, shoes, beer cans, or even balloons, solid wastes are always generated. It should not be surprising, then, that industrial solid wastes are more varied in their categories than are municipal wastes.

14.3
DISPOSAL ON LAND AND IN THE OCEAN

All too frequently garbage is simply deposited in an **open dump.** Waste is collected and, to save space and transportation costs, is compacted. The compacted waste is hauled to a suitable site and simply dumped on the ground. Organic matter rots or is consumed by insects, by rats, or if permitted, by hogs. Various salvaging operations may go on during the day. Bottles, rags, knickknacks, and especially metal scraps are collected by junk dealers or by individuals for their own use. In

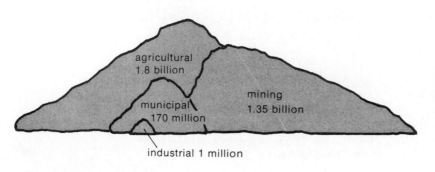

FIGURE 14–2. Sources and quantities of solid wastes in the United States (approximations expressed in metric tons per year).

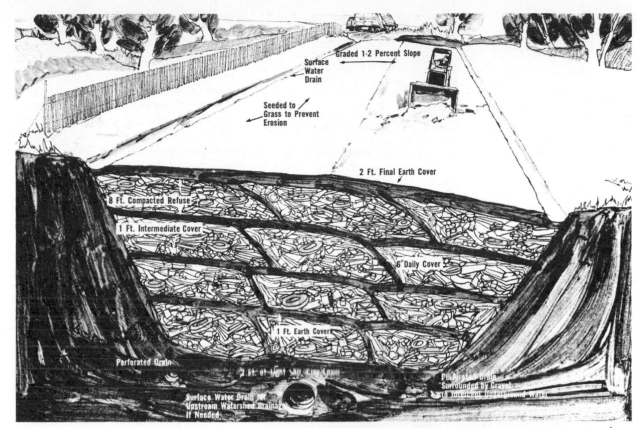

Surface Water Drain

Graded 1-2 Percent Slope

Seeded to Grass to Prevent Erosion

2 Ft. Final Earth Cover

8 Ft. Compacted Refuse

1 Ft. Intermediate Cover

6" Daily Cover

1 Ft. Earth Cover

Perforated Drain

1 Ft. of Light Silt Clay Drain

Perforated Drain Surrounded by Gravel to Intercept Underground Water

Surface Water Drain for Upstream Watershed Drainage If Needed

FIGURE 14–3. Sanitary landfill in a ravine or valley. Where the ravine is deep, refuse should be placed in lifts of 6 to 10 feet deep. Cover material may be obtained from the sides of the ravine. To minimize settlement problems, it is desirable to allow the first lift to settle about a year. This is not always necessary, however, if the refuse has been adequately compacted. Succeeding lifts are constructed by trucking refuse over the first one to the head of the ravine. Surface and groundwater pollution can be avoided by intercepting and diverting water away from the fill area through diversion trenches or pipes, or by placing a layer of highly-permeable soil beneath the refuse to intercept water before it reaches the refuse. It is important to maintain the surface of completed lifts to prevent ponding and water seepage. (Courtesy of New York State Department of Environmental Conservation)

some communities the pile is set afire in the evening to reduce the total volume and to expose the metal scraps for possible salvage. Of course, the organic decay, the burning, and the salvaging are recycling operations. However, there are serious ecological problems with the open dump. The dump is a potential source of disease. The fires are uncontrolled and therefore are always smoky and polluting. Rain erodes the dump and the polluted water flows into nearby rivers and streams; and, of course, the dumps are ugly.

Ocean dumping is practiced by many coastal cities. Barges carrying the refuse travel some distance from the harbor and dump their loads into a natural trench or canyon on the ocean floor. In this way most of the trash is removed from sight, though not from the biosphere. As would be expected, ocean dumping upsets the ecological balance of regions of the sea. Many organisms are killed outright. Although certain plankton and fish may survive in these areas, they are affected by the unusual environment. For example, flounder caught in the former New York City dump region have had an off-taste. Biologists have found old adhesive bandages and cigarette butts in fish stomachs. Therefore, it is not surprising that the flesh had a foul flavor.

The **sanitary landfill** is far less disruptive to the environment than the uncontrolled dumping on land or into the ocean. A properly engineered landfill

A botanical garden became the final layer of this completed landfill (From *Sanitary Landfill Facts.* U.S. Department of Health, Education and Welfare, PHS, 1970)

On the other hand, several serious problems are associated with sanitary landfills. First, land conversion is not always desirable. As we saw in Chapter 2, marshes and swamps house many valuable species of plants and animals. To destroy them is to destroy valuable ecosystems. Second, many large metropolitan areas have used up their available sites for landfills. They are now being forced to transport their trash further into the countryside. In these instances, transportation costs are high. Third, and perhaps most serious, such disposal represents a depletion of resources. Food wastes and sewage sludge which could be used as fertilizers are buried deep underground. Paper and wood scraps which could be recycled are lost, and nonrenewable supplies of metals are dissipated.

14.4
ENERGY FROM REFUSE

Many metropolitan areas no longer simply dump their garbage. They burn it. The process, as applied to waste disposal, is more complex than simply setting fire to a mass of garbage in an open dump. A modern incinerator unit is currently used in Montreal, Canada. The trash is burned in a carefully engineered furnace. The heat from the fire is used to boil water and produce steam. Then the steam is sold for industrial use. Thus, the trash is used as an energy source.

When the Montreal plant was first built, the money received from the sale of steam did not pay the cost of operating the incinerator. The economic problems arose because trash is not an ideal fuel. Municipal wastes contain food scraps and wet garbage that are difficult to burn efficiently. Also, the incineration of certain waste products produces acidic gases which corrode furnace walls. Particularly notorious is polyvinyl chloride (PVC), a plastic used in the manufacture of rainwear, toys, containers, garden hoses, and records. The burning of PVC produces hydrogen chloride gas. This gas reacts with water to produce a strongly corrosive liquid, hydrochloric acid. Even more threatening is the fact that some of the PVC decomposes before it burns

should be located on a site where rainwater will not flow through the trash and pollute nearby ecosystems. After waste is brought to a landfill, it is further compacted with bulldozers or other heavy machinery. Each day 15 to 30 centimeters (6 inches to 1 foot) of soil is pushed over the trash to exclude air, rodents, or vermin (see Fig. 14–3). In practice, however, the distinction between a sanitary landfill and the open dump is not always sharp. For example, a thin layer of earth may be an ineffective barrier against burrowing rats, flies, or gases evolving from decomposition.

The sanitary landfill is an effective way to dispose of trash. It does not produce much pollution, disease or unsightliness. With proper ingenuity, landfills may even reclaim spoiled land. In some cases swamps and marshes have been filled with garbage and used as building sites for apartment complexes or parks. In other cases, landfill mountains have been constructed and used as ski slopes, and in one case as a "soapbox derby" raceway.

and releases vinyl chloride, a known carcinogen.

In spite of the difficulties, incineration may become profitable in the future because:

(a) Increasingly large quantities of dry paper and cardboard have appeared in refuse, thereby increasing the fuel content of trash.

(b) The price of fuel has skyrocketed since 1973, and therefore the value of steam has also skyrocketed. On the other hand, costs of handling and burning trash have increased only moderately.

(c) The rising cost of land has made it harder to find adequate sites for landfills. If valuable land is used for dumping, or trash is hauled long distances to less expensive sites, landfills become more expensive. A high cost of discarding trash makes alternative solutions more desirable.

At the present time, many industries and municipalities across the globe are burning trash as fuel. As mentioned in Chapter 8, large scale incineration is practiced in France. In the United States 30 cities and towns are expected to be incinerating garbage by 1980 and using the heat to produce steam for the generation of electricity.

An ice cream cone is an ideal, biodegradable container.

14.5
RECYCLING — AN INTRODUCTION

Most refuse contains a wealth of raw materials that can easily be reused or recycled. There are many kinds of recycling paths. Consider, for example, the element phosphorus. Much of the phosphate ore now being mined is used as a fertilizer. The phosphorus is absorbed by plants, eaten by people, and excreted into sewage systems. In the United States most of the sewage is concentrated into sludge and dumped into the ocean or into sanitary landfills. Little is returned to the land as fertilizer. "So what," you may say, "can't we eventually mine the oceans or the old dumps?" The problem here, of course, is that we must think of the total environmental costs of our waste disposal practices. Mining old dumps is so expensive and uses so much fuel that it is impractical. The phosphorus would be recycled most efficiently if sewage sludge were used directly as a fertilizer or soil conditioner.

Think about the articles that people normally throw away. What would be the best way to minimize such waste? Obviously, there would be less junk if fewer items were discarded in the first place. That peanut butter jar in your

Much energy could be conserved if glass jars were refilled rather than discarded. In this store, cooking oil is dispensed from drums into reusable jars. (Photo by Marion Mackay)

Abandoned automobile, pre-World War II model.

garbage could be used over and over again if there were a large tub of peanut butter at the grocery store. Then you could simply take your jar to the store and refill it. The paper towel that you threw away need never have been purchased, for a cloth towel would have worked just as well. Or what about that old automobile that was carted off to the dump? If an automobile consumes too much oil and runs inefficiently, the engine can probably be rebuilt; perhaps it could be used for years to come. Similarly, when a refrigerator no longer cools efficiently, the fault generally lies with the compressor or the thermostat, not with frame or the food compartment. If new parts were installed, many valuable raw materials would be conserved.

But not all items are repairable or reusable. A car that has been in a bad accident often cannot be rebuilt. A tire can be recapped only once with safety. Week-old newspapers and spoiled meat are useless to most people. When an item cannot be used in its present condition, it must be destroyed and treated somehow to extract its useful raw materials. For example, used tires can be shredded and converted to raw rubber. Old newspapers can be repulped and converted to new paper. Spoiled meat can be rendered and converted to tallow and animal feed. Discarded metals can be melted and reused. In general, recycling conserves not only material resources but fuel reserves as well. For example, nearly twenty times as much fuel is needed to produce aluminum from virgin ore as from scrap aluminum. Over twice as much energy is needed to manufacture steel from ore as from scrap. It is also twice as costly in energy to make paper from trees as it is to recycle used paper.

In many cases, recycling operations also emit less pollution than the original processes. Significant quantities of pollutants are released when paper is manufactured from wood pulp or when metal is refined from ore. The Environmental Protection Agency recently es-

One evenin' he decided t'go out an' still hunt; try t'kill 'im a bear or somethin'. Sittin' at th' head a'th' swamp an' there's three bear come walkin' out. A small little bear in front, and they's a big he bear in th' center, an' they's a little cub behind this'um. An' he waited 'til this big bear got betwixt him an' a tree t'shoot it wi' his hog rifle so he could save his bullet — go cut it out of th' tree. And he shot this big bear.... An' he went home an' took a axe an' cut th' bullet out. He'd take it back an' remold it. Lead was hard t'get, so that's th' way they'd try t'save their bullets." (From Eliot Wigginton, Ed.: The Foxfire Book. Garden City, N.Y., Anchor Books, Doubleday & Co., 1972. 380 pp.)

"Admit it. Now that they're starting to recycle this stuff, aren't you glad I didn't throw it out?" (© 1971—reprinted by permission of *Saturday Review* and Joseph Farris)

timated that recycling all the metal and papers in municipal trash in the United States would prevent the release of over 2000 metric tons of air pollutants and 700 metric tons of water pollutants every year.

One can easily envision an ideal sequence in which durable goods are used for a long time, then repaired or patched to prolong their lives still further. Finally they could be recycled for use as raw materials. Unfortunately, no such plan has operated effectively in modern society. Complex social and economic factors have interfered.

14.6
RECYCLING TECHNIQUES

Much municipal, industrial, and agricultural trash can neither be reused nor repaired, and consequently must be reduced to raw materials suitable for remanufacture. Several techniques are available for this type of recycling.

Many materials such as metal, glass, and some plastics can be **melted,** purified, and reused.

Any material containing natural cel-

At best, recycling processes conserve both energy and materials. However, not every item can be recycled efficiently. For example, imagine that a careless person throws a soft drink bottle onto a roadway in some isolated rural region, and that it falls under the wheels of a trailer truck and is crushed. Crushed glass is a recyclable item, but the collection and concentration of all the crushed glass along rural roadways would consume so much energy as to be grossly inefficient. Therefore, recycling operations, despite their theoretical appeal, lose their effectiveness if the waste is widely dispersed.

Wastepaper cycle: Scrap paper is collected, pressed into bales, pulped into a slurry, and reformed into new paper. (Courtesy of Container Corporation)

Industrial composting operation.

lulose fiber such as wood, cloth, paper, sugar cane stalks, and marsh reeds can be beaten, **pulped,** and made into paper. Thus, for example, scrap newspaper can easily be repulped to manufacture recycled paper. Agricultural wastes can be reused as well. When sugar is extracted from cane, the remaining fibrous stalks are well suited to paper production. These stalks currently contribute 60 million metric tons of solid waste annually, and could easily be converted into paper.

If organic wastes are partially decomposed by bacteria, worms, and other living organisms, a valuable fertilizer and soil conditioner can be produced. This process is called **composting.** Almost any plant or animal matter, such as food scraps, old newspaper, straw, sawdust, leaves, or grass clippings, will form a satisfactory base for a composting operation.

When sewage sludge and animal manure are composted, large quantities of methane gas are released. Methane is an excellent fuel. Some farmers have collected methane from cow manure,

used the fuel to drive their tractors, and then recycled the compost as fertilizer. Although many such small-scale methane generators have been used throughout the world, few large-scale units are in operation.

If animal wastes such as fat, bones, feathers, or blood are cooked (**rendered**) several valuable products can be produced. These include a fatty product called tallow which is the raw material for soap, and a nonfatty product that is high in protein and can be used as an ingredient in animal feed. The raw material for a rendering plant contains wastes from farms, slaughterhouses, retail butcher shops, fish processing plants, poultry plants, and canneries. If there were no rendering plants, these wastes would impose a heavy burden on sewage treatment plants. They would add pollutants to streams and lakes and nourish disease organisms. At the rendering plant, the waste materials are sterilized and con-

TABLE 14-1. VARIOUS RECYCLING ROUTES OF SOME COMMON WASTES

Waste	Recycling Possibilities
Paper	Use the backs of business letters for scrap paper or personal stationery; lend magazines and newspapers to friends, etc. Repulp to reclaim fiber. Compost. Incinerate for heat.
Glass	Purchase drinks in deposit bottles and return them; use other bottles as storage bins in the home. Crush and remelt for glass manufacture. Crush and use as aggregate for building material or anti-skid additive for road surface.
Tires	Recap usable casings. Use for swings, crash guards, boat bumpers, etc. Shred and use for manufacture of new tires. Grind and use as additive in road construction.
Manure	Compost or spread directly on fields. Ferment to yield methane, use residue as compost. Convert to oil by chemical treatment. Treat chemically and reuse as animal feed.
Food scraps	Save for meals of left-overs. Sterilize and use as hog food. Compost. Use as culture for yeast for food production.
Slaughterhouse and butcher shop wastes	Sterilize and use as animal feed. Render. Compost.

verted to useful products such as tallow and chicken feed. But the rendering process generates odors. Although these odors can be controlled most of the time, they do get away now and then.

If people would separate their household trash into piles of aluminum, steel, paper, plastics, and food scraps, then perhaps all these materials could be recycled. But homemakers have not been willing to do this. Some people have suggested that governments pass laws requiring residents to separate their trash. But such a law could hardly be enforced. Policemen would have to examine garbage cans and try to discover who put the chicken bone in with the beer bottle, and that just wouldn't work. It is possible to separate municipal garbage automatically in large recycling plants. Several garbage separating factories have been designed. The trash is dumped into a huge bin and carried into the separator by means of a conveyor belt. Useful materials are removed and recycled. Unfortunately, although the technology is available, few automated recycling centers have been built.

14.7
THE FUTURE OF RECYCLING

Techniques are available for the recycling of most solid wastes. But the social attitude is not favorable for immediate large-scale recycling operations (see Table 14–2). In 1975, nearly 70 percent of the municipal trash in the United States was discarded in the ocean or in open dumps. Approximately 22 percent was deposited in landfills. About 8 percent was incinerated, but not all of the heat produced was used industrially. And only 1 percent was recycled.

Recycling is not popular today because, in general, it costs more to manufacture items from recycled goods than from raw materials. For example, in 1976, it cost $2.75 more to produce a metric ton of iron from scrap than from ore. Let us compare the two processes.

In an iron mine, huge power shovels dig the ore from the gound and load it into waiting freight trains. The ore is transported to a mill where it is crushed, converted to the metal, and refined. The entire process is performed using large, semi-automated machinery. Comparatively little labor is needed, but large amounts of energy are used.

In recycling operations, the scrap that is scattered about the countryside must first be collected. In many ways, the problem of scrap collection is central to the whole recycling issue. Obviously, if a person drove around the back roads of Montana looking for discarded metal cans and lost hub caps, many hours of labor and many gallons of gasoline would be consumed to collect a ton of scrap. Therefore, not all iron can be recycled practically. On the other hand, if people saved metal scraps and dropped them off at the recycling center on their way to work, profitable recycling could be realized. Similarly, it usually pays to recycle junk cars because there is a lot of iron concentrated in one item. Under the most favorable circumstances, collection of scrap requires large amounts of labor but little energy.

Since the scrap is already highly refined, comparatively little energy is needed to remelt and reuse it. The entire process uses comparatively little energy, but large amounts of labor.

TABLE 14–2. SOME FACTORS THAT CURRENTLY DISCOURAGE RECYCLING IN THE UNITED STATES

1. Public apathy toward bringing waste to recycling centers.
2. Scarcity of recycling centers.
3. Poor recyclability of many manufactured goods
4. Tax incentives that favor mining and logging over recycling.
5. Freight rates that favor mining and logging over recycling.
6. Lack of markets for recycled goods owing to
   (a) farmer apathy toward compost, and
   (b) reluctance of many industries to try recycled goods.
7. Poor funding of research on recycling products from trees.
8. Lack of consumer pressure for recycled goods.
9. Zoning laws that discriminate against recycling.

Production from virgin ore depletes resources and uses large quantities of energy.

Depletion of
resources

Large Consumption
of Energy

Collection and transportation of scrap use relatively large amounts of labor,
but little energy. Energy is cheap in the United States because economic
externalities are not considered, so recycling is often more expensive than
production from virgin ores.

The Franklin, Ohio, recycling center. (Courtesy of Black Clawson Fibre-claim, Inc.)

In our society, energy is relatively cheap and labor is expensive. Therefore, it is cheaper to mine iron than to recycle it. However, this cost analysis does not include economic externalities. If fuels were taxed to conserve dwindling supplies, if steel companies were charged for depleting iron ore reserves, and if mills were forced to meet strict water and air pollution standards, then iron from recycled scrap would be cheaper than iron from ores. But these costs are not all counted. If we don't recycle materials now, fuel and ore prices will rise sooner than they would otherwise. The costs of pollution damage will rise. There are real and tangible costs of wasteful practices. Recycling would conserve resources and in the long run save money.

Is it possible, then, to increase recycling now? Industry responds to a variety of pressures, not only economic but also social, political, and legal. A combination of these can be very effective. For example, there is a large automated garbage recycling plant in Franklin, Ohio. The Franklin reclamation plant is slightly more expensive than a landfill. But the citizens of that city decided to pay a marginal increase in cost, originally about 25 cents per family per month, because they believe in recycling. Also, this plant was financed partly with federal grants. Therefore the choice to enter an "unprofitable" business was also stimulated by governmental policy. The benefits of such a policy are illustrated by the fact that the Franklin operation has served as a model for similar, larger plants in Miami (Florida), Tokyo (Japan), and elsewhere.

Other examples of this sort are abundant. The General Services Administration of the United States Government, the City of New York, the Bank of America, Coca-Cola, and Canada Dry have all ordered large quantities of recycled paper for stationery or annual reports. At present there is no economic advantage to the use of recycled paper. But these companies have responded to public pressure to reuse scrap.

Another way to accelerate recycling rates is to pass laws that encourage reuse. For example, the states of Vermont, Oregon, Maine, and Michigan

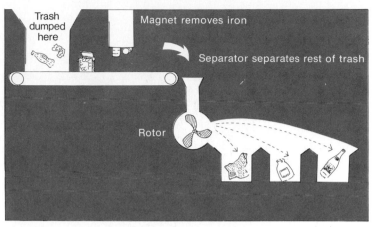

FIGURE 14–4.   Schematic view of one type of automated trash separator. Separators of this type can be used to automatically classify municipal trash for recycling.

have banned all nonreturnable beverage containers. However, the 2- to 5-cent deposit required may be too low. The total environmental cost of a discarded bottle includes the energy and raw materials of manufacture as well as the litter and waste disposal problem. Therefore, a bottle's true cost in 1978 was close to 25 cents. Perhaps a law requiring a 25-cent deposit on glass bottles would be an effective inducement for returning them.

The world cannot afford to wait to recycle a material until a resource is on the brink of exhaustion. Some supply of raw materials will always be needed, and our natural resources must be conserved now to insure adequate supplies of ore, fertilizers, and fiber for future generations.

In Europe, fuel prices are generally higher than they are in North America, and recycling is more profitable.

In Ecuador, South America, a bottle of beer costs approximately 40 cents in U.S. dollars. The deposit on the bottle, however, is three times the price of the beer. Thus an empty bottle is worth $1.20 and people simply do not discard them.

14.8
CASE HISTORY: RECYCLING OF BEVERAGE CONTAINERS

A century ago, milk, beer, cooking oil, and other liquids were shipped to general stores in large barrels. Customers then filled their own reusable jars. Many solid items such as coffee and nuts that are now available in glass containers were scooped from kegs into cloth sacks. These sacks were used over and over again. In time, glass packaging became increasingly popular, espe-

Bottling plant capable of filling 1000 bottles per minute.

In 1972, the energy used to manufacture nonreturnable bottles in the United States was 62 billion kilowatt hours more than that which would have been needed if returnables were used. That difference represents enough energy to supply almost ten million Americans with electric power for a year.

cially for foods. Later the soft drink, dairy, and beer industries added a new approach to glass packaging — the deposit bottle. These generally well-constructed bottles cost more to manufacture than the two- or five-cent deposit values assigned to them. But the small deposit was sufficient incentive for the customer to return the bottles. This system worked so well that at first a deposit bottle averaged about 30 round trips. Some were broken, lost, neglected, or used as flower vases and therefore never returned. These were, of course, replaced with new bottles. Gradually, consumers became so indifferent to the deposit that by 1960 the average number of round trips had declined to four. Moreover, the cost of washing bottles increased, so bottling companies initiated a shift to no-deposit, no-return containers.

This shift to throwaway glass has led to a tripling of energy consumption in the bottling industry. Today, 1.7 kilowatt hours of electricity, enough to light a 100-watt bulb for 17 hours, is needed to manufacture and transport *one* 12-ounce throwaway bottle.* Glass from a no-deposit, no-return bottle can be recycled. But it must be collected, transported to the recycling plant,

*Bruce Hanon: *System Energy and Recycling, A Study of the Beverage Industry.* Center for Advanced Computation Document #23, University of Illinois Press, 1973.

crushed, sorted according to color, remelted, and reprocessed. Once again, a key problem here is collection. In a recycling conscious society people would ideally sort their waste glass and deposit it in convenient centers. Under these conditions, recycling of glass would be profitable and conservative of energy. In our present society, the collection process is so inefficient that overall, more energy is needed to recycle glass from a no-deposit bottle than is needed to manufacture one from virgin material. It is, therefore, obvious that it would be environmentally more sound to sell only returnable bottles. The present use of throwaways is wasteful of resources.

In the 1940's some beer and soft drink manufacturers shifted partially to the use of steel cans as containers. Cans are lighter, easier to ship, and less susceptible to breakage than glass. But steel cans are hard to open, so in the late 1950's aluminum cans appeared on the market. Soon thereafter "flip-tops" were added. Aluminum cans require

even more energy to manufacture than glass bottles do. But they are cheaper to handle *for the beverage bottlers and shippers*. Therefore, they have been marketed extensively during the past decade. Much less energy is needed to remelt an aluminum can than to mine and purify aluminum ore. Therefore, even in our present society, considerable quantities of energy are saved whenever aluminum is recycled. However, despite active recycling campaigns, only about 20 percent of the aluminum cans sold in the United States are returned.

Many environmental groups have mounted advertising campaigns to educate people about the problems of beverage containers. Numerous recycling centers have been established throughout North America. These campaigns depend mainly on voluntary cooperation. They have met with only limited success. In 1972 the state of Oregon initiated a more comprehensive program to outlaw throwaway containers. The Oregon bottle bill banned flip-top cans and placed a mandatory two- to five-cent deposit on all beverage containers. Within the first year, the total quantity of roadside litter was reduced by 23 percent. The energy conserved amounted to enough fuel to heat 11,000

Nonbiodegradable containers.

homes for an entire year. Public enthusiasm later declined somewhat, so that in 1974 and 1975 more returnable bottles were discarded. But considerable savings were still realized. Bottle bills have been enacted in Vermont, Maine, and Michigan, but similar laws have been defeated by voters in Colorado and Massachusetts.

Energy is conserved when aluminum is recycled.

CHAPTER SUMMARY

14.1　SOLID WASTE CYCLES

Recycling paths do not operate efficiently in modern society, so the movement of materials through industrial paths generates an ever-increasing quantity of solid wastes.

14.2　SOURCES AND QUANTITIES

Municipal refuse is composed largely of unnecessary packaging materials and of items that have been discarded because they weren't built to last in the first place. Industrial, agricultural, and mining operations also contribute significantly to the solid waste problem.

14.3　DISPOSAL ON LAND AND IN THE OCEAN

Much of the solid waste in the United States is deposited in unsightly, uncontrolled, smelly, polluting open dumps. Some wastes are dumped untreated into the ocean where they are removed from view, but the trash destroys natural ecosystems. A sanitary landfill is an area where trash is deposited, compacted, and covered with soil daily to reduce pollution.

14.4　ENERGY FROM REFUSE

Many metropolitan areas incinerate their garbage and use the heat to produce steam. In recent years the fuel content of trash, the value of the steam, and the operating costs of alternative methods of disposal have all increased, and thus incineration has become more desirable. However, trash contains some materials that release noxious pollutants when burned, thus causing problems.

14.5　RECYCLING — AN INTRODUCTION

The real crux of the recycling issue is to guarantee the greatest possible use of a material in the most efficient manner. If an item cannot be reused or repaired, it may be practical to recover and reuse the materials of which it is made. However, such recycling will operate effectively only if it is compatible with present social and/or economic systems. In general, recycling conserves not only material resources but fuel reserves as well.

14.6　RECYCLING TECHNIQUES

Many recycling processes are available:
Melting: Metals and glass are re-formed to new metals and glass products.
Repulping: Paper, wood, sugar cane stalks, etc., are re-formed to make paper.
Composting: Organic trash is converted to fertilizers. Methane fuel can also be recovered.
Rendering: Slaughterhouse wastes are used to make soap and animal feed.
Industrial salvage: Heterogeneous mixtures of trash are separated and recycled.

14.7　THE FUTURE OF RECYCLING

Recycling has stagnated in our society mainly because the economic externalities of unecological waste practices are generally ignored. For example, it is often cheaper to manufacture steel from ore than from scrap, but the depletion of fuel and metal reserves and the pollution from mining operations are not considered in the present economic evaluation.

KEY WORDS

Open dump
Ocean dump
Sanitary landfill
Incineration

Recycling
Melting
Pulping
Composting
Rendering

TAKE-HOME EXPERIMENTS

1. **Solid waste disposal.**　Sort through your waste and garbage containers at home. List the items that (a) need not have been purchased in the first place, (b) could have been reused, and (c) could be easily recycled. Weigh your garbage daily for a week. Then start a program to reduce the solid waste in your household and repeat the weighings. How much difference has your program made? How have your personal habits been changed? Locate the recycling centers in your area. Compile data on (a) the time required to prepare cans, bottles, and paper for recycling and to bring them to the

center, and (b) the cost of transporting them.

2. **Recycled paper.** In this experiment you will manufacture recycled paper from old newspaper. Cut a square of newspaper approximately 30 cm on a side and shred it into small pieces. Fill a large bowl ¼ full of water, add the shredded paper, and let the mixture soak for an hour or two. Beat the mixture vigorously with an electric mixer or hand-operated egg beater until the paper breaks up into fibers and the mixture appears creamy and homogeneous. Next, dissolve two heaping tablespoons of starch or wallpaper paste in ½ liter of warm water, add this solution to the creamy slurry, and stir. Take a piece of fine window screen and dip it into the solution. Lift the screen out horizontally, as shown in the sketch. If you have done this correctly there will be a fine layer of paper fibers on the screen. Stir up the contents of the bowl and redip the screen carefully as many times as is needed to build up a layer about ¼ cm thick.

This layer must now be pressed and dried. To do this, place some cloth towels on a tabletop and lay the screen over the

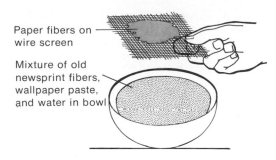

Paper fibers on wire screen

Mixture of old newsprint fibers, wallpaper paste, and water in bowl

Recycling: Making New Paper from Old

towels so that the fiber layer is facing upward. Cover the fibers with a piece of thin plastic (a plastic bag is fine). Now squeeze down evenly on the plastic, using a block of wood for a press. Most of the water should be squeezed from between the fibers, through the screen, and onto the cloth towel. Set the screen out to dry for a day or two, peel off the newly fabricated recycled paper, and write a letter to a friend!

PROBLEMS

1. **Disposal of manufactured products.** Which of the following manufactured products will undergo biological recycling, industrial recycling, or accumulation as a solid waste in our society? (a) a polyethylene squeeze bottle used as a container for mustard; (b) a copper drainpipe from a wrecked house that is sold to a dealer in scrap metals; (c) a woolen sweater; (d) the gold filling in an automobile body that is returned as scrap to the mill; (f) a "no-deposit" soda bottle; (g) the coffee grounds you used to make this morning's coffee.

2. **Mine wastes.** Explain how mine wastes differ from municipal, industrial, or agricultural wastes.

3. **Types of wastes.** Give an example of a biodegradable, a combustible, a poisonous, and an odorous solid waste.

4. **Land disposal.** Explain the difference between an open dump and a sanitary landfill.

5. **Incineration.** Discuss the differences between burning trash in an open dump and in a modern incinerator.

6. **Incineration.** Discuss the advantages and disadvantages of incineration as a method of waste disposal.

*7. **Incineration.** Plastics, made from coal and oil, have a high heat content. If garbage incinerators were commonly used in the United States, and the resulting heat of combustion were used industrially, would you feel that plastic packaging would be an advantageous way to use fossils fuels twice? Defend your answer.

8. **Recycling.** As mentioned in the text, broken or obsolete items can be repaired, broken down for the extraction of materials, or discarded. Which route is more conservative of raw materials and energy for each of the following items? (a) a 1948-model passenger car that doesn't run; (b) a 1977-model passenger car that doesn't run; (c) an ocean liner grounded on a sandbar and broken in two; (d) an ocean liner sunk in the central ocean; (e) last year's telephone directory; (f) an automobile battery that won't produce current because the owner of the car left the lights on all day; (g) an empty ink cartridge from a fountain pen?

*9. **Planned obsolescence.** An old-timer complains that years ago a man could store canned milk in the creek for three years before the can would rust through, but now a can will only last one year in the creek. Would you agree with the old man that cans should be made to be more durable? What

---

*Indicates more difficult problem.

about automobiles? Explain any differences.

10. **Recycling.** Sand and bauxite, which are the raw materials for glass and aluminum, respectively, are plentiful in the Earth's crust. Since we are in no danger of depleting these resources in the near future, why should we concern ourselves with recycling glass bottles and aluminum cans?

*11. **Recycling.** List the most efficient recycling technique and the resultant products for each of the following: (a) steer manure; (b) old clothes; (c) scrap lumber; (d) aluminum foil used to wrap your lunch; (e) a broken piece of pottery; (f) old bottle caps; (g) stale beer; (h) old eggshells; (i) a burnt-

*Indicates more difficult problem

out power saw; (j) worn-out furniture; (k) old garden tools; (l) tin cans; (m) disposable diapers. How does your choice of technique depend on your location?

12. **Composting.** What is composting? Name three materials that can be recycled by composting. What is the resulting product?

13. **Rendering.** What is rendering? Name three materials that can be recycled by rendering. What are the resulting products?

14. **Recycling.** Explain why recycling is not common practice in the United States today. Would an energy tax and stricter pollution laws encourage recycling? Discuss.

15. **Recycling.** Compare the cost, energy consumption, pollution, and other environmental factors of steel production from ore versus steel production from recycled scrap.

QUESTIONS FOR CLASS DISCUSSION

1. At the present time, recycling of paper is a marginally profitable business, and many repulping mills have gone bankrupt in recent years. List some of the economic externalities associated with the production of paper from raw materials. Who bears these costs? How could the burden be shifted to encourage paper recycling?

2. A family lives in a sparsely populated canyon in the northern Rocky Mountains. Their household trash is disposed of in the following manner: Papers are used to start the morning fire in the potbelly stove; food wastes are either fed to livestock or composted. Ashes are incorporated into the compost mixture; metal cans are cleaned, cut open, and used to line storage bins to make them rodent-resistant; glass bottles are saved to store food; miscellaneous refuse is hauled to a sanitary landfill. Comment on this system. Can you think of situations where this system would be undesirable? Do you think that it is likely that many people will adopt this system?

3. Rendering plants recycle various slaughterhouse and cannery wastes. Despite careful controls these factories sometimes emit foul odors. Comment on the overall environmental impact of a rendering plant.

4. Environmental organizations have been active in establishing collection centers for old newspapers, cans, and so on. Can you think of other activities that these groups might engage in which would produce increased recycling?

5. Consider the possible reasons why the average number of round trips of a returnable bottle, originally about 30, fell to 4 by 1960. Can you think of other reasons besides the fact that a two-cent or five-cent deposit is not worth as much now as it used to be? Ask your parents or grandparents to recall what was actually involved in returning bottles for deposit. How would you feel about depositing bottles today? Also, interview some store managers, including some older ones, and get their answers to these questions.

6. In 1976 voters in Colorado were asked to decide whether or not to levy a mandatory tax on nonreturnable cans and bottles. The beverage industry strongly opposed the law. In one brochure published by a major beer manufacturer it was stated:

   Claim: Amendment #8 [the proposed bottle law] would conserve energy and resources.
   Fact: Any savings in coal consumption resulting from the law would be offset by an

increase in the consumption of gasoline, natural gas and water.

Returnable bottles are heavier than nonreturnable cans. Manufacturing their heavy, durable carrying cases would require an increase in energy consumption.

Bottles would also require twice as much space as cans. Trucks would have to make at least twice as many trips to haul refillable containers, to say nothing of the extra trips to pick up empties, resulting in increased gasoline consumption.

Washing re-usable bottles requires five times more water than cans. Heating the water for sterilization and removing the detergents that are used means increased energy consumption.

Examine these statements critically. Have all the facts been presented fully and accurately, or do you feel that the brochure states the case incompletely? Is it reasonable or misleading? Defend your position.

BIBLIOGRAPHY

A compilation of valuable articles on solid wastes is: Solid Wastes. Environmental Science and Technology Reprint Book. Washington, D.C., American Chemical Society, 1971.

Most of the popular books on the environment that are cited in the bibliography of Chapter 3 include some discussion of solid wastes. For more technical information, refer to the following:

Andrew W. Breidenbach: *Composting of Municipal Solid Wastes in the United States.* Environment Protection Agency Publication, Washington, D.C., U.S. Government Printing Office, Stock number 5502–0033, 1971.

Richard B. Engdahl: *Solid Waste Processing.* Public Health Service Publication No. 1856. Washington, D.C., U.S. Government Printing Office, 1969.

Bruce Hanon: *System Energy and Recycling, A Study of the Beverage Industry.* Center for Advanced Computation Document #23, University of Illinois Press, 1973.

N. Y. Kirov: *Solid Waste Treatment and Disposal* Ann Arbor, Michigan, Ann Arbor Science Publishers, 1972.

Fred C. Price. Steven Ross, and Robert L. Davidson (eds.); *McGraw-Hill's 1972 Report on Business and the Environment.* New York, McGraw-Hill Publications, 1972.

United States Bureau of Mines: *Automobile Disposal—A National Problem.* Washington, D.C., U.S. Government Printing Office.

Noise is unwanted sound.

15
NOISE

15.1
SOUND AND NOISE

Sound

To understand how sound is produced, just consider its opposite — silence. Think of the quietest occasions of your life. Perhaps standing in the woods after a snow on a windless winter day was one such instance. Certainly, you would not think of an occasion in which you were moving, whether walking, running, or riding. We realize that motion itself produces sound. Thus a bell makes a sound when it is struck and starts to vibrate. A violin string can only be heard after it has been set into motion.

If a person throws a stone into a pond of water, the falling stone generates a series of waves as shown in Figure 15–1A. Similarly, if a person strikes a bell, the vibrations produced generate air

FIGURE 15–1. *A,* Small water waves.

FIGURE 15–1. *B*, Schematic drawing of sound waves in air.

waves as shown in Figure 15–1B. These sound waves travel through the air and eventually beat against a person's eardrum. The signal is then transmitted to the brain and the person hears the bell.

Noise

Noise can be defined, simply, as unwanted sound. This concept is straightforward enough, but it does not teach us how to predict which sounds will be disliked. A given sound may be music to one person but noise to another. It may be pleasant when soft but noise when loud, acceptable for a short time but noise when prolonged. Many sounds are reasonable when you make them but noise when someone else makes them. Most sounds become irri-

tating and unwanted when they are very loud. There is ample evidence that exposure to loud sounds is harmful in various ways, and, in any event, loudness tends to be annoying. Therefore, the louder a sound is, the more likely it is to be considered noise.

Noise is not a material like sulfuric acid or sewage sludge. It does not accumulate in the environment. Once its echo dies away, it is gone. But sounds can affect human beings, and these effects do not disappear as easily as the sounds themselves.

15.2
LOUDNESS AND THE DECIBEL SCALE

We have noted that, other things being equal, the louder a sound, the more likely it is to be considered noisy. And with good reason — loud sounds tend to interfere with our activities and, as we shall see, they can be physically harmful.

The human ear is sensitive to a wide range of sound intensities. The loudest sounds that we can hear, such as a rocket taking off or the noise of battle, are billions of times more powerful than the softest sounds, such as the patter of raindrops on soft earth or a child's whisper. Sound intensity is measured on a scale of values called a **decibel** (abbreviated as dB) **scale**. The decibel scale is set up as follows:

(a) The softest sound that can be heard by humans is called 0 dB.

(b) Each *tenfold increase* in sound intensity is represented by an *additional* 10 dB. Thus, a 10-dB sound is 10 times as intense as the faintest audible sound. (That still isn't very much.) The

sound level in a quiet library is about 1000 times as intense as the faintest audible sound. Therefore the sound level in the library is 10 + 10 + 10 or 30 dB. To summarize:

Sound	Sound Level in Decibels
Faintest audible sound (threshold of hearing).	0
A leaf rustling, at 10 times the intensity of the threshold of hearing.	10
Sound level in a broadcasting studio, at 10 times the level of the rustling leaf. (This is also 100 times the level of the faintest audible sound.)	20
Sound level in a quiet library, at 10 times the level of the broadcasting studio. (We are now at 1000 times the threshold of hearing.)	30

15.3
THE EFFECTS OF NOISE

Noise can interfere with our communication, diminish our hearing, and affect our health and our behavior.

Interference with Communication

We have defined noise as unwanted sound. Let us think for a moment about the sounds that we do want. Some of these are speech or music, danger warnings such as the cry of a baby or the rattle of a rattlesnake, pleasant natural sounds such as the chirp of a bird or the rustle of leaves in a gentle breeze. We want to hear these sounds at the right level — not too loud or too quiet. Noise keeps us from hearing what we want to hear.

If you wish to observe how noise interferes with communication, perform the following simple experiment. Venture out some quiet winter day into a snowy woodland and hold a muted conversation there, or sing, or listen to the little natural sounds that noisier environments extinguish. Then take a ride on a noisy subway with your girl or boyfriend and yell sweet nothings back and forth.

Loss of Hearing

An occasional loud noise interferes with sounds that we wish to hear, but

Decibel meter. (Photo courtesy of GenRad, Inc. [formerly General Radio Company], Concord, Mass.)

Snowy woods, a silent environment.

Combat noise has partially deafened many soldiers. (Photo of action in Vietnam, courtesy of the U.S. Army)

first day in a noisy factory. Of course, he recognizes the noisiness, and may even feel the effect as a "ringing in the ears." He will have suffered a temporary hearing loss. As he walks out of the factory, then, most sounds will seem softer. His car will seem to be better insulated, because he will not hear the rattles and squeaks so well. He will judge people's voices to be just as loud as usual, but they will seem to be speaking through a blanket. He will also feel rather tired.

By morning he will be rested. The ringing in his ears will have stopped, and his hearing will be partly but not completely restored. The factory will therefore not seem quite so noisy as it was on the first day. As the months go by he will become more and more accustomed to his condition, but his condition will be getting worse and worse. Can he recover if he is removed from his noisy environment? That will depend on how noisy it has been and how long he has been exposed. In many cases, his chances for almost complete recovery will be pretty good for about a year or so. However, if the person continues to work in the same noisy place for many years, hearing loss becomes irreversible, and eventually he will become deaf. Occupational noise such as that produced by bulldozers, jackhammers, diesel trucks, and aircraft is deafening many millions of workers. In technologically developed societies, women hear better than men, because they are generally less exposed to occupational noise. In undeveloped, quiet societies, women and men hear equally well. This difference between the sexes in developed societies may diminish as more effort is exerted to make industry quiet and as more women work in in-

we recover when quiet is restored. However, if a person is exposed to loud noises for long periods of time, then there may be significant permanent loss of hearing.

The general level of city noise, for example, is high enough to deafen us gradually as we grow older. In the absence of such noise, hearing ability need not deteriorate with advancing age. Thus, inhabitants of quiet societies, such as tribespeople in southeastern Sudan, hear as well in their seventies as New Yorkers do in their twenties.

It is important to understand that most instances of loss of hearing that result from environmental noise are not immediately noticeable. The victim is often unaware that he or she is slowly losing the ability to hear well. Let us picture a worker who completes his

(FUNKY WINKERBEAN by Tom Batiuk. © Field Enterprises, Inc., 1972. Courtesy of Field Newspaper Syndicate.)

dustry. Battle noises, such as those made by tanks, helicopters, jets, and artillery, are so deafening that many American soldiers who undergo combat training suffer severe hearing loss.

What about extremely loud noises? There has been recent concern that rock-and-roll music in night clubs is often indeed very loud. Sound levels of 125 decibels have been recorded in some discothèques. Such noise is at the edge of pain and is unquestionably deafening. Noise levels as high as 135 dB should never be experienced, even for a brief period, because the effects can be instantaneously damaging. If the noise level exceeds about 150 or 160 dB, the eardrum might be ruptured beyond repair.

Other Effects on Health and Behavior

As we have already discussed in many contexts, a living organism, such as a human being, is a very complicated system. The effects of a stress or a disturbance are often difficult to predict. Having read this book thus far, you should realize that a disturbance great enough to deafen you will have other effects as well. Indeed, many investigators believe that loss of hearing is not the most serious consequence of excess noise. The first effects are anxiety and stress or, in extreme cases, fright. These reactions produce body changes such as increased rate of heart beat, constriction of blood vessels, digestive spasms, and dilation of the pupils of the eyes. The long-term effects of such overstimulation are difficult to assess, but we do know that in animals it damages the heart, brain, and liver and produces emotional disturbances. The emotional effects on people are, of course, also difficult to measure. We do know that work efficiency goes down when noise goes up.

Is some exposure to noise unavoidable if we are to live in a developed, technological society? This question is similar to questions we have posed in earlier chapters. Are air pollution, water pollution, and other environmental disruptions necessary? It all depends on the cost of pollution control and how much people are willing to pay for a healthy environment. In the

next two sections we will discuss some technical, legal, and economic problems of noise control.

15.4
NOISE CONTROL

Noise is produced by some source such as a truck engine or a jackhammer. It then travels — is transmitted — through some medium such as air. Finally, it is received by a person's ear. To control noise, therefore, we can reduce the source, interrupt the path of transmission, or protect the receiver.

Reducing the Source

The most obvious way to reduce the problem is to limit the number or intensity of various sources of noise. Don't beat the drum so hard, or ring the bell so loud, or run so many trucks or motorcycles, or mow the lawn with a power mower so often. There are obvious limitations to this type of solution. For example, if we run fewer trucks, we will have less food and other essentials delivered to us.

Even if we do not reduce the sound power, we may be able to reduce the noise production by changing the source in some way. Our purpose in pushing a squeaky baby carriage is to move the baby, not to make noise. Therefore, we can oil the wheels to reduce the squeaking. Thus, machinery should be designed so that parts do not needlessly hit or rub against each other (see Fig. 15–2). It is possible to design machines that work quietly. Rotary saws can be used to break up street pavement. They do the job perfectly well and are much quieter than jackhammers.

We could also change our procedures. If a suburban sidewalk must be broken up by jackhammers, it would be better not to start early in the morning, when many people are asleep. Construction crews could work in the middle of the day when many residents have left for work. Similarly, aircraft takeoffs could be routed over less densely inhabited areas.

FIGURE 15–2. Hydraulically operated shear has less impact noise than
mechanically driven shear. (Courtesy
of Pacific Industrial Manufacturing
Company)

All too often, machines are built to perform a task most efficiently without consideration of how much noise is produced. If machines were originally designed properly, noise levels could be substantially reduced.

FIGURE 15–3.   Man wearing acoustical earmuffs while using chainsaw.

Interrupting the Path

Sound waves travel through air. They also travel through several other types of media such as water or wood. If you put your head underwater in the bathtub and bang two rocks together, you can hear this noise transmitted through the water. Similarly, noise travels through wooden walls. Many people who live in apartment buildings hear the sounds their neighbors make. Some materials absorb sound waves and do not transmit them efficiently. Sound-absorbing media are called **acoustical materials.** These can be used as barriers to sound transmission. Certain sound-absorbing media can also be built directly into machinery. One example of this type of approach is the use of mufflers in automobile exhaust systems. Similarly, acoustical tiles and wallboard can be used in house and apartment construction to reduce noise levels.

Protecting the Receiver

The final line of defense is strictly personal. We protect ourselves instinctively when we hold our hands over our ears. Alternatively, we can use ear plugs or muffs as shown in Figure 15–3. (Stuffing in a bit of cotton does very little good.) A combination of ear plugs and muff can reduce noise by 40 or 50 decibels, which could make a jet plane sound no louder than a vacuum cleaner. Such protection could prevent the deafness caused by combat training, and should also be worn for recreational shooting.

We can also protect ourselves from a noise source by going away from it. Unfortunately, many people cannot simply move away from big cities or quit a job in a noisy factory.

15.5
LEGISLATION AND PUBLIC POLICY CONCERNING NOISE CONTROL

Noise is hardly a new phenomenon, and the annoyance it produces has

been recognized since ancient times. Julius Caesar banned chariots from Rome because their wheels clattered on the stone streets. English Common Law recognized that freedom from noise is essential to the full enjoyment of a private dwelling place. Literature contains references to the effects of noise, especially from the loudest source, which is battle. For example, records show that Admiral Lord Rodney was deafened for 14 days following the firing of 80 broadsides from his ship, H.M.S. *Formidable,* in the year 1782.

Despite these early beginnings, federal noise legislation in the United States is a relatively recent development. Before the 1960's, the federal government left noise control in the hands of state and local governments. But as people became more and more environmentally aware, Congress started to act on this matter. The Noise Control Act was passed in 1972. This law was written "to promote an environment for all Americans free from noise that jeopardizes their health or welfare." Under this law, the U.S. Environmental Protection Agency (EPA) coordinates all federal programs relating to noise research and noise control. Some of these programs deal with the following areas:

Laws have been established limiting the amount of noise permissible in factories and other job sites.

EPA has the authority to prescribe noise standards for trucks, buses, transportation and construction equipment, engines, and electronic devices.

EPA conducts and finances research on the effects of noise (including sonic booms) on humans, animals, wildlife, and property. The Agency also conducts research to determine acceptable levels of noise, and develops improved methods for measuring and controlling this environmental problem.

If we have learned anything in this book it is the lesson that solutions to environmental problems are not simple and straightforward. Consider the example of a new slum housing project that was proposed in New York City. The federal government supports the construction of new housing for slum regions, provided the new housing

(From *Saturday Review,* May 13, 1972, by permission of Mort Gerberg)

The Concorde SST.

meets government standards. In this case the noise levels in the apartment complex were expected to exceed federal standards. Therefore, the government refused to support the project.

New York officials replied that the government position amounted to saying "that the poor people can live in their old houses in rundown neighborhoods, but not in new houses," and that "to be told that you have to be worried about noise levels of brand new, safe, sanitary housing in Harlem, when hundreds of thousands live in slums, is patently ridiculous." People were living in noisy, dirty, unsanitary housing. The proposed new project would build houses that were still noisy but at least they would be clean and sanitary.

Local contractors then estimated the cost of adding soundproofing to the building. The expense was so great that the federal government then refused to support the project because it would cost too much. Thus, without soundproofing the government refused support because of too much noise. With the sound insulation the government refused support because the project would be too expensive. This situation poses a dilemma. Strict application of noise regulations will make it more costly to built new houses for the poor. But relaxation of the standards will make the new housing less comfortable to live in.

15.6
CASE HISTORY: SUPERSONIC TRANSPORT (SST)

The SST is a passenger aircraft that travels faster than sound and at much higher altitudes than subsonic airplanes. Higher speed requires more power, and more power makes more noise. Near airports the planes travel at subsonic speed. But the SST's are still noisier than conventional aircraft because of the design of the engines and the rapid climb shortly after takeoff.

When the SST reaches supersonic speed in flight, another effect, a **sonic boom**, is heard. This, too, creates problems.

To be struck unexpectedly by a sonic boom can be quite unnerving. It sounds like a loud, close thunderclap, which can seem quite eerie when it comes from a cloudless sky. Depending on the power it generates, the sonic boom can rattle windows or shatter them or even destroy buildings. It is important to avoid the misconception that the sonic boom occurs only when the aircraft "breaks the sound barrier," that is, passes from subsonic to supersonic

Supersonic is faster than the speed of sound, and **subsonic** is slower than the speed of sound. The speed of sound is 331 meters/sec, or 727 miles/hr.

"At least they have agreed to show only silent movies." (From *Industrial Research*, Nov., 1976)

speed. On the contrary, the sonic boom is continuous and, like the wake of a speedboat, trails the aircraft all during the time that its speed is supersonic. Furthermore, the power of the sonic boom increases as the supersonic speed of the aircraft increases.

During the late 1960's an American SST was being developed by the Boeing Aircraft Corporation under financial support by the U.S. government. This project provided a great many jobs for the people living in the Seattle, Washington, area (Boeing's home). Thus it was a significant asset to the economy of the region. However, as it became evident that the SST aircraft could give rise to environmental problems, significant opposition to the project was mounted. A conflict arose between environmentalists and industrial and labor groups. One fundamental question that arose was, "How annoying would noise from the SST actually be?"

It has been pointed that annoyance is a subjective experience. A sound may bother one person and be barely noticeable to another. People have argued that if SST's were used solely for flights over the ocean, the sonic booms wouldn't bother anyone. Furthermore, the extra noise over airports is a small price to pay for human progress and technological advances. Opponents argue that SST's cause increased pollution in the upper level of the atmosphere, and the noise pollution really is bothersome. Whenever such differences of opinion arise it is important to look closely at the balance between the expected gain and the environmental disruptions. Examine the SST balance sheet below.

SST BALANCE SHEET

Expected Gains From SST	*Environmental Problems*
Faster travel for people who can afford the added cost	Increased noise Increased air pollution

In 1971, the U.S. Congress stopped the funding of the SST project in this country. Meanwhile, a British and French team developed an independent supersonic airliner called the Concorde that now carries transatlantic passengers commercially.

In 1978, an SST flight from New York to Paris required 5 hours' flying time and cost $862. A conventional flight travels from New York to Paris in 8 hours for a total cost of $325. It is hard to imagine that three hours saved by a few wealthy people riding in a supersonic flight will really improve the human condition appreciably. Perhaps the single most important problem for our generation is to understand the benefits and potential problems of technological developments. Then that understanding can serve as a basis for the regulation of technology to improve life on Earth.

CHAPTER SUMMARY

15.1 SOUND AND NOISE

Sound is produced by motion. Sound waves travel through the air and are detected by the ear.

Noise is unwanted sound.

15.2 LOUDNESS AND THE DECIBEL SCALE

Sound intensity is measured on a scale of values called a **decibel scale.** The decibel scale represents each tenfold increase in sound intensity as an additional ten decibels.

15.3 THE EFFECTS OF NOISE

Noise interferes with communication. Excessive noise causes gradual loss of hearing and eventually deafness, which is irreparable. Excessive noise can cause anxiety, fright, and manifestations of other psychological or physical strains.

15.4   NOISE CONTROL

Noise can be controlled by (a) reducing the noise at the source by proper design of machinery; (b) interrupting the path of the noise through the use of mufflers and other acoustical devices or materials; and (c) protecting the receiver with ear plugs or muffs.

15.5   LEGISLATION AND PUBLIC POLICY ON NOISE CONTROL

Although legal recognition of noise problems is old, the important modern U.S. legislation is the Noise Control Act of 1972. This law authorizes the U.S. Environmental Protection Agency (EPA) to set standards limiting noise generations. Nonetheless, conflicting interests often make it difficult to administer and enforce the regulations.

KEY WORDS

Sound
Noise
Decibel scale
Acoustical materials

Supersonic
Subsonic
SST

TAKE-HOME EXPERIMENTS

1. **Loudness.** Carry this book around with you for a few days so you can refer to Table 15–1 (page 315). Make a diary of various sounds you hear, recording your information as shown below:

7:00 A.M.	Baby crying.	84 dB
7:30	Dishwasher in kitchen.	70 dB
7:45	Garbage truck, 150 feet away	90
8:00	Traffic noise while waiting for bus.	81
8:45	Arrived at entrance to office. Noise of jackhammer on sidewalk.	106
9:00-12:00 Noon	Average sound in office.	45
12:00-1:00 P.M.	Noise in restaurant—dishes, etc.	45
5:00-5:30	Rode home on subway (windows open).	90-111
6:00	Mowed lawn with power mower.	93

Estimate the decibel levels as well as you can from Table 15–1. Try to include the following sources: (a) a television or radio at normal listening volume in your room or apartment; (b) the central study area in the school library; (c) your environmental science classroom; (d) the street outside your classroom; (e) the background noise in your room at night; (f) the school cafeteria; (g) a local factory or construction job site. Which sounds are annoying? Offer suggestions for reducing the perceived loudness in each of the instances where the sound is annoying.

2. **Sound.** Select a convenient constant source of sound, such as a ringing alarm clock, and try to reduce the loudness you hear from it by the following means: (a) stuff some cotton loosely into your ears; (b) hold your hands over your ears; (c) use both the cotton and your hands; (d) submerse your head in the bathtub (face up) until your ears are underwater; (e) if you can borrow a pair of earmuffs of the kind used in factories or at airports, try them. (*Safety Note:* Don't try to stuff any small hard objects in your ears; the results could be harmful.)

Inaudible (zero loudness)	Loudness to naked ear

Draw a straight line (of any length) in your notebook, labeling one end "inaudible" and the other end "loudness to the naked ear." Mark the positions of each of the sound-reducing methods on the line at a point that corresponds to the loudness you heard. For example, if you think that one of the methods reduced the loudness by half, mark its position halfway along the line. If you think it reduced it by only 25 per cent, mark it at ¼ of the length away from the "naked ear" end. Discuss the reasons for your findings.

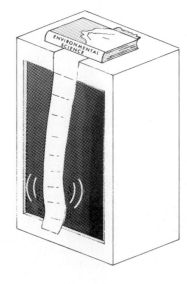

3. **Sound power.** Hang a strip of tissue paper in front of the loudspeaker of a sound system (see sketch). Shut off the sound, close the windows, and turn off any fan or air conditioner in the room. The paper should hang motionless. Now switch the sound on and turn up the volume. Describe the effect on the paper and explain your observations.

TABLE 15-1. SOUND LEVELS AND HUMAN RESPONSES

Sound Intensity Factor	Sound Level, dB	Sound Sources	Effects		
			Perceived Loudness	Damage to Hearing	Community Reaction to Outdoor Noise
1,000,000,000,000,000,000	180 —	• Rocket engine			
100,000,000,000,000,000	170 —				
10,000,000,000,000,000	160 —				
1,000,000,000,000,000	150 —	• Jet plane at takeoff	Painful	Traumatic injury	
100,000,000,000,000	140 —			Injurious range; Irreversible damage	
10,000,000,000,000	130 —	• Maximum recorded rock music			
1,000,000,000,000	120 —	• Thunderclap • Textile loom • Auto horn, 1 meter away	Uncomfortably loud		
100,000,000,000	110 —	• Riveter		Danger zone; progressive loss of hearing	
10,000,000,000	100 —	• Jet fly-over at 300 meters • Newspaper press			
1,000,000,000	90 —	• Motorcycle, 8 meters away • Food blender	Very loud		Vigorous action
100,000,000	80 —	• Diesel truck, 80 km/hr, 15 m away • Garbage disposal		Damage begins after long exposure	
10,000,000	70 —	• Vacuum cleaner	Moderately loud		Threats
1,000,000	60 —	• Ordinary conversation • Air conditioning unit, 6 meters away			Widespread complaints
100,000	50 —	• Light traffic noise, 30 meters away			Occasional complaints
10,000	40 —	• Average living room • Bedroom	Quiet		No action
1000	30 —	• Library • Soft whisper			
100	20 —	• Broadcasting studio	Very quiet		
10	10 —	• Rustling leaf	Barely audible		
1	0 —	• Threshold of hearing			

PROBLEMS

1. **Noise.** What is noise? Do you think it would be feasible to develop an instrument that would indicate how noisy a given sound is? Defend your answer.

2. **Noise.** Which of the following phenomena would you be willing to classify as noise? Defend your answers. (a) Your neighbor's dog barking while you are listening to music; (b) your own dog barking to warn you of a smell of smoke.

3. **Decibel scale.** What is the noise level in decibels of a sound that is: (a) 100, (b) 10,000, (c) 1,000,000, and (d) 1,000,000,000 times the faintest audible sound?

4. **Decibels.** A person hears a cry in the woods that is 1000 times the intensity of the faintest audible sound. What is the sound level in decibels?

5. **Interference with communication.** Give three examples from your daily experience where noise interferes with communication.

6. **Loss of hearing.** Describe what happens to a person's hearing ability after spending one day in a noisy environment. What happens after one year, after ten years?

7. **Noise levels.** Imagine that you owned a factory. It would cost a certain amount of money to reduce the noise level in the plant. Would there be an economic advantage to this reduction of noise? Discuss.

8. **Noise control.** There are three categories of noise control: reducing the source, interrupting the path, or protecting the receiver. Which category would each of the following be listed under: (a) closing a window while a jackhammer is working outside; (b) wearing ear plugs; (c) oiling a squeaking door; (d) having an old, noisy, automobile engine rebuilt; (e) putting a new muffler on a noisy car.

9. **Legislation.** Identify the legislation that establishes national policy in the United States for promoting a quiet environment. What are the methods that are used to achieve this objective?

*10. **SST.** In an effort to keep noise to a minimum, SST's are now permitted to fly only on trans-ocean flights. Do planes flying over the ocean produce less sound? If so, why? If not, why are transcontinental flights not allowed?

*Indicates more difficult problem.

QUESTIONS FOR CLASS DISCUSSION

1. It has often been stated that some damage to hearing may be "inevitable" to people living in a developed society. Do you agree? If so, defend your answer. If not, describe some ways that you could live which would prevent damage to hearing over a lifetime.

2. People have suggested that noise standards should be reduced in some specific cases. Compare and contrast the problems and benefits of reducing noise pollution standards from SST's or slum housing.

BIBLIOGRAPHY

A delightfully written, non-mathematical, yet authoritative paperback book that covers the entire field very well is:
Rupert Taylor: Noise. Baltimore, Penguin Books, 1970, 268 pp.

Two books that emphasize the effects of noise are:
Karl D. Kryter: The Effects of Noise on Man. New York, Academic Press, 1970. 632 pp.
William Burns: Noise and Man. 2nd Ed. Philadelphia, J. B. Lippincott, 1973, 459 pp.

For a basic text on noise control, refer to:
Leo L. Beranek: Noise Reduction. New York, McGraw-Hill, 1960. 753 pp.

Four popular books that take up the environmental aspects of noise are:
Theodore Berland: The Fight For Quiet. New York, Prentice-Hall, 1970. 370 pp.
Robert Alex Baron: The Tyranny of Noise. New York, St. Martin's Press, 1970, 294 pp.
Henry Still: In Quest of Quiet. Harrisburg, Pa., Stackpole Books, 1970. 220 pp.
Clifford R. Bragdon: Noise Pollution: The Unquiet Crisis. Philadelphia, University of Pennsylvania Press, 1971. 280 pp.

GLOSSARY

acoustical materials—Materials that absorb sound energy. They are used to reduce noise levels in buildings and machines.

acid mine drainage — The pollution of waterways that occurs when chemicals from mines react with air and water to form acids. These acids upset the plant and animal life of streams and rivers.

adaptation — The process of accommodation to change; the process by which the characteristics of an organism become suited to the environment in which the organism lives.

age-sex distribution — A graph showing the numbers of males and females of each age group that live in a population.

air — A mixture of oxygen, nitrogen, and other gases, and small solid particles which surround the Earth and form the atmosphere.

air pollution — The deterioration of the quality of air that results from the addition of impurities.

altruism — Devotion to the interests of others.

aquaculture — The science and practice of raising fish in artificially controlled ponds or pools.

asbestos — A fibrous mineral that has many industrial uses. It has also been shown to be carcinogenic.

atmosphere — The predominantly gaseous envelope that surrounds the Earth.

atmospheric inversion — See *inversion.*

atom — The fundamental unit of an element. An atom consists of a nucleus surrounded by an electron cloud.

atomic nucleus — The small positive central portion of the atom that contains its protons and neutrons.

background radiation — The level of radiation on Earth from natural sources.

balance of trade — The balance between the total value of all of a nation's imports compared with the total value of all of its exports.

baleen — A set of elastic, horny plates that form a sieve-like region in the mouths of certain whales. Plankton-eating whales have no true teeth, only baleen.

biodegradable — Refers to substances that can readily be decomposed by living organisms.

biosphere — That part of the Earth and its atmosphere which can support life.

birth rate — The number of individuals born during some time period, usually a year, divided by the midyear population of that year.

black lung disease — A series of debilitating and often fatal diseases that affect the lungs of miners who work in underground coal mines.

breeder reactor — A nuclear reactor that produces more fissionable material than it consumes.

broad spectrum poison — A poison that kills many species of animals (and/or plants) indiscriminately.

calorie — A unit of energy used to express quantities of heat. When calorie is spelled with a small c, it refers to the quantity of heat required to heat 1 gram of water 1° C.

317

(This definition is not precise, because the quantity depends slightly on the particular temperature range chosen.) When Calorie is spelled with a capital C, it means 1000 small calories, or one kilocalorie, the quantity of heat required to heat 1000 grams (1 kilogram) of water 1° C. Food energies for nutrition are always expressed in Calories. The exact conversions are:

$$1 \text{ cal} = 4.184 \text{ joules}$$
$$1 \text{ kcal} = 4184 \text{ joules}$$

carnivore — An animal that eats the flesh of other animals.

carrying capacity — The maximum number of individuals of a given species that can be supported by a particular environment.

census — A count of a population.

chain reaction — A reaction that proceeds in a series of steps, each step being made possible by the preceding one.

clearcutting — The process by which loggers cut *all* the trees from a given region. Clearcutting is usually cheaper than other logging practices, but it is disruptive of natural ecosystems.

climax system — A natural system that represents the end, or apex, of an ecological succession.

competition — An interaction in which two or more organisms try to gain control of a limited resource.

cooling pond — A large pond or lake used to cool water from an electric generating station or any other industrial facility.

cooling tower — A large towerlike structure used to cool water from an electric generating station or any other industrial facility.

critical level — In general, a critical condition relates to a point at which a system changes very abruptly in response to a small change in some part of the system. In ecology, a population is said to be reduced to its critical level when its numbers are so few that it is in acute danger of extinction.

DDT — An organochloride insecticide that was banned in 1973 because of its harmful effect on the environment.

death rate — The number of individuals dying during some time period, usually a year, divided by the midyear population of that year.

decay organisms — Plants and animals that consume the tissue of dead organisms.

decibel (dB) — A unit of sound intensity equal to 1/10 of a Bel. The decibel scale is a logarithmic scale used in measuring sound intensities relative to the intensity of the faintest audible sound.

demographic transition — The pattern of change in birth and death rates typical of a developing society. The process can be outlined briefly as follows. Both birth and death rates in preindustrial society are typically very high. Therefore, population growth is very slow. Introduction, or development, of modern medicine causes a decline in death rates, but birth rates remain high. Hence, there is a rapid increase in population growth. Finally, birth rates fall, and the population grows slowly once more.

demography — The branch of sociology or anthropology which deals with the statistical study of human populations. There is emphasis on the total size, density, number of deaths, births, migrations, marriages, prevalence of disease, and so forth.

desert — A climax system in which rainfall is less than 25 centimeters per year. These systems are barren and support relatively little plant or animal life.

ecological niche — The description of the unique functions and habitats of an organism in an ecosystem.

ecology — The study of the interrelationships among plants and animals and the interactions between living organisms and their physical environment.

economic externality — That portion of the cost of a product which is not accounted for by the manufacturer but is borne by some other sector of society. An example is the cost of environmental degradation that results from a manufacturing operation.

ecosystem — A group of plants and animals occurring together plus that part of the physical environment with which they interact. An ecosystem is defined to be nearly self-contained, so that the matter which flows into and out of it is small compared with the quantities which are internally recycled in a continuous exchange of the essentials of life.

ecosystem homeostasis — The control mechanisms within an ecosystem that act to maintain constancy by opposing external stresses.

electron — The fundamental atomic unit of negative electricity.

energy — The capacity to perform work or to transfer heat.

environmental degradation — Any action that makes the environment less fit for plant or animal life.

Environmental Protection Agency (EPA) — A branch of the United States government devoted to protection of the environment.

environmental resistance — The sum of various pressures, such as predation, competition, adverse weather, etc., which collectively inhibit the potential growth of every species.

estuary — A partially enclosed shallow body of water with access to the open sea and usually a supply of fresh water from the land. Estuaries are less salty than the open ocean. They are affected by tides and, to a lesser extent, by wave action of the sea.

eutrophication — The enrichment of a body of water with nutrients, with the consequent deterioration of its quality for human purposes.

externality — See *economic externality*.

First Law of Thermodynamics — See *thermodynamics*.

fission (of atomic nuclei) — The splitting of atomic nuclei into approximately equal fragments.

fixed nitrogen — Nitrogen that is found in a chemical form readily usable by most plants. Atmospheric nitrogen, N_2, is not usable and must be converted to "fixed" forms such as ammonia.

fly ash — Mineral material released when coal is burned.

food chain — An idealized pattern of flow of energy in a natural ecosystem. In the classical food chain, plants are eaten only by primary consumers; primary consumers are eaten only by secondary consumers; secondary consumers only by tertiary consumers, and so forth. (See *food web*.)

food pyramid — A relationship that demonstrates that the energy available to an ecosystem decreases from plants to carnivores to higher level carnivores.

food web — The actual patterns of food consumption in a natural ecosystem. A given organism may obtain nourishment from many different types of organisms, and thus give rise to a complex, interwoven series of energy transfers.

Freon — A trade name of the Dupont Company that refers to the class of chlorofluorocarbons. The compounds that may be implicated in stratospheric pollution are Freon-11 ($CFCl_3$), and Freon-12 (CCl_2F_2).

fusion (of atomic nuclei) — The combination of nuclei of light elements (particularly hydrogen) to form heavier nuclei.

Gaia — The ancient Greek goddess of the Earth. This word has recently been used to describe the biosphere and to emphasize the interdependence of the Earth's ecosystems by likening the entire biosphere to a single living organism.

gas — A state of matter that consists of molecules that are moving independently of each other in random patterns.

geometric growth — In population studies, growth such that in each unit of time, the population increases by a constant multiple. An example would be a population that grew by multiples of 2: 2, 4, 8, 16, 32, 64, 128, and so on.

geothermal energy — Energy derived from the heat of the Earth's interior.

Green Revolution — The realization of increased crop yields in many areas due to the development of new high-yielding strains of wheat, rice, and other grains in the 1960's.

gross national product (GNP) — The total value of all goods and services produced by the economy in a given year.

half-life (of a radioactive substance) — The time required for half of a sample of radioactive matter to decompose.

heat — A form of energy. Every object contains heat energy in an amount that depends on its mass, its temperature, and the specific heat of the materials of which it consists.

heat engine — A mechanical device that converts heat to work.

hectare — A metric measure of surface area. One hectare is equal to 10,000 square meters or 2.47 acres.

herbicide — A chemical used to control unwanted plants.

hormone — A chemical secreted by an organism. Hormones regulate and control various life processes.

humus — The complex mixture of decayed organic matter that is an essential part of healthy natural soil.

hydroelectric power — Power derived from the energy of falling water.

incineration — A process by which solid wastes are burned. Sometimes the heat produced is utilized for space heating or to produce steam to generate electricity.

inbreeding — The mating of closely related individuals. Inbreeding generally weakens a population.

inversion — A meteorological condition in which the lower layers of air are cooler than those at higher altitudes. This cool air remains relatively stagnant and causes a concentration of air pollutants and unhealthy conditions in congested urban regions.

ion — An electrically charged atom or group of atoms.

isotopes — Atoms of the same element that have different mass numbers.

joule — A fundamental unit of energy. 4.184 joules = 1 calorie.

juvenile hormone — A chemical naturally secreted by an insect while it is a larva. When the flow of juvenile hormone stops, the insect metamorphoses to become an adult. These compounds can be used as insecticides because if sprayed at critical times they will interrupt the natural metamorphoses and eventually kill specific insect pests.

krill — Small, shrimp-like crustacea that grow in large numbers in the cold waters of the southern oceans. Krill represent a primary food supply for many species of whales.

land use — Issues associated with how a particular area of land will be used. Particular problems might include: should a farm be dug up to make way for a coal mine? Should a climax forest be cut for timber?

legal standing — The necessary requirements to bring a case to court.

legume — Any plant of the family *Leguminosae,* such as peas, beans, or alfalfa. Bacteria living on the roots of legumes change atmospheric nitrogen, N_2, to nitrogen-containing salts which can be readily assimilated by most plants.

loss of cooling accident (LOCA) — An accident in a nuclear reactor that may occur if the cooling system fails. Heat would build up rapidly and a major leak of radioactivity could occur.

mass transit — Transportation systems that utilize vehicles that carry large numbers of people — such as buses, trains, and trolleys.

meteorology — The science of the Earth's atmosphere.

mine spoil — The earth and rock removed from a mine that has no commercial value.

mineral reserves — The estimated supply of ore in the ground.

mutation — An alteration in the genes of a parent that leads to significant changes in newborn infants.

natural succession — The sequence of changes through which an ecosystem passes during the course of time.

neutron — A fundamental particle of the atomic nucleus that is electrically neutral.

noise — Unwanted sound.

nuclear fission — See *fission.*

nuclear fusion — See *fusion.*

nuclear reactor — A device that utilizes nuclear reactions to produce useful energy.

nucleus — See *atomic nucleus.*

ocean dump — A site where solid waste is deposited on the ocean floor with little or no treatment.

omnivore — An organism that eats both plant and animal tissue. Common omnivores include bears, pigs, rats, chickens, and people.

open dump — A site where solid waste is deposited on a land surface with little or no treatment.

open pit mine — See *strip mine.*

ore — A rock mixture that contains enough valuable minerals to be mined profitably with currently available technology.

organochlorides — A class of organic chemicals that contain chlorine bonded within the molecule. Some organochlorides, such as DDT, are effective pesticides. They are generally broad-spectrum, and long-lived in the environment.

organophosphates — A class of organic compounds that contain phosphorus and oxygen bonded within the molecule. Some organophosphates, used as pesticides, are broad-spectrum and extremely poisonous, although they are not long-lasting in the environment.

ozone—Triatomic oxygen, O_3.

parasitism—A special case of predation in which the predator is much smaller than the victim and obtains its nourishment by consuming the tissue or food supply of a larger living organism.

perpetual motion machine — A machine that will run forever and perform work without the use of an external energy supply. Such a machine is impossible to build.

photosynthesis — The process by which chlorophyll-bearing plants use energy from the sun to convert carbon dioxide and water to sugars.

phytoplankton — Any microscopic, or nearly microscopic, free-floating autotrophic plant in a body of water. There are a great many different species which exist in a community of phytoplankton; these plants occur in large numbers and account for most of the primary production in deep bodies of water.

plankton — Any small, free-floating organism living in a body of water. See *Phytoplankton* and *Zooplankton*.

pollution — The contamination of the quality of some portion of the environment by the addition of harmful impurities.

pollution tax — A tax on a polluter that is determined by the quantity of pollutants emitted.

power — The amount of energy delivered in a given time interval.

prairie — An extensive area of fairly level, predominantly treeless land. Prairies are characterized by an abundance of various types of grasses.

predation — An interaction in which some individuals eat others.

predator — An animal that attacks, kills, and eats other animals. More broadly, an organism that eats other organisms.

proton — A fundamental particle of the atomic nucleus that bears a unit positive charge.

radioactivity — The emission of radiation by atomic nuclei.

rationing — A method of restricting consumption by allowing individuals to purchase only a fixed amount of some item such as gasoline.

recycling — The process whereby waste materials are reused for the manufacture of new materials and goods.

rendering — The cooking of animal wastes such as fat, bones, feathers, and blood to yield tallow (used in the manufacture of soap) and high-protein animal feed.

replacement level — The average number of children that each woman must bear to maintain a constant population — that is, to achieve *zero population growth*.

respiration — The process by which plants and animals combine oxygen with sugar and other plant matter to produce energy and maintain body functions. Carbon dioxide and water are released as by-products.

sanitary landfill — A site where solid waste is deposited on a land surface, compacted, and covered with dirt to reduce odors, and prevent disease and fire.

Second Law of Thermodynamics — See *thermodynamics*.

slurry — A mixture of a solid and a liquid. Coal is mixed with water to form a slurry that can be transported in a pipeline. Slurry transportation is cheaper than transportation by rail.

smog — Smoky fog. The word is used loosely to describe visible air pollution.

smog precursor — An organic gaseous air pollutant that can undergo chemical reaction in the presence of sunlight to produce smog.

smoke — An aerosol that is usually produced by combustion or decomposition processes.

solar cell — A device that converts sunlight directly into electrical energy.

solar collector — A device designed to concentrate solar energy for a useful purpose.

solar energy — Energy derived from the sun.

sonic boom — The sharp disturbance of air pressure caused by the reinforcing waves that trail an object moving at supersonic speed.

sonic speed — The speed of sound.

soot—Particles, composed mostly of carbon, that are released from most flame sources.

species—A group of organisms that interbreed with other members of the group but not with individuals outside the group.

SST—Supersonic transport.

steam turbine—A machine consisting of a shaft equipped with a series of blades that spin when hot steam is passed through them. The shaft is generally connected to a generator to produce electricity.

sterilization — A procedure that renders a person or animal incapable of fathering or conceiving a child.

strip mining — Any mining operation that operates by removing the surface layers of soil and rock, thereby exposing the deposits of ore or coal to be removed.

subsonic speed — Less than the speed of sound.

supersonic speed — Greater than the speed of sound.

synergism — A condition in which a whole effect is greater than the sum of its parts.

system — An assemblage of objects united by some form of regular interaction or interdependence; an organic or organized whole.

thermal pollution — A change in the quality of an environment (usually an aquatic environment) caused by raising its temperature.

thermocline — Middle waters of a lake, where temperature and oxygen content fall off rapidly with depth.

thermodynamics — The science concerned with heat and work and the relationships between them.

   First Law of Thermodynamics — Energy cannot be created or destroyed.

   Second Law of Thermodynamics — It is impossible to derive mechanical work from any portion of matter by cooling it below the temperature of the coldest surrounding object.

tundra — Arctic or mountainous areas that are too cold to support trees and are characterized by low mosses and grasses.

tunnel mine — A mining operation that operates by digging tunnels underground to extract veins of ore or coal.

urban sprawl — A condition where a central city is surrounded by a vast suburban region.

urbanization — A process characterized by movement of people from rural to urban settlements, from small towns to large cities, and from large cities to their suburbs.

use tax — A tax placed on a product such as gasoline. A use tax forces consumers to pay a tax when they use a certain product.

work — The energy expended when something is forced to move.

zero population growth — A condition where the birth and death rates in a population are equal. Therefore, the size of the population remains constant.

zooplankton — Microscopic or nearly microscopic free-floating aquatic animals that feed on other forms of plankton. Some zooplankton are larvae of larger animals, while others remain as zooplankton during their entire life cycle.

APPENDIX

A THE METRIC SYSTEM (INTERNATIONAL SYSTEM OF UNITS)

The metric system (Système International d'Unités, abbreviated SI) is internationally recognized and is now used in nearly all the nations in the world. Terminology in environmental science has often used a mixture of metric and other units. More recently, however, the shift has been toward exclusive use of the metric system. This book is concerned with four of the fundamental metric units. These are:

Quantity	Unit	Abbreviation
length	meter	m
mass	kilogram	kg
time	second	s or sec
temperature	degree Celsius	°C

Larger or smaller units in the metric system are expressed by the following prefixes:

Multiple or Fraction	Prefix	Symbol
1000	kilo	k
1/100	centi	c
1/1000	milli	m
1/1,000,000	micro	μ

Length

The **meter** was once defined in terms of the length of a standard bar; it is now defined in terms of wavelengths of light. A meter is about 1.1 yards, or slightly greater than 3 feet.

One **centimeter,** cm, is 1/100 of a meter, or about 0.4 inch. There are 2.54 cm in an inch.

Area

The metric unit is the **hectare** (abbreviated ha), which equals 10,000 m^2, or about 2.47 acres.

Mass

The **kilogram,** kg, is the mass of a piece of plantinum-iridium metal called the Prototype Kilogram Number 1, kept at the International Bureau of Weights and Measures, in France. It is equal to about 2.2 pounds.

One gram, g, is 1/1000 kg. There are approximately 28 grams in an ounce. One metric ton is 1000 kg = 2204.6 lb.

Volume

Volume is not a fundamental quantity; it is derived from length.
One **cubic centimeter,** cm³, is the volume of a cube whose edge is 1 cm.
One **liter** = 1000 cm³. One liter equals about 1.1 quarts.

Temperature

If two bodies, A and B, are in contact, and if there is a spontaneous transfer of heat from A to B, then A is said to be *hotter* or at a higher temperature than B. Thus, the greater the tendency for heat to flow away from a body, the higher its temperature is.

The Celsius (formerly called Centigrade) temperature scale is defined by several fixed points. The most commonly used of these are the freezing point of water, 0°C, and the boiling point of water, 100°C.

The Fahrenheit scale, commonly used in medicine and engineering in England and the United States, designates the freezing point of water as 32°F and the boiling point of water as 212°F.

Energy

The metric unit of energy is the **joule.**

The unit commonly used to express heat energy, or the energies involved in chemical changes, is the **calorie,** cal. One calorie is 4.184 joules. The energy of one calorie is sufficient to warm one gram of water 1°C.

The **kilocalorie,** kcal, is 1000 calories. This unit is also designated Calorie (capital C), especially when it is used to express food energies for nutrition.

Some other units of energy are:

1 erg = one-ten-millionth of a joule.

1 watt-hour (the energy released when one watt of power is delivered for one hour) = 860 cal.

1 kilowatt-hour (kW-hr) = 1000 W-hr = 860 kcal.

1 Btu (British thermal unit) is the heat required to raise the temperature of one pound of water 1°F = 252 cal.

Power

Power is energy per unit time. The metric unit is the **watt.**
One watt is equal to one joule per second.
One kilowatt = 1000 watts.
One horsepower = 745 watts.

B CHEMICAL SYMBOLS, FORMULAS, AND EQUATIONS

Atoms or elements are denoted by symbols of one or two letters, like H, U, W, Ba, and Zn.

Compounds or molecules are represented by formulas that consist of symbols and subscripts, sometimes with parentheses. The subscript denotes the number of atoms of the element represented by the symbol to which it is attached. Thus H_2SO_4 is a formula that represents a molecule of sulfuric acid, or the substance sulfuric acid. The molecule consists of two atoms of hydrogen, one atom of sulfur, and four atoms of oxygen. The substance consists of matter that is an aggregate of such molecules. The formula for oxygen gas is O_2; this tells us that the molecules consist of two atoms each.

Chemical transformations are represented by chemical equations, which tell us the molecules or substances that react and the ones that are produced, and the molecular ratios of these reactions. The equation for the burning of methane in oxygen to produce carbon dioxide and water is:

$$CH_4 + 2O_2 \rightarrow CO_2 + 2H_2O$$

Each coefficient applies to the entire formula that follows it. Thus $2H_2O$ means $2(H_2O)$. This gives the following molecular ratios: reacting materials, two molecules of oxygen to one of methane; products, two molecules of water to one of carbon dioxide. The above equation is balanced because the same number and kinds of atoms, one of carbon and four each of hydrogen and oxygen, appear on each side of the arrow.

C CHEMICAL FORMULAS OF PESTICIDES

Organochlorides

Organophosphates

A Carbamate

INDEX

327